AN ELEMENTARY TREATISE ON STATICS

AN ELEMENTARY TREATISE

ON

STATICS

BY

S. L. LONEY

Cambridge
at the University Press
1956

CAMBRIDGE
UNIVERSITY PRESS

University Printing House, Cambridge CB2 8BS, United Kingdom

Cambridge University Press is part of the University of Cambridge.

It furthers the University's mission by disseminating knowledge in the pursuit of education, learning and research at the highest international levels of excellence.

www.cambridge.org
Information on this title: www.cambridge.org/9781316603819

Published March 1912
Second edition 1917
Reprinted 1920, 1924, 1927, 1930, 1934, 1942, 1944, 1946, 1951, 1956
First paperback edition 2016

A catalogue record for this publication is available from the British Library

ISBN 978-1-316-60381-9 Paperback

PREFACE

THE present work is a companion book to my *Dynamics of a Particle and of Rigid Bodies.* It is meant to cover the usual course of Statics for Students who are reading for a Degree in Science or Engineering, and for Junior Students for Mathematical Honours.

The book starts with the elementary Principles of the subject, but a Student would profit more by its use if he had previously read some elementary work, such as my *Elements of Statics.* A knowledge of the ordinary processes of the Differential and Integral Calculus is assumed, and also, in some articles, of the notions of Solid Geometry.

It will be evident that, in a book of this size, many parts of the subject must be quite untouched, but I have some hopes that, within the limits I have set to myself, the book is fairly complete.

The number of examples is large, and is intended to be useful for Students of very varying capacity. I have verified most of the questions, and hope that the number of important errors will be found to be small.

For any corrections, or suggestions for improvement, I shall be grateful.

S. L. LONEY.

ROYAL HOLLOWAY COLLEGE,
ENGLEFIELD GREEN, SURREY.
January 25, 1912.

CONTENTS

STATICS

CHAPTER I

INTRODUCTION. COMPOSITION AND RESOLUTION OF FORCES ACTING AT ONE POINT

1. A BODY is a portion of matter limited in every direction.

FORCE is anything which changes, or tends to change, the state of rest, or uniform motion, of a body.

A body is said to be at rest when it does not change its position with respect to surrounding objects.

STATICS is the science which treats of the action of forces on bodies, the forces being so arranged that the bodies are at rest.

The science which treats of the action of force on bodies in motion is called DYNAMICS.

2. A PARTICLE is a portion of matter which is indefinitely small in size, or which, for the purpose of our investigations, is so small that the distances between its different parts may be neglected.

A body may be regarded as an indefinitely large number of indefinitely small portions, or as a conglomeration of particles.

A RIGID BODY is a body whose parts always preserve an invariable position with respect to one another.

This conception, like that of a particle, is idealistic. In nature no body is perfectly rigid. Every body yields, perhaps only very slightly, if force be applied to it. If a rod, made of wood, have one end firmly fixed and the other end be pulled, the wood stretches slightly; if the rod be made of iron the deformation is very much less.

To simplify our enquiry we shall assume, unless it be otherwise stated, that all the bodies with which we have to deal are perfectly rigid.

3. EQUAL FORCES. Two forces are said to be equal when, if they act on a particle in opposite directions, the particle remains at rest.

4. MASS. The mass of a body is the quantity of matter in the body. The unit of mass used in England is a pound and is defined to be the mass of a certain piece of platinum kept in the Exchequer Office.

In France, and other foreign countries, the theoretical unit of mass used is a gramme, which is equal to about 15·432 grains. The practical unit is a kilogramme (1000 grammes), which is equal to about 2·2046 lbs.

WEIGHT. The idea of weight is one with which everyone is familiar. We all know that a certain amount of exertion is required to prevent any body from falling to the ground. The earth attracts every body to itself with a force which is called the weight of the body.

5. *Measurement of force.* We shall choose, as our unit of force in Statics, the weight of one pound. The unit of force is therefore equal to the force which would just support a mass of one pound when hanging freely.

It is found in Dynamics that the weight of one pound is not quite the same at different points of the earth's surface. In Statics, however, we shall not have to compare forces at different points of the earth's surface, so that this variation in the weight of a pound is of no practical importance; we shall therefore neglect this variation and assume the weight of a pound to be constant.

In practice the expression "weight of one pound" is, in Statics, often shortened into "one pound." The student will therefore understand that "a force of 10 lbs." means "a force equal to the weight of 10 lbs."

6. *Forces represented by straight lines.* A force will be completely known when we know (i) its magnitude, (ii) its direction, and (iii) its point of application, *i.e.* the point of the body at which the force acts.

Hence we can conveniently represent a force by a straight line drawn through its point of application; for a straight line has both magnitude and direction.

7. *Subdivisions of Force.* There are three different forms under which a force may appear when applied to a mass, *viz.* as (i) an attraction, (ii) a tension, and (iii) a reaction.

8. *Attraction.* An attraction is a force exerted by one body on another without the intervention of any visible instrument and without the bodies being necessarily in contact. The most common example is the attraction which the earth has for every body; this attraction is (Art. 4) called its weight.

9. *Tension.* If we tie one end of a string to any point of a body and pull at the other end of the string, we exert a force on the body; such a force, exerted by means of a string or rod, is called a tension.

If the string be light [*i.e.* one whose weight is so small that it may be neglected] the force exerted by the string is the same throughout its length.

For example, if a weight W be supported by means of a light string passing over the smooth edge of a table, it is found that the same force must be applied to the string whatever be the point, A, B, or C, of the string at which the force is applied.

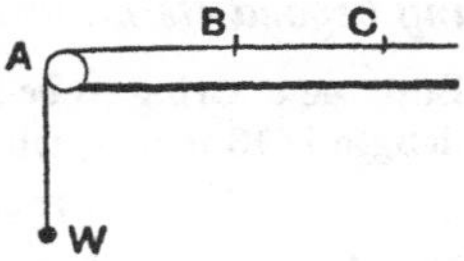

Now the force at A required to support the weight is the same in each case; hence it is clear that the effect at A is the same whatever be the point of the string to which the tension is applied, and that the tension of the string is therefore the same throughout its length.

Again, if the weight W be supported by a light string passing round a smooth peg A, it is found that the same force must be exerted at the other end of the string whatever be the direction (AB, AC, or AD) in which the string is pulled and that this force is equal to the weight W.

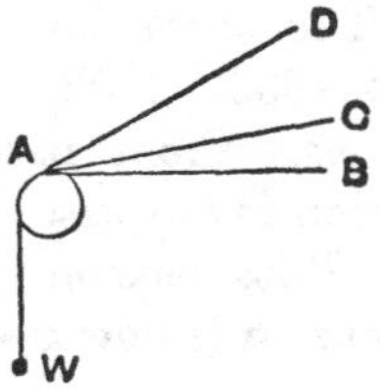

[These forces may be measured by attaching the free end of the string to a spring balance.]

Hence *the tension of a light string passing round a smooth peg is the same throughout its length.*

If two or more strings be knotted together the tensions are not necessarily the same in each string.

10. *Reaction.* If one body lean, or be pressed, against another body, each body experiences a force at the point of contact; such a force is called a reaction.

The force, or action, that one body exerts on a second body is equal and opposite to the force, or reaction, that the second body exerts on the first.

This statement will be found to be included in Newton's Third Law of Motion.

11. Tensions of Elastic Strings. All strings are extensible, although the extensibility is in many cases extremely small, and practically negligible. When the extensibility of the string cannot be neglected, there is a simple experimental law connecting the tension of the string with the amount of extension of the string. It may be expressed in the form

The tension of an elastic string varies as the extension of the string beyond its natural length.

Suppose a string to be naturally of length one foot; its tension, when the length is 13 inches, will be to its tension, when of length 15 inches, as

$$13-12 : 15-12, \text{ i.e. as } 1 : 3.$$

This law may be verified experimentally thus; take a spiral spring, or an india-rubber band. Attach one end A to a fixed point and at the other end B attach weights, and observe the amount of the extensions produced by the weights. These extensions will be found to be approximately proportional to the weights. The amount of the weights used must depend on the strength of the spring or of the rubber band; the heaviest must not be large enough to injure or permanently deform the spring or band.

The above law was published in the year 1676 by Hooke (A.D. 1635—1703), and enunciated by him in the form *Ut tensio, sic vis.* From it we easily obtain a formula giving us the tension in any case. Let a be the unstretched length of a string, and T its tension when it is stretched to be of length x. The extension is now $x-a$, and the law states that $T \propto x-a$.

This is generally expressed in the form $T = \lambda . \dfrac{x-a}{a}$.

The quantity λ depends only on the thickness of the string and on the material of which it is made, and is called the

Modulus of Elasticity of the String. It is equal to the force which would stretch the string, if placed on a smooth horizontal table, to twice its natural length; for, when $x = 2a$, we have the tension $= \lambda$. No elastic string will however bear an unlimited stretching; when the string, through being stretched, is on the point of breaking, its tension then is called the *breaking tension.*

Hooke's Law holds also for steel and other bars, but the extensions for which it is true in these cases are extremely small. We cannot stretch a bar to twice its natural length; but λ will be 100 times the force which will extend the bar by $\frac{1}{100}$th of its natural length. For if $x - a = \frac{a}{100}$, then $T = \frac{\lambda}{100}$.

The value of T will depend also on the thickness of the bar, and the bar is usually taken as one square inch section. Thus the modulus of elasticity of a steel bar is about 13500 tons per square inch.

12. Equilibrium. When two or more forces act upon a body and are so arranged that the body remains at rest, the forces are said to be in equilibrium.

We shall assume that if at any point of a rigid body we apply two equal and opposite forces, they will have no effect on the equilibrium of the body; similarly, that if at any point of a body two equal and opposite forces are acting they may be removed.

13. *Principle of the Transmissibility of Force. If a force act at any point of a rigid body, it may be considered to act at any other point in its line of action provided that this latter point be rigidly connected with the body.*

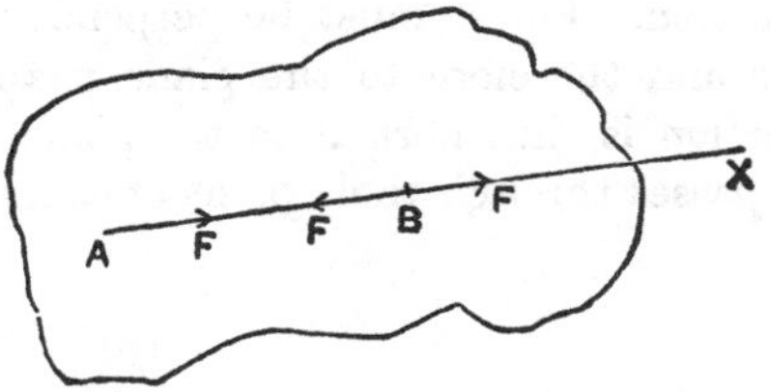

Let a force F act at a point A of a body in a direction AX. Take any point B in AX and at B introduce two equal and opposite forces, each equal to F, acting in the directions BA and BX; these will have no effect on the equilibrium of the body.

The forces F acting at A in the direction AB, and F at B in the direction BA, are equal and opposite; we shall assume that they neutralise one another and hence that they may be removed. We have thus left the force F at B acting in the direction BX and its effect is the same as that of the original force F at A. The internal forces in the above body would be different according as the force F is supposed applied at A or B.

14. *Smooth bodies.* If we place a piece of smooth polished wood, having a plane face, upon a table whose top is made as smooth as possible we shall find that, if we attempt to move the block along the surface of the table, some resistance is experienced. There is always some force, however small, between the wood and the surface of the table. If the bodies were perfectly smooth there would be no force, parallel to the surface of the table, between the block and the table; the only force between them would be perpendicular to the table.

When two bodies, which are in contact, are perfectly smooth the force, or reaction, between them is perpendicular to their common tangent plane at the point of contact.

In the case of an ordinary curved surface this direction is therefore along the normal to the surface at the point of contact, whose direction is definite.

If one of the bodies be in the shape of a thin wire, or edge, then at any point P there are an infinite number of lines perpendicular to its surface; for any line through P in a plane perpendicular to the tangent line at P satisfies this condition; but, if we have two edges in contact, the common perpendicular is a definite direction. For it must be perpendicular to each of the two edges and therefore to the plane passing through them, *i.e.* its direction is that normal to the plane through the two edges which passes through their point of contact.

Composition and Resolution of Forces

15. Suppose a flat piece of wood is resting on a smooth table and that it is pulled by means of three strings attached to three of its corners, the forces exerted by the strings being horizontal; if the tensions of the strings be so adjusted that the wood remains at rest it follows that the three forces are in equilibrium.

Hence two of the forces must together exert a force equal and opposite to the third. This force, equal and opposite to the third, is called the resultant of the first two.

Resultant. Def. *If two or more forces $P, Q, S \ldots$ act upon a rigid body and if a single force, R, can be found whose effect upon the body is the same as that of the forces $P, Q, S \ldots$ this single force R is called the resultant of the other forces and the forces $P, Q, S \ldots$ are called the components of R.*

It follows from the definition that if a force be applied to the body equal and opposite to the force R, then the forces acting on the body will balance and it be in equilibrium; conversely, if the forces acting on a body balance then either of them is equal and opposite to the resultant of the others.

16. If two forces act on a body in the same direction their resultant is clearly equal to their sum; and if they act on the body in opposite directions their resultant is equal to their difference and acts in the direction of the greater.

When two forces act at a point of a rigid body in different directions their resultant may be obtained by means of the following

Theorem. Parallelogram of Forces. *If two forces, acting at a point, be represented in magnitude and direction by the two sides of a parallelogram drawn from one of its angular points, their resultant is represented both in magnitude and direction by the diagonal of the parallelogram passing through that angular point.*

This fundamental theorem of Statics, or rather another form of it, *viz.* the Triangle of Forces (Art. 21), was first enunciated by Stevinus of Bruges in the year 1586. Before his time the science of Statics rested on the Principle of the Lever as its basis.

17. Experimental Proof. Take three light strings and knot an end of each together at a point O. Let two of the strings pass over light pulleys free to turn in any manner which are attached to fixed supports, and at the other ends of these strings let there be attached weights equal to P and Q lbs. respectively. To the end of the third string let there be attached a weight R lbs. and let this string hang vertically and freely. Then provided that neither of these three weights is greater than the sum of the other two, the system will take up some position of equilibrium. In this position of equilibrium mark off lengths OA, OB, OC along the three strings proportional to P, Q, and R respectively and complete the parallelogram $OADB$; then it will be found that OC will be equal and opposite to OD. But since P, Q, and R balance therefore R must be equal and opposite to the resultant of P and Q, *i.e.* OD must represent the resultant of forces which are represented by OA and OB.

The pulleys and weights of the foregoing experiment may be replaced by three Salter's Spring Balances. Each of these balances shews, by a pointer which travels up and down a graduated face, what force is applied to the hook at its end.

Three light strings are knotted at O and attached to the ends of the spring balances. The three balances are then drawn out to shew any convenient tensions, and are laid on a horizontal table and fixed to it by any convenient hooks or nails. The readings of the balances then give the tensions P, Q, R of the three strings. Just as in the preceding experiment we then verify the truth of the parallelogram of forces.

Dynamical Proof. A proof may also be deduced from the Parallelogram of Accelerations and Newton's Laws of Motion.

If a particle, of mass m, have accelerations f_1 and f_2 represented both in magnitude and direction by the straight lines OA and OB, its resultant acceleration, f_3, is represented by the diagonal OC of the parallelogram $OACB$.

Since the particle has an acceleration f_1 in the direction OA there must be a force $P\,(=mf_1)$ in that direction and similarly a force $Q\,(=mf_2)$ in the direction OB. Let OA_1, OB_1 represent these forces in magnitude and direction.

Then $$\frac{OA_1}{OB_1}=\frac{P}{Q}=\frac{f_1}{f_2}=\frac{OA}{OB}.$$

Complete the parallelogram $OA_1C_1B_1$; then, by simple geometry, we see that O, C, C_1 are in a straight line, and

$$\frac{OC_1}{OC}=\frac{OA_1}{OA}.$$

Hence OC_1 represents the force which produces the acceleration represented by OC, and hence it is the force which is equivalent to the forces represented by OA_1 and OB_1.

18. The magnitude and direction of the resultant, R, of two forces P and Q acting at an angle α may be easily obtained. For let OA and OB represent the forces P and Q acting at an angle α. Complete the parallelogram $OACB$ and draw CD perpendicular to OA, produced if necessary.

Then $OD=OA+AC\cos DAC=P+Q\cos BOD=P+Q\cos\alpha$.

[If D fall between O and A, as in the second figure, we have
$OD=OA-AC\cos DAC=P-Q\cos(180°-\alpha)=P+Q\cos\alpha$.]

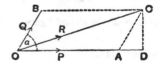

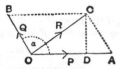

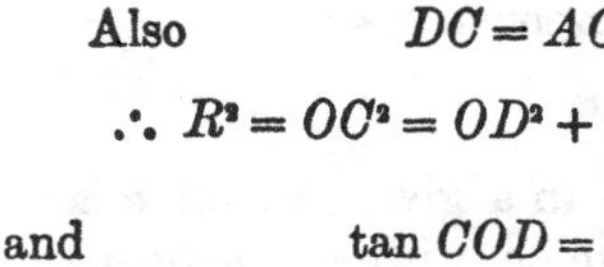

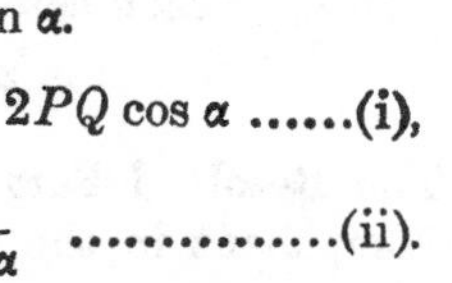

Also $$DC=AC\sin DAC=Q\sin\alpha.$$

$$\therefore\ R^2=OC^2=OD^2+CD^2=P^2+Q^2+2PQ\cos\alpha \quad\text{......(i)},$$

and $$\tan COD=\frac{DC}{OD}=\frac{Q\sin\alpha}{P+Q\cos\alpha} \quad\text{...............(ii)}.$$

These two equations give the required magnitude and direction of the resultant.

Cor. If the forces be at right angles, we have $\alpha=90°$, so that $R=\sqrt{P^2+Q^2}$, and $\tan COA=\frac{Q}{P}$.

19. A force may be resolved into two components in an infinite number of ways; for an infinite number of parallelograms can be constructed having OC as a diagonal and each of these parallelograms would give a pair of such components.

The most important case occurs when we resolve a force into two components at right angles to one another.

Suppose we wish to resolve a force F, represented by OC, into two components, one of which is in the direction OA and the other is perpendicular to OA.

Draw CM perpendicular to OA and complete the parallelogram $OMCN$. The forces represented by OM and ON are the required components.

Let the angle AOC be α.

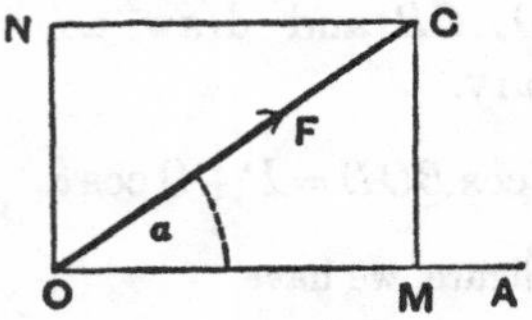

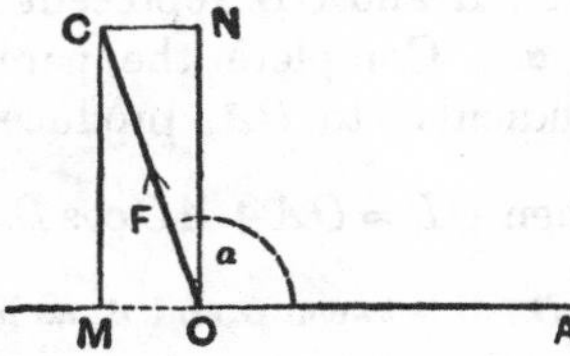

Then $OM = OC\cos\alpha = F\cos\alpha$, and $ON = OC\sin\alpha = F\sin\alpha$.

[If the point M lie in OA produced backwards, as in the second figure, the component of F in the direction OA

$$= -OM = -OC\cos COM = OC\cos\alpha = F\cos\alpha.$$

Also the component perpendicular to $OA = ON = MC = OC\sin COM = F\sin\alpha.$]

Hence, in each case, the required components are

$$F\cos\alpha \text{ and } F\sin\alpha.$$

The Resolved Part of a given force in a given direction is the component in the given direction which, with a component in a direction perpendicular to the given direction, is equivalent to the given force.

Thus the resolved part of the force F in the direction OA is $F\cos\alpha$. Hence *the Resolved Part of a given force in a given direction is obtained by multiplying the given force by the cosine of the angle between the given force and the given direction.*

20. A force may be resolved into two components in any two assigned directions.

Let the components of a force F, represented by OC, in the directions OA and OB be required and let the angles AOC and COB be α and β respectively.

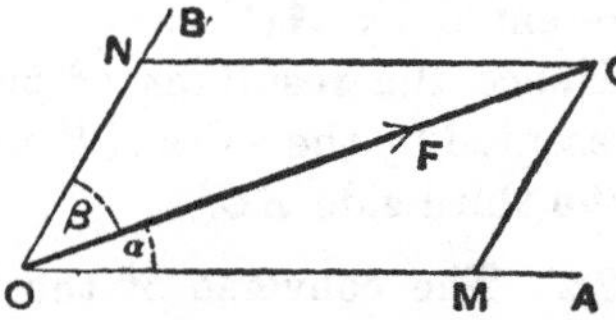

Draw CM parallel to OB to meet OA in M and complete the parallelogram $OMCN$. Then OM and ON are the required components.

Since the sides of the triangle OMC are proportional to the sines of the opposite angles, we have

$$\frac{OM}{\sin\beta} = \frac{MC}{\sin\alpha} = \frac{F}{\sin(\alpha+\beta)}.$$

Hence the required components are $F\dfrac{\sin\beta}{\sin(\alpha+\beta)}$ and $F\dfrac{\sin\alpha}{\sin(\alpha+\beta)}$.

The student must carefully notice that the components of a force in two assigned directions are not the same as the resolved parts of the forces in these directions. For example, the resolved part of F in the direction OA is, by Art. 19, $F\cos\alpha$.

21. Triangle of Forces. *If three forces, acting at a point, be represented in magnitude and direction by the sides of a triangle, taken in order, they will be in equilibrium.*

Let the forces P, Q, and R acting at the point O be represented in magnitude and direction by the sides AB, BC, and CA of the triangle ABC. Complete the parallelogram $ABCD$

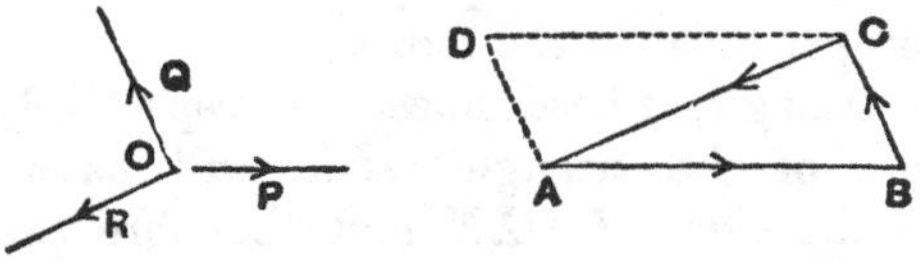

The forces represented by BC and AD are the same, since BC and AD are equal and parallel.

Now the resultant of the forces AB and AD is, by the parallelogram of forces, represented by AC. Hence the resultant of AB, BC, and CA is equal to the resultant of forces AC and CA, and is therefore zero.

Hence the three forces P, Q, and R are in equilibrium.

Cor. Since forces, acting at a point and represented by AB, BC, and CA, are in equilibrium, and since, when three forces are in equilibrium, each is equal and opposite to the resultant of the other two, it follows that the resultant of AB and BC is equal and opposite to CA, *i.e.* their resultant is represented by AC.

Hence the resultant of two forces, acting at a point and represented by the sides AB and BC of a triangle, is represented by the third side AC.

22. The converse of the Triangle of Forces is also true, *viz.* that *If three forces P, Q, and R acting at a point O be in equilibrium they can be represented in magnitude and direction by the sides of any triangle which is drawn so as to have its sides respectively parallel to the directions of the forces.*

Measure off lengths OL and OM along the directions of P and Q to represent these forces respectively. Complete the parallelogram $OLNM$ and join ON.

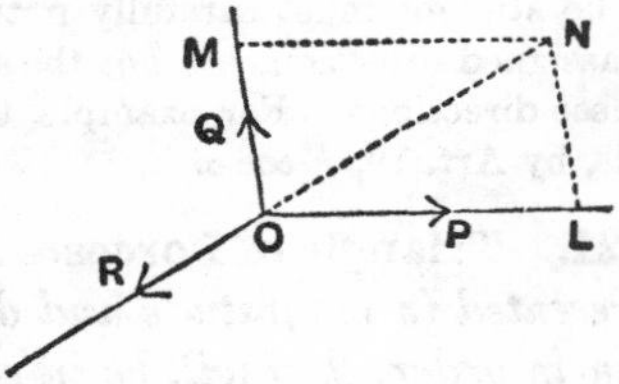

Since the three forces P, Q, and R are in equilibrium, R must be equal and opposite to the resultant of P and Q, and must therefore be represented by NO. Hence the three forces P, Q, and R are parallel and proportional to the sides OL, LN, and NO of the triangle OLN.

Any other triangle, whose sides are parallel to those of the triangle OLN, will have its sides proportional to those of OLN and therefore proportional to the forces.

Again, any triangle, whose sides are respectively perpendicular to those of the triangle OLN, will have its sides proportional to the sides of OLN and therefore proportional to the forces.

Hence we have an easy graphic method of determining the relative directions of three forces which are in equilibrium and whose magnitudes are known. We have to construct a triangle whose sides are proportional to the forces, and this can always be done unless two of the forces added together are less than the third.

23. Lami's Theorem. *If three forces acting on a particle keep it in equilibrium, each is proportional to the sine of the angle between the other two.*

For, in the previous article, since the sides of the triangle OLN are proportional to the sines of the opposite angles, we have

$$\frac{OL}{\sin LNO} = \frac{LN}{\sin LON} = \frac{NO}{\sin OLN}$$

i.e.

$$\frac{P}{\sin QOR} = \frac{Q}{\sin ROP} = \frac{R}{\sin POQ}.$$

24. Polygon of Forces. *If any number of forces, acting on a particle, be represented, in magnitude and direction, by the sides of a polygon, taken in order, the forces shall be in equilibrium.*

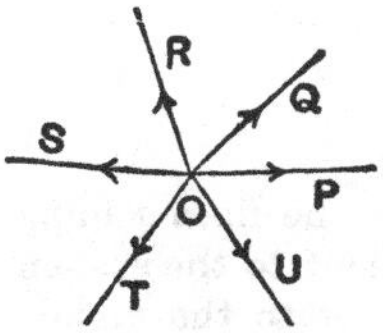

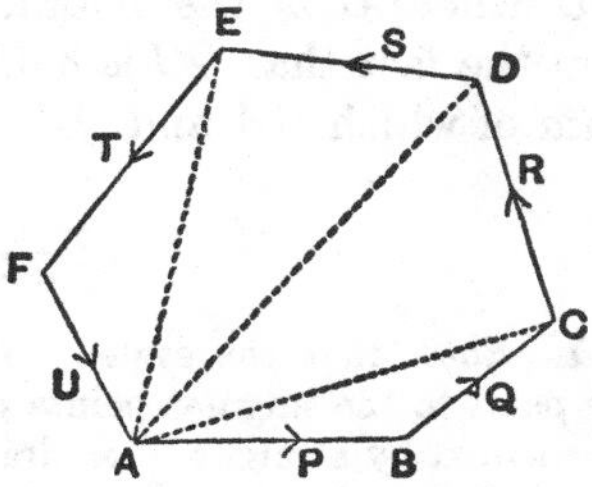

Let the sides AB, BC, CD, DE, EF, and FA of the polygon $ABCDEF$ represent the forces acting on a particle O. Join AC, AD, and AE.

By the corollary to Art. 21, the resultant of forces AB and BC is represented by AC. Similarly the resultant of forces AC and CD is represented by AD; the resultant of forces AD and DE by AE; and the resultant of forces AE and EF by AF.

Hence the resultant of all the forces is equal to the resultant of AF and FA, *i.e.* the resultant vanishes, and the forces are in equilibrium.

A similar method of proof will apply whatever be the number of forces. It is also clear from the proof that the sides of the polygon need not be in the same plane.

The converse of the Polygon of Forces is not true; for the ratios of the sides of a polygon are not known when the directions of the sides are known. For example, in the above figure, we might take any point A' on AB and draw $A'F'$ parallel to AF to meet EF in F'; the new polygon $A'BCDEF'$ has its sides respectively parallel to those of the polygon $ABCDEF$ but the corresponding sides are clearly not proportional.

25. *The resultant of two forces, acting at a point O in directions OA and OB and represented in magnitude by $\lambda . OA$ and $\mu . OB$, is represented by $(\lambda + \mu) . OC$, where C is a point in AB such that $\lambda . CA = \mu . CB$.*

For by the corollary to Art. 21, the force $\lambda . OA$ is equivalent to forces represented by $\lambda . OC$ and $\lambda . CA$, and the force $\mu . OB$ to forces $\mu . OC$ and $\mu . CB$.

Hence the given forces are together equivalent to forces $(\lambda + \mu) . OC$ together with forces $\lambda . CA$ and $\mu . CB$. Also the two latter balance.

Cor. The resultant of forces represented by OA and OB is $2OC$, where C is the middle point of AB. This is also clear from the fact that OC is half the diagonal OD of the parallelogram of which OA and OB are adjacent sides.

EXAMPLES

1. Shew that the system of forces represented by the lines joining any point to the angular points of a triangle is equivalent to the system represented by straight lines drawn from the same point to the middle points of the sides of the triangle.

2. Find a point within a quadrilateral such that, if it be acted on by forces represented by straight lines joining it to the angular points of the quadrilateral, it will be in equilibrium.

3. Four forces act along and are proportional to the sides of the quadrilateral $ABCD$; three act in the directions AB, BC, and CD and the fourth acts from A to D; find the magnitude and direction of their resultant, and determine the point in which it meets CD.

4. The sides BC and DA of a quadrilateral $ABCD$ are bisected in F and H respectively; shew that if two forces parallel and equal to AB and DC act on a particle, then the resultant is parallel to HF and equal to $2 . HF$.

5. The sides AB, BC, CD, and DA of a quadrilateral $ABCD$ are bisected at E, F, G, and H respectively. Shew that the resultant of the forces acting at a point, which are represented in magnitude and direction by EG and HF, is represented in magnitude and direction by AC.

6. From a point P, within a circle whose centre is fixed, straight lines PA_1, PA_2, PA_3, and PA_4 are drawn to meet the circumference, all being equally inclined to the radius through P; shew that, if these lines represent forces radiating from P, their resultant is independent of the magnitude of the radius of the circle.

7. Two constant equal forces act at the centre C of an ellipse parallel to SP and PH, where P is any point on the ellipse and S and H are the foci; shew that the end of the straight line which represents their resultant lies on a circle which passes through C.

8. Explain how a vessel is enabled to sail in a direction nearly opposite to that of the wind. If its sail be considered as a rigid plane, shew that it should be set so as to bisect the angle between the keel and the apparent direction of the wind in order that the force to urge the vessel on may be as great as possible.

26. Parallelopiped of Forces. Three forces acting at a point O and represented by straight lines OA, OB, OC are equivalent to a force represented by OD, the diagonal of the parallelopiped whose edges are OA, OB, and OC.

For the two forces OA, OB are equivalent to a force OE, where $OAEB$ is a parallelogram. Also forces OE, OC are equivalent to a force OD, since $OEDC$ is a parallelogram.

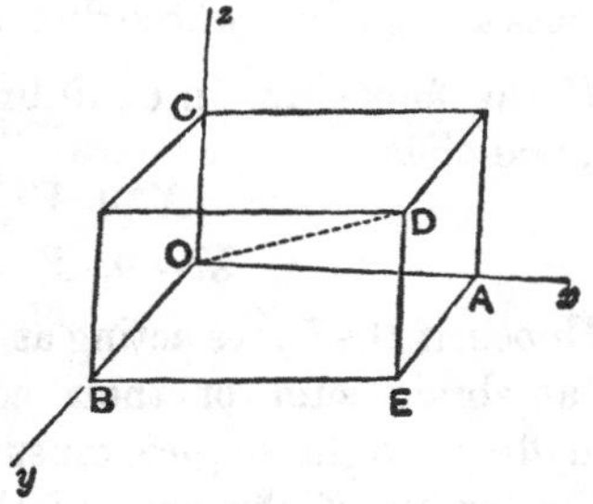

If the parallelopiped is rectangular, so that OA, OB, OC may be taken to be along rectangular axes of coordinates, and if the forces OA, OB, OC be X, Y, and Z, their resultant $R = \sqrt{X^2 + Y^2 + Z^2}$, and acts along a line whose direction cosines are cos AOD, cos BOD, and cos COD,

$$i.e.\ \frac{OA}{OD},\ \frac{OB}{OD},\ \frac{OC}{OD},\quad i.e.\ \frac{X}{R},\ \frac{Y}{R},\ \frac{Z}{R}.$$

Conversely, a force R acting at the origin O along a line whose direction cosines are (l, m, n) has as components along the axes of coordinates $X\ (= lR)$, $Y\ (= mR)$, and $Z\ (= nR)$.

27. *The sum of the resolved parts of two forces in a given direction is equal to the resolved part of their resultant in the same direction.*

For it is easily seen that if OA, OB represent the two forces, and if $OACB$ be a parallelogram, then the sum of the projections of OA and OB on *any* line OX is equal to the sum of the projections of OA, AC on the same line and is therefore equal to the projection of OC on the same line. Hence the result.

28. ***To find the resultant of any number of forces acting at a given point O, and the conditions that they may be in equilibrium.***

Take any three mutually perpendicular axes Ox, Oy, Oz through O. Let the given forces be R_1 acting along a line whose direction cosines are (l_1, m_1, n_1), R_2 acting along a line (l_2, m_2, n_2), etc.

By Art. 26, R_1 is equal to the components l_1R_1, m_1R_1, n_1R_1 along the three axes, and similarly for the other forces.

If the total components along these axes be X, Y, Z, then

$$X = l_1R_1 + l_2R_2 + l_3R_3 + \ldots,$$
$$Y = m_1R_1 + m_2R_2 + m_3R_3 + \ldots,$$
$$Z = n_1R_1 + n_2R_2 + n_3R_3 + \ldots.$$

Hence, by Art. 26, the resultant force $R = \sqrt{X^2 + Y^2 + Z^2}$, and acts along a line whose direction cosines are $\left(\frac{X}{R}, \frac{Y}{R}, \frac{Z}{R}\right)$.

If the forces are in equilibrium their resultant R must be zero, and thus

$$X^2 + Y^2 + Z^2 = 0.$$
$$\therefore\ X = 0,\ Y = 0, \text{ and } Z = 0.$$

Hence, if the forces acting at a given point are in equilibrium, the algebraic sum of their components in three directions mutually at right angles must separately vanish.

Conversely, if the sum of the components along three such directions separately vanish, the forces are in equilibrium.

If the forces of the previous article are coplanar, we need only resolve along two straight lines in their plane.

When there are only three coplanar forces acting at a point the conditions of equilibrium are often most easily found by Lami's Theorem (Art. 23).

29. ***Equilibrium of a particle at rest in contact with a smooth material curve or surface.***

Let the particle be at rest at a point P of the curve whose coordinates are (x, y, z) and let s be the length of the arc OP measured from a fixed point O.

The direction cosines of the tangent to the curve are

$$\frac{dx}{ds}, \frac{dy}{ds}, \frac{dz}{ds}$$

and are known if the form of the curve is known.

Since the curve is smooth, the only action it can exert on the particle is in a direction normal to the curve; hence the resultant force tangential to the curve must vanish. If therefore X, Y, Z be the components parallel to the axes of the forces acting on the particle, we have

$$X\frac{dx}{ds}+Y\frac{dy}{ds}+Z\frac{dz}{ds}=0.$$

If the curve be in one plane this condition becomes

$$X\frac{dx}{ds}+Y\frac{dy}{ds}=0, \qquad i.e.\ X+Y\frac{dy}{dx}=0.$$

If the particle be in contact with a smooth surface $f(x, y, z)=0$ at a point P, the resultant force along *any* tangent line at P must clearly vanish. Hence the resultant force of the components X, Y, Z (which acts along a line whose direction cosines are proportional to X, Y, Z) must coincide with the normal at P $\left(\text{whose direction cosines are proportional to } \frac{df}{dx}, \frac{df}{dy}, \frac{df}{dz}\right)$.

Hence we must have $\dfrac{\frac{df}{dx}}{X}=\dfrac{\frac{df}{dy}}{Y}=\dfrac{\frac{df}{dz}}{Z}$.

Also the normal reaction of the surface must be equal to the resultant force $\sqrt{X^2+Y^2+Z^2}$.

30. Ex. 1. *A bead rests on a smooth wire in the form of the ellipse* $\frac{x^2}{a^2}+\frac{y^2}{b^2}=1$, *being acted on by forces* λx^n, μy^n *parallel to the axes. Find its position of equilibrium and consider the case when n is unity.*

The position is given by $\lambda x^n\frac{dx}{ds}+\mu y^n\frac{dy}{ds}=0$,

i.e. $$\lambda x^n=-\mu y^n\frac{dy}{dx}=\mu y^n\frac{xb^2}{ya^2} \quad \text{.......................(1)},$$

from the equation to the ellipse.

$$\therefore \frac{x}{(\mu b^2)^{\frac{1}{n-1}}}=\frac{y}{(\lambda a^2)^{\frac{1}{n-1}}}=\frac{1}{\sqrt{\frac{1}{a^2}(\mu b^2)^{\frac{2}{n-1}}+\frac{1}{b^2}(\lambda a^2)^{\frac{2}{n-1}}}},$$

giving the position of equilibrium.

If n be unity then (1) gives $a^2\lambda=b^2\mu$ as the only condition of equilibrium, and hence, if this condition holds, the particle will rest at any point of the curve. Under forces proportional to $\frac{x}{a^2}$, $\frac{y}{b^2}$ parallel to the axes the bead will therefore rest anywhere.

Ex. 2. *A particle P is acted upon towards two centres at A and B by forces $\frac{\mu}{AP^2}$ and $\frac{\mu_1}{BP^2}$ respectively. Shew that it will rest on any smooth curve whose equation is of the form*

$$(r-\mu c)(r_1-\mu_1 c)=\mu\mu_1 c^2,$$

where $AP=r$, $BP=r_1$ and c is any constant.

If s be the length of the arc OP measured from any fixed point O on the arc of the required smooth curve, and PT be the tangent at P, we have by resolving the forces along PT,

$$\frac{\mu}{r^2}\cos APT+\frac{\mu_1}{r_1^2}\cos BPT=0,$$

i.e.
$$\frac{\mu}{r^2}\frac{dr}{ds}+\frac{\mu_1}{r_1^2}\frac{dr_1}{ds}=0.$$

Hence, by integration,

$$\frac{\mu}{r}+\frac{\mu_1}{r_1}=\text{const.}=\frac{1}{c}.$$

$$\therefore\ (r-\mu c)(r_1-\mu_1 c)=\mu\mu_1 c^2.$$

EXAMPLES

1. A small bead P can slide on a smooth elliptic wire; it is attracted towards the foci S and H by forces proportional to SP^m and HP^n respectively; find the position of equilibrium.

2. A particle P is acted upon by forces $\frac{\mu}{r_1}$ and $\frac{\mu}{r_2}$ respectively towards two fixed points O_1 and O_2; shew that it will rest at any point of a smooth groove whose equation is $r_1r_2=$constant, where

$$O_1P=r_1 \text{ and } O_2P=r_2.$$

If it be acted on by constant forces P_1 and P_2 towards the same points, the equation to the corresponding groove would be $P_1r_1+P_2r_2=$constant.

3. A particle P is acted upon by an attractive force, $\frac{\mu}{OP^2}$, towards a fixed point O and a repulsive force, $\frac{\mu}{O'P^2}$, from a fixed point O'. Shew that the equation of the curve on which P lies, if the attraction on it is always along the tangent at P, is given by $\cos\theta-\cos\theta'=$constant, where θ and θ' are the angles that OP and $O'P$ make with $O'O$ produced.

If the forces be $\frac{\mu}{OP}$ and $\frac{\mu}{O'P}$, shew that the curve is given by $\theta-\theta'=$const., so that it is an arc of a circle.

4. A tube in the form of a parabola is placed with its axis vertical and vertex downwards, and a heavy particle is placed within it; shew that the particle can be kept at rest by a force along an ordinate, and outwards, which varies as the ordinate, and that the corresponding reaction of the tube varies as the square root of its distance from the focus of the parabola.

5. Shew that the point on the smooth surface $\frac{x^3}{a^3}+\frac{y^3}{b^3}+\frac{z^3}{c^3}=1$, where a particle would rest if acted on by any force towards the origin, is given by

$$\frac{x}{a^3}=\frac{y}{b^3}=\frac{z}{c^3}=\frac{1}{\sqrt[3]{a^6+b^6+c^6}}.$$

6. A framework consisting of eight equal light rods, jointed so as to have the appearance of half a regular octahedron, is placed with its square base on a horizontal plane. When a weight W is suspended from the vertex, shew that the stress in the slant rods is $\frac{1}{4}W\sqrt{2}$, and that in the horizontal rods is $\frac{1}{8}W\sqrt{2}$.

7. Twelve equal light rods are smoothly hinged together so as to form a regular octahedron. One corner is fixed and the framework hangs freely with equal weights attached to each of the remaining corners. Shew that the tensions in the lower rods, the horizontal rods, and the upper rods are in the ratios 1 : 3 : 5.

8. Three poles, each 9 feet long, form a tripod from the vertex of which a weight W is hung; the feet of the poles rest on a horizontal plane, rough enough to prevent any sliding, and form a triangle the lengths of whose sides are 5, 5, and 6 feet; if T_1, T_1, and T_2 be the thrusts of the poles, shew that

$$\frac{T_1}{25}=\frac{T_2}{14}=\frac{9W}{8\sqrt{4559}}.$$

CHAPTER II

PARALLEL FORCES. MOMENTS. COUPLES

31. ***To find the resultant of two parallel forces acting upon a rigid body.***

Case I. *Let the forces be* **like,** *i.e.* **let them act in the same direction.**

Let P and Q be the forces acting at points A and B of the body, and represented by the lines AL and BM.

At A and B apply two equal and opposite forces, each equal to S, and acting in the directions BA and AB respectively and represented by AD and BE. These two forces balance one another and have no effect upon the equilibrium of the body.

Complete the parallelograms $ALFD$ and $BMGE$; let the diagonals FA and GB be produced to meet in O. Draw OC parallel to AL to meet AB in C.

The forces P and S at A have a resultant P_1, represented by AF. Let its point of application be removed to O. So the forces Q and S at B have a resultant Q_1 represented by BG. Let its point of application be transferred to O.

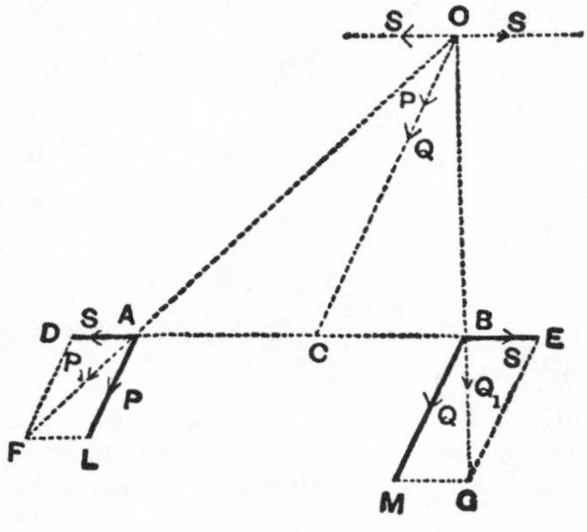

The force P_1 at O may be resolved into two forces, S parallel to AD, and P in the direction OC. So the force Q_1 at O may be resolved into two forces, S parallel to BE, and Q in the direction OC. Hence the original forces P and Q are equivalent to a force $(P+Q)$ acting along OC, *i.e.* acting at C parallel to the original directions of P and Q.

By construction, OCA and ALF are similar triangles;

$$\therefore \frac{OC}{CA}=\frac{AL}{LF}=\frac{P}{S}, \text{ so that } P.CA=S.OC\text{......(1)}.$$

So, since the triangles OCB and BMG are similar, we have

$$Q.CB=S.OC\text{......................(2)}.$$

Hence $\quad P.CA=Q.CB,$ so that $\dfrac{CA}{CB}=\dfrac{Q}{P}$,

i.e. C divides the line AB *internally* in the inverse ratio of the forces.

Case II. *Let the forces be* **unlike**, *i.e.* let them act in opposite directions.

Let P be the greater of the two forces. Making the same construction as before, the diagonals AF and BG always meet in a point O, unless they are parallel in which case the forces P and Q are equal.

As before, the original forces P and Q are now equivalent to a force $P-Q$ acting in the direction CO produced, *i.e.* acting at C in a direction parallel to that of P.

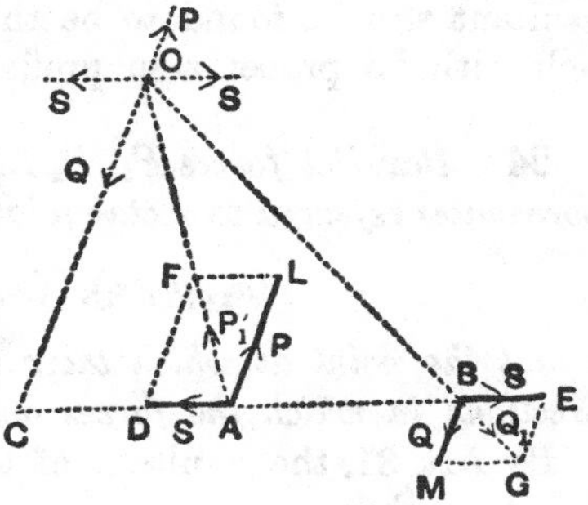

As in Case I we have $\dfrac{CA}{CB}=\dfrac{Q}{P}$, *i.e.* C divides the line AB *externally* in the inverse ratio of the forces.

To sum up; If two parallel forces, P and Q, act at points A and B of a rigid body,

(i) their resultant is a force whose line of action is parallel to the lines of action of the component forces; also, when the component forces are like, its direction is the same as that of the two forces, and, when the forces are unlike, its direction is the same as that of the greater component.

(ii) the point of application is a point C in AB, such that

$$P.AC = Q.BC.$$

(iii) the magnitude of the resultant is the sum of the two component forces when the forces are like, and the difference of the two component forces when they are unlike.

32. *Case of failure of the preceding construction.*

In the second figure of the last article, if the forces P and Q be equal, the triangles FDA and GEB are equal in all respects, and hence the angles DAF and EBG will be equal. In this case the lines AF and GB will be parallel and will not meet in any such point as O; hence the construction fails.

Hence there is no single force which is equivalent to two equal unlike parallel forces.

33. If we have a number of like parallel forces acting on a rigid body we can find their resultant by successive applications of Art. 31. We must find the resultant of the first and second, and then the resultant of this resultant and the third, and so on. The magnitude of the final resultant is the sum of the forces.

If the parallel forces be not all like, the magnitude of the resultant will be found to be the algebraic sum of the forces each with its proper sign prefixed.

34. *Parallel forces P_1, P_2 ... act at points A_1, A_2 ... whose coordinates referred to rectangular axes are*

$$(x_1, y_1, z_1), \ (x_2, y_2, z_2), \ \ldots;$$

to find the point at which their resultant acts whatever be the directions in which the forces act.

By Art. 31, the resultant of the forces at A_1, A_2 cuts A_1A_2 at G_1 such that

$$\frac{P_1}{P_2} = \frac{G_1A_2}{G_1A_1} = \frac{z_2 - G_1N_1}{G_1N_1 - z_1},$$

where G_1N_1 is perpendicular to the plane xOy.

$$\therefore G_1N_1 = \frac{P_1z_1 + P_2z_2}{P_1 + P_2},$$

The resultant of the force $P_1 + P_2$ at G_1 and P_3 at A_3 cuts G_1A_3 at G_2, such that

$$\frac{P_1+P_2}{P_3} = \frac{G_2A_3}{G_2G_1} = \frac{z_3 - G_2N_2}{G_2N_2 - G_1N_1},$$

so that

$$G_2N_2 = \frac{(P_1+P_2)\,G_1N_1 + P_3z_3}{P_1+P_2+P_3} = \frac{P_1z_1 + P_2z_2 + P_3z_3}{P_1+P_2+P_3},$$

and so on, whatever be the number of the forces.

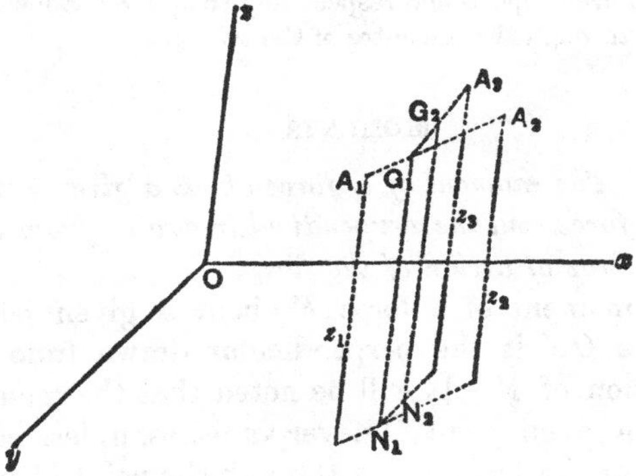

Hence, finally, the z-coordinate of the point of action of the parallel forces is given by

$$\bar{z} = \frac{P_1z_1 + P_2z_2 + \ldots}{P_1 + P_2 + \ldots} = \frac{\Sigma(Pz)}{\Sigma(P)}.$$

So the other coordinates of the point of action are

$$\bar{x} = \frac{\Sigma(Px)}{\Sigma(P)} \text{ and } \bar{y} = \frac{\Sigma(Py)}{\Sigma(P)}.$$

This point is called the centre of the System of Parallel Forces.

EXAMPLES

1. At the angular points of a square, taken in order, there act parallel forces in the ratio 1 : 3 : 5 : 7 ; find the distance from the centre of the square of the point at which their resultant acts.

2. ***A, B, C,*** **and *D* are the angles of a parallelogram taken in order, like parallel forces proportional to 6, 10, 14, and 10 respectively act at *A, B, C,* and *D*; shew that the centre and resultant of these parallel forces remain the same, if, instead of these forces, parallel forces, proportional to 8, 12, 16, and 4, act at the points of bisection of the sides *AB, BC, CD,* and *DA* respectively.**

3. Find the centre of parallel forces equal respectively to P, $2P$, $3P$, $4P$, $5P$, and $6P$, the points of application of the forces being at distances 1, 2, 3, 4, 5, and 6 inches respectively from a given point *A* measured along a given line *AB*.

4. Three parallel forces, P, Q, and R, act at the vertices *A, B,* and *C* of a triangle and are proportional respectively to a, b, c. Shew that their resultant passes through the in-centre of the triangle.

Moments

35. Def. *The moment of a force about a given point is the product of the force and the perpendicular drawn from the given point upon the line of action of the force.*

Thus the moment of a force F about a given point O is $F \times ON$, where ON is the perpendicular drawn from O upon the line of action of F. It will be noted that the moment of a force F about a given point O never vanishes, unless either the force vanishes or the force passes through the point about which the moment is taken.

Suppose the force F to be represented in magnitude, direction, and line of action by the line AB.

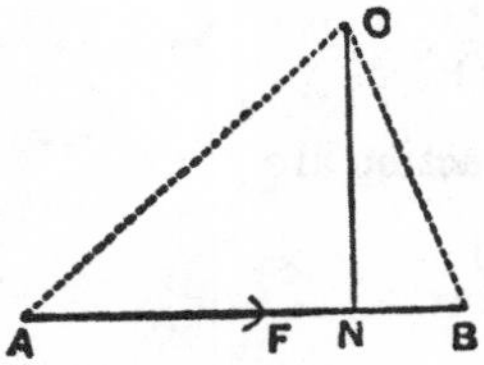

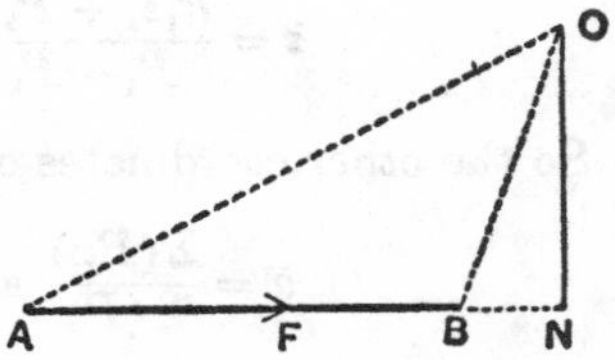

Join OA and OB.

The moment of F about O is $F \times ON$, *i.e.* $AB \times ON$. But $AB \times ON$ is equal to twice the area of the triangle OAB.

Hence *the moment of the force F about the point O is represented by twice the area of the triangle whose base is the line representing the force and whose vertex is the point about which the moment is taken.*

36. *Physical meaning of the moment of a force about a point.*

Suppose the body is a plane lamina resting on a smooth table and that the point O of the body is fixed. The effect of a force F acting on the body would be to cause it to turn about the point O as a centre, and this effect would not be zero unless (1) the force F were zero, or (2) the force F passed through O, in which case the distance ON would vanish. Hence the product $F \times ON$ would seem to be a fitting measure of the tendency of F to turn the body about O. This may be experimentally verified as follows:

Let the lamina be at rest under the action of two strings whose tensions are F and F_1, which are tied to fixed points of the lamina and whose lines of action lie in the plane of the lamina. Let ON and ON_1 be the perpendiculars drawn from the fixed point O upon the lines of action of F and F_1.

If we measure the lengths ON and ON_1 and also the forces F and F_1, it will be found that the product $F.ON$ is always equal to the product $F_1.ON_1$. Hence the two forces, F and F_1, will have equal but opposite tendencies to turn the body about O if their moments about O have the same magnitude.

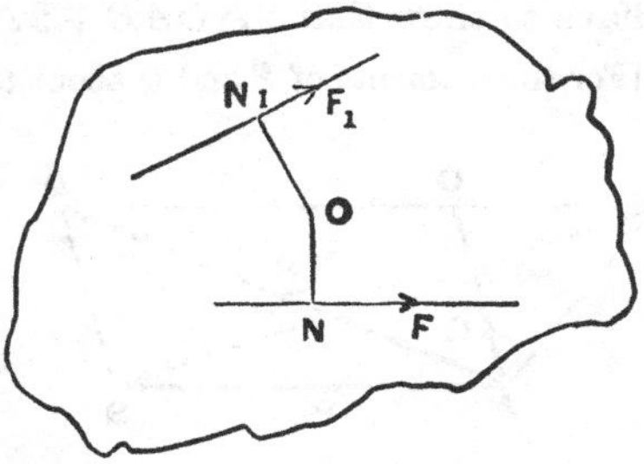

These forces F and F_1 may be measured by carrying the strings over light smooth pulleys and hanging weights at their ends sufficient to give equilibrium; or by tying the strings to the hooks of two spring balances and noting the readings of the balances, as in the cases of Art. 17.

37. *Positive and negative moments.* In Art. 36 the force F would, if it were the only force acting on the lamina, make it turn in a direction opposite to that in which the hands of a watch move, when the watch is laid on the table with its face upwards. The force F_1 would, if it were the only force acting on the lamina, make it turn in the same direction as that in which the hands of the watch move.

The moment of F about O, *i.e.* in a direction ↻, is said to be *positive*, and the moment of F_1 about O, *i.e.* in a direction ↺, is said to be *negative*.

The algebraic sum of the moments of a set of forces about a given point is the sum of the moments of the forces, each moment having its proper sign prefixed to it.

38. *The algebraic sum of the moments of any two forces about any point O in their plane is equal to the moment of their resultant about the same point.*

Case I. *Let the forces P and Q meet in a point A.*

From O draw OC parallel to the direction of P to meet the line of action of Q in the point C.

Let AC represent Q in magnitude and on the same scale let AB represent P; complete the parallelogram $ABDC$, and join OA and OB. Then AD represents the resultant, R, of P and Q.

(α) If O be without the angle DAC, as in the first figure, we have to shew that $2\triangle OAB + 2\triangle OAC = 2\triangle OAD$.

[For the moments of P and Q about O are in the same direction.]

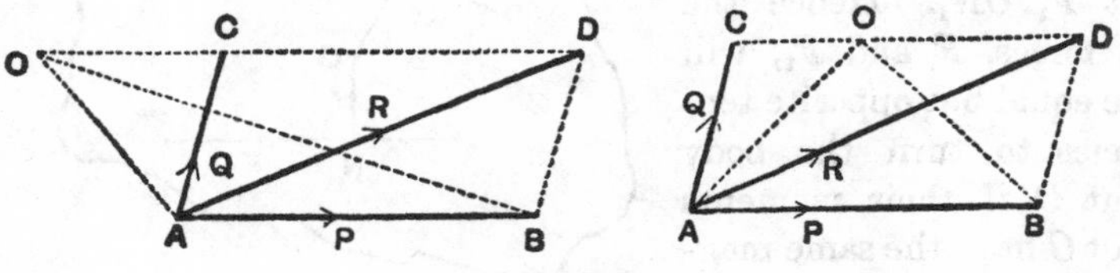

Since AB and OD are parallel, $\triangle OAB = \triangle DAB = \triangle ACD$.

$\therefore 2\triangle OAB + 2\triangle OAC = 2\triangle ACD + 2\triangle OAC = 2\triangle OAD$.

(β) If O be within the angle CAD, as in the second figure, we have to shew that $2\triangle AOB - 2\triangle AOC = 2\triangle AOD$.

[For the moments of P and Q about O are in opposite directions.]

As in (α), we have $\triangle AOB = \triangle DAB = \triangle ACD$.

hence $2\triangle AOB - 2\triangle AOC = 2\triangle ACD - 2\triangle OAC = 2\triangle OAD$.

Case II. *Let the forces P and Q be parallel.*

From O draw $OACB$ perpendicular to the forces and their resultant R $(= P + Q)$ to meet them in A, B, and C respectively.

By Art. 31 we have $P.AC = Q.CB$....................(1).

Hence the sum of the moments of P and Q about O

$$= Q.OB + P.OA = Q(OC + CB) + P(OC - AC)$$

$$= (P + Q).OC, \text{ by equation (1)},$$

$$= \text{moment of the resultant about } O.$$

The case when the point has any other position, as also the case when the forces have opposite parallel directions, are left for the student to prove for himself.

39. If the point O about which the moments are taken lie on the resultant, the moment of the resultant about the point vanishes. In this case the algebraic sum of the moments of the component forces about the given point vanishes, *i.e.* *The moments of two forces about any point on the line of action of their resultant are equal and of opposite sign.*

40. Generalised theorem of moments. *If any number of forces P, Q, R, S ... in one plane acting on a rigid body have a resultant, the algebraic sum of their moments about any point O in their plane is equal to the moment of their resultant.*

Let P_1 be the resultant of P and Q,

P_2 the resultant of P_1 and R,

P_3 the resultant of P_2 and S,

and so on till the final resultant is obtained.

Then the moment of P_1 about O = sum of the moments of P and Q (Art. 38);

Also the moment of P_2 about O = sum of the moments of P_1 and R = sum of the moments of P, Q, and R.

So the moment of P_3 about O

= sum of the moments of P_2 and S

= sum of the moments of P, Q, R, and S,

and so on until all the forces have been taken.

Hence the moment of the final resultant

= algebraic sum of the moments of the component forces.

41. It follows, similarly as in Art. 39, that the algebraic sum of the moments of any number of forces about a point on the line of action of their resultant is zero; so, conversely, if the algebraic sum of the moments of any number of forces about any point in their plane vanishes, then, *either* their resultant is zero (in which case the forces are in equilibrium), *or* the resultant passes through the point about which the moments are taken.

We can thus find points on the line of action of the resultant of a system of forces. For we have only to find a point about which the algebraic sum of the moments of the system of forces vanishes, and then the resultant must pass through that point.

If we have a system of parallel forces the resultant is known both in magnitude and direction when one such point is known.

Ex. *Forces equal to 3P, 7P, and 5P act along the sides AB, BC, and CA of an equilateral triangle ABC; find the magnitude, direction, and line of action of the resultant.*

Let the side of the triangle be a, and let the resultant force meet the side BC in Q. Then the sum of the moments of the forces about Q vanish.

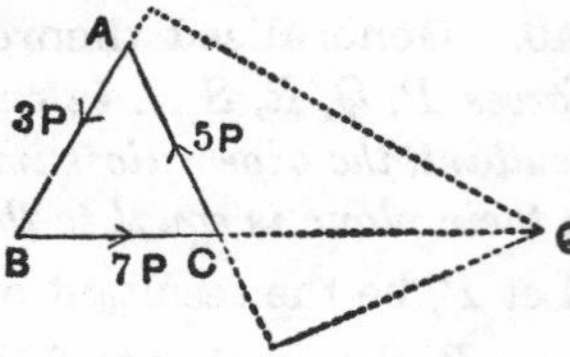

$$\therefore\ 3P\times(QC+a)\sin 60^\circ = 5P\times QC\sin 60^\circ.$$

$$\therefore\ QC=\frac{3a}{2}.$$

The sum of the components of the forces perpendicular to BC

$$=5P\sin 60^\circ - 3P\sin 60^\circ = P\sqrt{3}.$$

Also the sum of the components in the direction BC

$$=7P-5P\cos 60^\circ - 3P\cos 60^\circ = 3P.$$

Hence the resultant is $P\sqrt{12}$ inclined at an angle $\tan^{-1}\frac{\sqrt{3}}{3}$, *i.e.* 30°, to BC and passing through Q where $CQ=\frac{3}{2}BC$.

EXAMPLES

1. Forces proportional to AB, BC, and $2CA$ act along the sides of a triangle ABC taken in order; shew that the resultant is represented in magnitude and direction by CA and that its line of action meets BC at a point X where CX is equal to BC.

2. ABC is a triangle and D, E, and F are the middle points of the sides; forces represented by AD, $\frac{2}{3}BE$, and $\frac{1}{3}CF$ act on a particle at the point where AD and BE meet; shew that the resultant is represented in magnitude and direction by $\frac{1}{2}AC$ and that its line of action divides BC in the ratio $2:1$.

3. Three forces act along the sides of a triangle; shew that, if the sum of two of the forces be equal in magnitude but opposite in sense to the third force, then the resultant of the three forces passes through the centre of the inscribed circle of the triangle.

4. The wire passing round a telegraph pole is horizontal and the two portions attached to the pole are inclined at an angle of 60° to one another. The pole is supported by a wire attached to the middle point of the pole and inclined at 60° to the horizon; shew that the tension of this wire is $4\sqrt{3}$ times that of the telegraph wire.

5. At what height from the base of a pillar must the end of a rope of given length be fixed so that a man standing on the ground and pulling at its other end with a given force may have the greatest tendency to make the pillar overturn?

6. The magnitude of a force is known and also its moments about two given points A and B. Find, by a geometrical construction, its line of action.

7. Find the locus of all points in a plane such that two forces given in magnitude and position shall have equal moments, in the same sense, round any one of these points.

8. AB is a diameter of a circle and BP and BQ are chords at right angles to one another; shew that the moments of forces represented by BP and BQ about A are equal.

9. A man carries a bundle at the end of a stick which is placed over his shoulder; if the distance between his hand and his shoulder be changed how does the pressure on his shoulder change?

10. A cyclist, whose weight is 150 lbs., puts all his weight upon one pedal of his bicycle when the crank is horizontal and the bicycle is prevented from moving forwards. If the length of the crank is 6 inches and the radius of the chain-wheel is 4 inches, find the tension of the chain.

11. A letter-weigher consists of a uniform plate in the form of a right-angled isosceles triangle ABC, of mass 3 ozs., which is suspended by its right angle C from a fixed point to which a plumb-line is also attached. The letters are suspended from the angle A, and their weight read off by observing where the plumb-line intersects a scale engraved along AB, the divisions of which are marked 1 oz., 2 oz., 3 oz., etc. Shew that the distances from A of the divisions of the scale form a harmonic progression.

12. A pack of cards is laid on a table, and each card projects in the direction of the length of the pack beyond the one below it; if each project as far as possible, shew that the distances between the extremities of successive cards will form a harmonical progression.

13. A cylinder, whose length is *b* and the diameter of whose base is *c*, is open at the top and rests on a horizontal plane; a uniform rod rests partly within the cylinder and in contact with it at its upper and lower edges; supposing the weight of the cylinder to be *n* times that of the rod, find the length of the rod when the cylinder is on the point of falling over.

Couples

42. Def. Two equal unlike parallel forces, whose lines of action are not the same, form a Couple.

A couple is by some writers called a Torque; by others the word Torque is used to denote the Moment of the Couple.

Examples of a couple are the forces applied to the handle of a screw-press, or to the key of a clock in winding it up, or by the hands to the handle of a door in opening it.

The Arm of a couple is the perpendicular distance between the lines of action of the two forces which form the couple.

The Moment of a couple is the product of one of the forces forming the couple and the arm of the couple.

Thus the arm of the couple (P, P) is AB and its moment is $P \times AB$.

From any point O in the plane of the couple draw OAB perpendicular to the lines of action of the forces to meet them in A and B respectively.

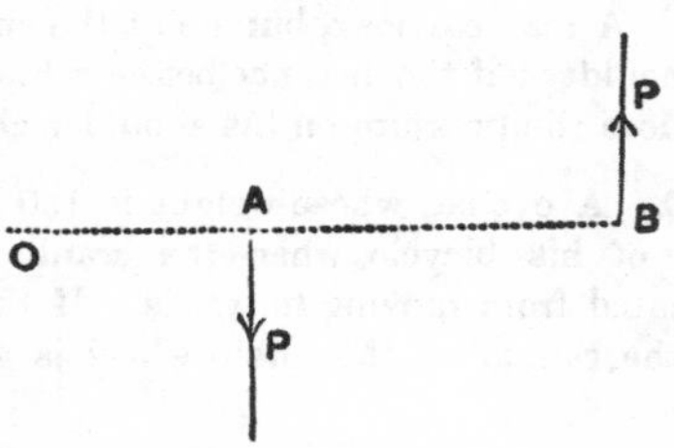

The algebraic sum of the moments of the forces about O

$= P.OB - P.OA = P.AB$

= the moment of the couple,

and is therefore the same whatever be the point O about which the moments are taken.

43. *Two couples, acting in one plane upon a rigid body, whose moments are equal and opposite, balance one another.*

Let one couple consist of two forces (P, P), acting at the ends of an arm p, and let the other couple consist of two forces (Q, Q), acting at the ends of an arm q.

Case I. Let one of the forces P meet one of the forces Q in a point O, and let the other two forces meet in O'. From O' draw perpendiculars, $O'M$ and $O'N$, upon the forces which do not pass through O', so that the lengths of these perpendiculars are p and q respectively.

Since the moments of the couples are equal in magnitude, we have $P.p = Q.q$.

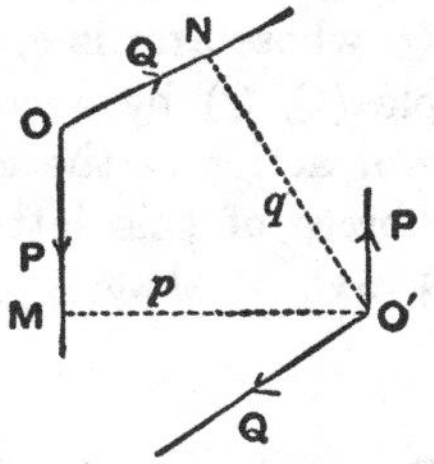

Hence, (Art. 41), O' is on the line of action of the resultant of P and Q acting at O, so that OO' is the direction of this resultant.

Similarly, the resultant of P and Q at O' is in the direction $O'O$.

Also these resultants are equal in magnitude; for the forces at O are respectively equal to, and act at the same angle as, the forces at O'.

Hence these two resultants destroy one another, and therefore the four forces composing the two couples are in equilibrium.

Case II. Let the forces composing the couples be all parallel, and let any straight line perpendicular to their directions meet them in the points A, B, C, and D, as in the figure, so that, since the moments are equal, we have

$$P.AB = Q.CD \text{..................(i).}$$

Let L be the point of application of the resultant of Q at C and P at B, so that

$$P.BL = Q.CL \text{...(ii)}$$

(Art. 31).

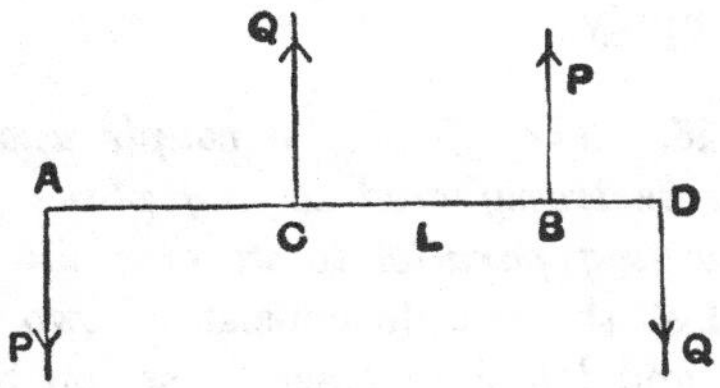

By subtracting (ii) from (i), we have $P.AL = Q.LD$, so that L is the point of application of the resultant of P at A, and Q at D.

But the magnitude of each of these resultants is $(P + Q)$, and they have opposite directions; hence they are in equilibrium and therefore the four forces composing the two couples balance.

44. Since two couples in the same plane, of equal but opposite moment, balance, it follows, by reversing the directions of the forces composing one of the couples, that

Any two couples of equal moment in the same plane are equivalent.

45. *Any number of couples in the same plane acting on a rigid body are equivalent to a single couple, whose moment is equal to the algebraic sum of the moments of the couples.*

For let the couples consist of forces (P, P) whose arm is p, (Q, Q) whose arm is q, (R, R) whose arm is r, etc. Replace the couple (Q, Q) by a couple whose components have the same lines of action as the forces (P, P). The magnitude of each of the forces of this latter couple will be X, where $X.p = Q.q$, (Art. 44), so that

$$X = Q\frac{q}{p}.$$

So let the couple (R, R) be replaced by a couple $\left(R\frac{r}{p}, R\frac{r}{p}\right)$, whose forces act in the same lines as the forces (P, P); similarly for the other couples.

Hence all the couples are equivalent to a couple, each of whose forces is $P + Q\frac{q}{p} + R\frac{r}{p} + \ldots$ acting at an arm p.

The moment of this couple is $P.p + Q.q + R.r + \ldots$.

Hence the original couples are equivalent to a single couple, whose moment is equal to the sum of their moments.

If all the component couples have not the same sign we must give to each moment its proper sign, and the same proof will apply.

46. *The effect of a couple upon a rigid body is unaltered if it be transferred to any plane parallel to its own, the arm remaining parallel to its original position.*

Let the couple consist of two forces (P, P), whose arm is AB, and let their lines of action be AC and BD.

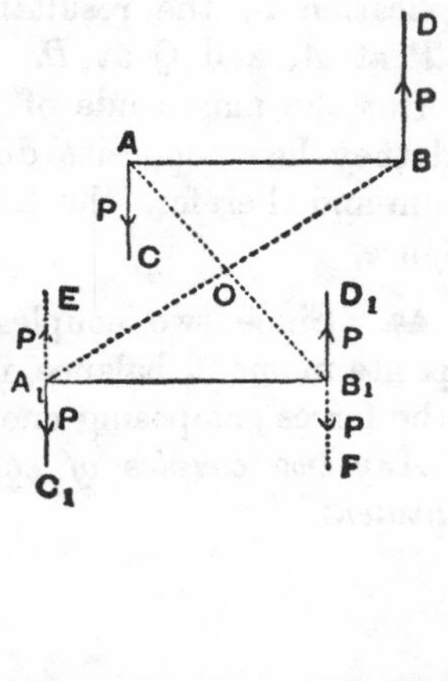

Let A_1B_1 be any line equal and parallel to AB. Draw A_1C_1 and B_1D_1 parallel to AC and BD respectively.

At A_1 introduce two equal and opposite forces, each equal to P, acting in the direction A_1C_1 and the opposite direction A_1E. At B_1 introduce, similarly, two equal and opposite forces, each equal to P, acting in the direction B_1D_1 and the opposite direction B_1F. These forces will have no effect on the equilibrium of the body.

Join AB_1 and A_1B, and let them meet in O; then O is the middle point of both AB_1 and A_1B.

The forces P at B and P acting along A_1E have a resultant $2P$ acting at O parallel to BD.

The forces P at A and P acting along B_1F have a resultant $2P$ acting at O parallel to AC.

These two resultants are equal and opposite, and therefore balance. Hence we have left the two forces (P, P) at A_1 and B_1 acting in the directions A_1C_1 and B_1D_1, *i.e.* parallel to the directions of the forces of the original couple.

Also the plane through A_1C_1 and B_1D_1 is parallel to the plane through AC and BD.

Hence the theorem is proved.

Cor. From this proposition and Art. 44 we conclude that *A couple may be replaced by any other couple acting in a parallel plane, provided that the moments of the two couples are the same.*

47. The *Axis* of a couple is a straight line OP drawn from a point O in the plane of a couple perpendicular to the plane of the couple and proportional to the moment of the couple. The direction in which this axis is to be drawn, *i.e.* its sense, is determined by the following convention: *Suppose a watch to be placed in the plane of a couple; if the couple would produce rotation in the direction of the motion of the hands of the watch, the axis is drawn upwards through the face of the watch; whilst if the couple would produce rotation in the opposite direction the axis is drawn through the back of the watch.*

In the case of the figure of Art. 34 an axis drawn along the positive direction of Ox would represent a couple in the plane of yz which would turn the body from Oy to Oz; whilst an axis drawn in the direction xO produced would represent a couple in the plane yz tending to turn the body from Oz towards Oy.

The directions of rotation of positive couples about the axes Ox, Oy, Oz thus follow the cyclical order 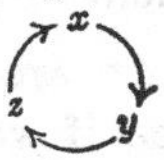

From Art. 44 it follows that the effect of a couple is known when its moment and its plane are known, *i.e.* when its moment and the direction of the normal to its plane are known, *i.e.* when its axis is given in magnitude and direction.

48. *Resultant couple of two couples whose planes are not parallel.*

Let the planes of the two couples meet in a line AB. If the arm of each couple is not AB, replace each by a couple in its own plane of suitable moment whose arm is AB, as in Art. 45.

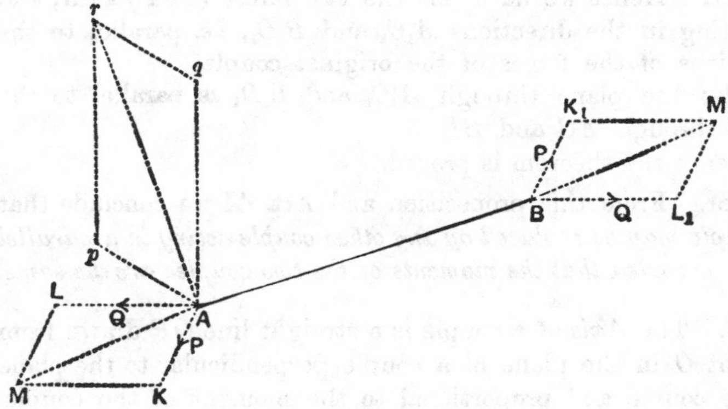

With this arm let the forces of the first couple be AK and BK_1, each equal to P, and those of the second couple AL and BL_1, each equal to Q. Complete the parallelograms $AKML$, $BK_1M_1L_1$. Then clearly the forces P, Q at A compound into a force represented by AM; and the forces P and Q at B into a force BM_1 which is equal, parallel, and in an opposite direction to AM.

Hence the two couples compound into a couple.

Draw Ap, Aq perpendicular to the planes $KABK_1$, $LABL_1$ to represent the axes of the couples, so that

$$\frac{Ap}{Aq}=\frac{\text{moment of the couple }(P, P)}{\text{moment of the couple }(Q, Q)}=\frac{P\,.\,AB}{Q\,.\,AB}=\frac{AK}{AL}.$$

Hence Ap, Aq are perpendicular and proportional to AK and AL. Hence if we complete the parallelogram $Aqrp$, Ar is perpendicular and proportional to AM, so that

$$\frac{Ap}{\text{moment of the couple } (P, P)} = \frac{Aq}{\text{moment of the couple } (Q, Q)}$$
$$= \frac{Ar}{\text{moment of the couple } (AM, BM_1)}.$$

Hence Ar is the axis of the resultant couple.

Therefore two given couples compound into a couple whose axis is obtained by compounding the axes of the given couples according to the parallelogram law.

49. The rule for the composition of couples is thus similar to that for the composition of forces, so that all theorems relating to the composition or resolution of forces apply to the composition or resolution of couples.

Thus (Fig., Art. 26) if we have three component couples about the axes of x, y and z whose moments L, M, N are represented by OA, OB, OC, they compound into a couple whose moment is $G\,(=\sqrt{L^2+M^2+N^2})$ about a line whose direction cosines are $\left(\frac{L}{G}, \frac{M}{G}, \frac{N}{G}\right)$.

Conversely, a couple G about a line OD whose direction cosines are (l, m, n) is equivalent to three couples about the axes whose moments are lG, mG, and nG.

50. We can now compound into one couple any system of couples acting in any planes whatever on a rigid body. For take any point O as origin and any three rectangular axes Ox, Oy, Oz. Any one of the given couples, if its plane does not pass through O, can, by Art. 46, be replaced by an equivalent couple in a parallel plane through O, and its axis may be taken to be a straight line through O perpendicular to this plane. Resolve this axis, by the parallelogram law, into axes along the axes of coordinates, and let the component couples be (L_1, M_1, N_1). Similarly for all the other of the given couples. We thus have a component couple

$$L = L_1 + L_2 + \ldots = \Sigma\,(L_1) \text{ about } Ox,$$

$$M = M_1 + M_2 + \ldots = \Sigma\,(M_1) \text{ about } Oy,$$

and $$N = N_1 + N_2 + \ldots = \Sigma\,(N_1) \text{ about } Oz.$$

These compound into a couple of moment

$$G\left(=\sqrt{L^2+M^2+N^2}\right)$$

about an axis whose direction cosines are $\frac{L}{G}, \frac{M}{G}, \frac{N}{G}$.

51. *A single force and a couple acting in the same plane upon a rigid body cannot produce equilibrium, but are equivalent to the single force acting in a direction parallel to its original direction.*

Let the couple consist of two forces, each equal to P, their lines of action being OB and O_1C respectively.

Let the single force be Q.

If Q be not parallel to the forces of the couple, let it be produced to meet one of them in O. Then P and Q, acting at O, are equivalent to some force R, acting in some direction OL which lies between OA and OB.

Let OL be produced (backwards if necessary) to meet the other force of the couple in O_1, and let the point of application of R be transferred to O_1. Draw O_1A_1 parallel to OA.

Then the force R may be resolved into two forces Q and P, the former acting in the direction O_1A_1, and the latter in the direction opposite to O_1C.

This latter force P is balanced by the second force P of the couple acting in the direction O_1C.

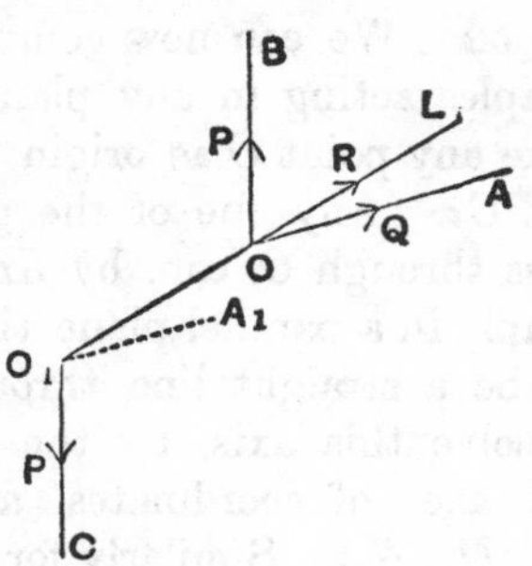

Hence we have left as the resultant of the system a force Q acting in the direction O_1A_1 parallel to its original direction OA.

When Q is parallel to each of the forces P, it is clear by Art. 31 that their resultant is parallel to P and equal to Q.

52. *If three forces, acting upon a rigid body, be represented in magnitude, direction, and line of action by the sides of a triangle taken in order, they are equivalent to a couple whose moment is represented by twice the area of the triangle.*

Let ABC be the triangle and P, Q, and R the forces, so that P, Q, and R are represented by the sides BC, CA, and AB of the triangle.

Through B draw LBM parallel to the side AC, and introduce two equal and opposite forces, equal to Q, at B, acting in the directions BL and BM respectively. By the triangle of forces (Art. 21) the forces P, R, and Q acting in the straight line BL are in equilibrium.

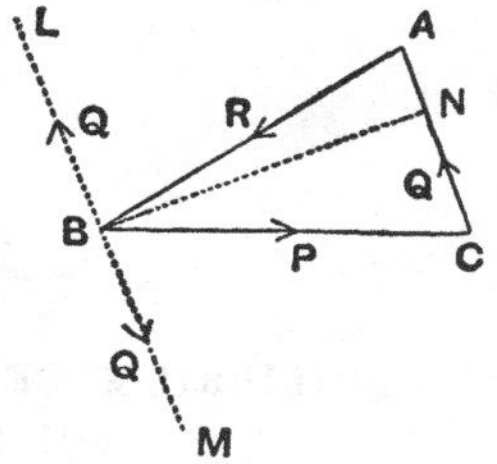

Hence we are left with the two forces, each equal to Q, acting in the directions CA and BM respectively.

These form a couple whose moment is $Q \times BN$, where BN is drawn perpendicular to CA. Also $Q \times BN = CA \times BN =$ twice the area of the triangle ABC.

Cor. In a similar manner it may be shewn that if a system of forces acting on one plane on a rigid body be represented in magnitude, direction, and line of action by the sides of the polygon, they are equivalent to a couple whose moment is represented by twice the area of the polygon.

53. *A couple and a force which does not lie in its plane cannot be in equilibrium.*

For let the force R meet the plane of the couple in O and replace the couple if necessary, by Art. 44, by a couple, one of whose forces P passes through O. Then R and this force P compound into a force acting through O which does not meet the other force P of the couple. Hence they cannot be in equilibrium.

CHAPTER III

EQUILIBRIUM OF A RIGID BODY ACTED ON BY FORCES IN ONE PLANE

54. IN the present chapter we shall discuss the equilibrium of a rigid body acted upon by forces lying in a plane.

By the help of the following theorem we shall find that, when the forces are only three in number, the conditions of equilibrium reduce to those of a single particle.

If three forces, acting in one plane upon a rigid body, keep it in equilibrium, they must either meet in a point or be parallel.

If the forces are not all parallel, at least two of them, P and Q, must meet in a point O, and their resultant must be a force passing through O. But, since the forces P, Q, and R are in equilibrium, this resultant must balance R. Also two forces cannot balance unless they have the same line of action.

Hence the line of action of R must pass through O.

By the preceding theorem we see that the conditions of equilibrium of three forces, acting in one plane, are easily obtained. For the three forces must meet in a point; and by using Lami's Theorem, (Art. 23), or by resolving the forces in two directions at right angles, (Art. 28), or by a graphic construction, we can obtain the required conditions.

55. Trigonometrical Theorems. There are two trigonometrical theorems which are useful in Statical Problems, *viz.* *If P be any point in the base AB of a triangle ABC, and if CP divides AB into two parts m and n, and the angle C into two parts α and β, and if the angle CPB be θ, then*

$$(m+n)\cot\theta = m\cot\alpha - n\cot\beta \;\;\ldots\ldots\ldots\ldots(1),$$

and

$$(m+n)\cot\theta = n\cot A - m\cot B \;\;\ldots\ldots\ldots\ldots(2).$$

For

$$\frac{m}{n} = \frac{AP}{PB} = \frac{AP}{PC} \cdot \frac{PC}{PB} = \frac{\sin ACP}{\sin PAC} \cdot \frac{\sin PBC}{\sin PCB}$$

$$= \frac{\sin \alpha}{\sin (\theta - \alpha)} \cdot \frac{\sin (\theta + \beta)}{\sin \beta} = \frac{\cot \beta + \cot \theta}{\cot \alpha - \cot \theta}.$$

$$\therefore \ m \cot \alpha - n \cot \beta = (m + n) \cot \theta.$$

Again

$$\frac{m}{n} = \frac{\sin ACP}{\sin PAC} \cdot \frac{\sin PBC}{\sin PCB}$$

$$= \frac{\sin (\theta - A)}{\sin A} \cdot \frac{\sin B}{\sin (\theta + B)}$$

$$= \frac{\cot A - \cot \theta}{\cot B + \cot \theta}.$$

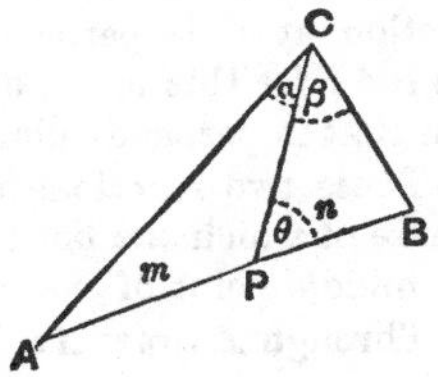

$$\therefore \ (m + n) \cot \theta = n \cot A - m \cot B.$$

56. Ex. 1. *A beam whose centre of gravity divides it into two portions, a and b, is placed inside a smooth sphere ; shew that, if θ be its inclination to the horizon in the position of equilibrium and 2α be the angle subtended by the beam at the centre of the sphere, then*

$$\tan \theta = \frac{b - a}{b + a} \tan \alpha.$$

In this case both the reactions, R and S, at the ends of the rod pass through the centre, O, of the sphere. Hence the centre of gravity, G, of the rod must be vertically below O. Let OG meet the horizontal line through A in N. Draw OD perpendicular to AB.

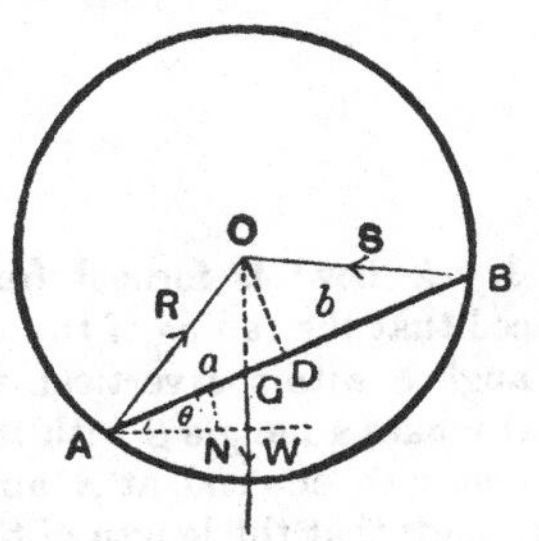

Then $\angle AOD = \angle BOD = \alpha,$

and $\angle DOG = 90° - \angle DGO = \angle DAN = \theta.$

The second relation of Art. 55 then gives

$$(a + b) \cot OGB = b \cot OAB - a \cot OBA,$$

i.e. $(a + b) \tan \theta = (b - a) \tan \alpha.$

Also, by Lami's Theorem, $\dfrac{R}{\sin BOG} = \dfrac{S}{\sin AOG} = \dfrac{W}{\sin AOB}.$

$$\therefore \ \frac{R}{\sin (\alpha + \theta)} = \frac{S}{\sin (\alpha - \theta)} = \frac{W}{\sin 2\alpha}, \ \text{giving the reactions.}$$

Ex. 2. *A heavy uniform rod, of length 2a, rests partly within and partly without a fixed smooth hemispherical bowl, of radius r ; the rim of the bowl is horizontal, and one point of the rod is in contact with the rim ; if θ be the inclination of the rod to the horizon, shew that*

$$2r \cos 2\theta = a \cos \theta.$$

Let the figure represent that vertical section of the hemisphere which passes through the rod. Let AB be the rod, G its centre of gravity, and C the point where the rod meets the edge of the bowl.

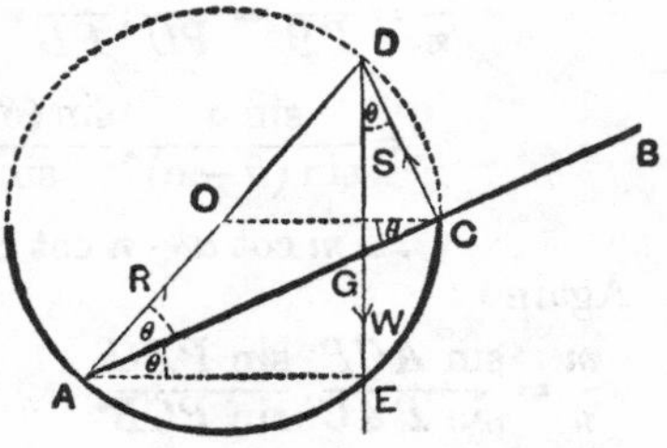

The reaction at A is along the line to the centre, O, of the bowl; for AO is the only line through A which is perpendicular to the surface of the bowl at A. Also the reaction at C is perpendicular to the rod; for this is the only direction that is perpendicular to both the rod and the rim of the bowl.

These two reactions meet in a point D which lies on the geometrical sphere of which the bowl is a portion. Hence the vertical line through G, the middle point of the rod, must pass through D.

Through A draw AE horizontal to meet DG in E and join OC.

Then $$\angle OAC = \angle OCA = \angle CAE = \theta.$$

$$\therefore a\cos\theta = AE = AD\cos 2\theta = 2r\cos 2\theta.$$

Also, by Lami's Theorem, if R and S be the reactions at A and C, we have

$$\frac{R}{\sin\theta} = \frac{S}{\sin ADG} = \frac{W}{\sin ADC},$$

i.e. $$\frac{R}{\sin\theta} = \frac{S}{\cos 2\theta} = \frac{W}{\cos\theta}.$$

EXAMPLES

1. A bowl is formed from a hollow sphere, of radius a, and is so placed that the radius of the sphere drawn to each point in the rim makes an angle α with the vertical, whilst the radius drawn to a point A of the bowl makes an angle β with the vertical; if a smooth uniform rod remain at rest with one end at A and a point of its length in contact with the rim, shew that the length of the rod is

$$4a\sin\beta\sec\frac{\alpha-\beta}{2}.$$

2. A cylinder, of radius r, whose axis is fixed horizontally, touches a vertical wall along a generating line. A flat beam of uniform material, of length $2l$ and weight W, rests with its extremities in contact with the wall and the cylinder, making an angle of 45° with the vertical. Shew that, in the absence of friction, $\frac{l}{r} = \frac{\sqrt{5}-1}{\sqrt{10}}$, that the pressure on the wall is $\frac{1}{2}W$, and that the reaction of the cylinder is $\frac{1}{2}\sqrt{5}\,W$.

3. A hemispherical bowl, of radius r, rests on a smooth horizontal table and partly inside it rests a rod, of length $2l$ and of weight equal to that of the bowl. Shew that the position of equilibrium is given by the equation

$$l \sin(\alpha+\beta) = r \sin \alpha = -2r \cos(\alpha+2\beta),$$

where α is the inclination of the base of the hemisphere to the horizon, and 2β is the angle subtended at the centre by the part of the rod within the bowl.

4. A smooth rod, of length $2a$, has one end resting on a plane of inclination α to the horizon, and is supported by a horizontal rail which is parallel to the plane and at a distance c from it. Shew that the inclination θ of the rod to the inclined plane is given by the equation

$$c \sin \alpha = a \sin^2 \theta \cos(\theta - \alpha).$$

5. A solid cone, of height h and semi-vertical angle α, is placed with its base against a smooth vertical wall and is supported by a string attached to its vertex and to a point in the wall; shew that the greatest possible length of the string is $h\sqrt{1+\frac{16}{9}\tan^2\alpha}$.

6. The altitude of a cone is h and the radius of its base is r; a string is fastened to the vertex and to a point on the circumference of the circular base, and is then put over a smooth peg; shew that, if the cone rest with its axis horizontal, the length of the string must be $\sqrt{h^2+4r^2}$.

7. Two equal circular discs of radius r, with smooth edges, are placed on their flat sides in the corner between two smooth vertical planes inclined at an angle 2α, and touch each other in the line bisecting the angle. Shew that the radius of the smallest disc that can be pressed between them, without causing them to separate, is $r(\sec\alpha - 1)$.

8. A picture frame, rectangular in shape, rests against a smooth vertical wall, from two points in which it is suspended by parallel strings attached to two points in the upper edge of the back of the frame, the length of each string being equal to the height of the frame. Shew that, if the centre of gravity of the frame coincide with its centre of figure, the picture will hang against the wall at an angle $\tan^{-1}\frac{b}{3a}$ to the vertical, where a is the height and b the thickness of the picture.

9. It is required to hang a picture on a vertical wall so that it may rest at a given inclination, α, to the wall and be supported by a cord attached to a point in the wall at a given height h above the lowest edge of the picture; determine, by a geometrical construction, the point on the back of the picture to which the cord is to be attached and find the length of the cord that will be required.

57. *Any system of forces, acting in one plane upon a rigid body, can be reduced to either a single force or a single couple.*

By the parallelogram of forces any two forces, whose

directions meet, can be compounded into one force; also, by Art. 31, two parallel forces can be compounded into one force provided they are not equal and unlike.

First compound together all the parallel forces, or sets of parallel forces, of the given system.

Of the resulting system take any two forces, not forming a couple, and find their resultant R_1; next find the resultant R_2 of R_1 and a suitable third force of the system; then determine the resultant of R_2 and a suitable fourth force of the system; and so on until all the forces have been exhausted.

Finally, we must either arrive at a single force, or we shall have two equal parallel unlike forces forming a couple.

58. *If a system of forces act in one plane upon a rigid body, and if the algebraic sum of their moments about each of three points in the plane (not lying in the same straight line) vanish separately, the system of forces is in equilibrium.*

For any such system of forces, by the last article, reduces to either a single force or a single couple.

In our case they cannot reduce to a single couple; for, if they did, the sum of their moments about any point in their plane would, by Art. 42, be equal to a constant which is not zero, and this is contrary to our hypothesis.

Hence the system of forces cannot reduce to a single couple. The system must therefore either be in equilibrium or reduce to a single force F.

Let the three points about which the moments are taken be A, B, C. Since the algebraic sum of the moments of a system of forces is equal to that of their resultant (Art. 40), therefore the moment of F about the point A must be zero. Hence F is either zero, or passes through A.

Similarly, since the moment of F about B vanishes, F must be either zero or must pass through B, *i.e.* F is either zero or acts in the line AB.

Finally, since the moment about C vanishes, F must be either zero or pass through C.

But (since the points A, B, C are not in the same straight line) the force cannot act along AB and also pass through C. Hence the only admissible case is that F should be zero, *i.e.* that the forces should be in equilibrium.

The system will also be in equilibrium if (1) the sum of the moments about each of two points, A and B, separately vanish, and if (2) the sum of the forces resolved along AB be zero. For, if (1) holds, the resultant, by the foregoing article, is either zero or acts along AB. Also, if (2) be true, there is no resultant in the direction AB; hence the resultant force is zero. Also, as in the foregoing article, there is no resultant couple. Hence the system is in equilibrium.

59 *Any system of forces, acting in one plane upon a rigid body, is equivalent to a force acting at an arbitrary point of the body together with a couple.*

Let P be any force of the system acting at a point A of the body, and let O be any arbitrary point. At O introduce two equal and opposite forces, the magnitude of each being P, and let their line of action be parallel to that of P. These do not alter the state of equilibrium of the body.

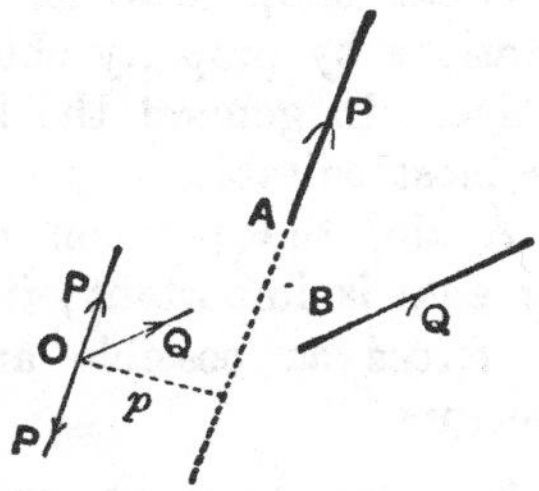

The force P at A and the opposite parallel force P at O form a couple of moment $P.p$, where p is the perpendicular from O upon the line of action of the original force P. Hence the force P at A is equivalent to a parallel force P at O and a couple of moment $P.p$.

So the force Q at B is equivalent to a parallel force Q at O and to a couple of moment $Q.q$, where q is the perpendicular from O on the line of action of Q. The same holds for each of the system of forces.

Hence the original system of forces is equivalent to forces $P, Q, R, \ldots$ acting at O, parallel to their original directions, and a number of couples; these are equivalent to a single resultant force at O, and a single resultant couple of moment

$$P.p + Q.q + \ldots.$$

60. Let the forces of the previous article be in equilibrium. By Art. 51 a force and a couple cannot balance unless each is zero.

Hence the resultant of $P, Q, R, \ldots$ at O must be zero, and therefore, by Art. 28, *the sum of their resolved parts in two directions must separately vanish.*

Also the moment $Pp + Qq + \ldots$ must be zero, *i.e. the algebraic sum of the moments of the forces about an arbitrary point O must vanish also.*

The first condition ensures that there shall be no motion of the body as a whole; the second ensures that there shall be no motion of rotation about any point.

The above three statical relations, together with the geometrical relations holding between the component portions of a system, will, in general, be sufficient to determine the equilibrium of any system acted on by forces which are in one plane.

Great simplifications can often be introduced into the equations by properly choosing the directions along which we resolve. In general, the horizontal and vertical directions are the most suitable.

Again, the position of the point about which we take moments is important; it should be chosen so that as few of the forces as possible are introduced into the equation of moments.

61. We have shewn that the conditions given in the previous article are sufficient for the equilibrium of the system of forces; they are also necessary.

Suppose we knew only that the first two conditions were satisfied. The system of forces might then reduce to a single couple; for the forces of this couple, being equal and opposite, are such that their components in any direction would vanish. Hence, resolving in any third direction would give us no additional condition. In this case the forces would not be in equilibrium unless the third condition were satisfied.

Suppose, again, that we knew only that the components of the system along one given line vanished and that the moments about a given point vanished also; in this case the forces might reduce to a single force through the given point perpendicular to the given line; hence we see that it is necessary to have the sum of the components parallel to another line zero also.

62. In solving any statical problem the student should proceed as follows:

(1) Draw the figure according to the conditions given.

(2) Mark all the forces acting on the body or bodies,

taking care to assume an unknown reaction (to be determined) wherever one body presses against another, and to mark a tension along any supporting string, and to assume a reaction wherever the body is hinged to any other body or fixed point.

(3) For each body, or system of bodies, involved in the problem, equate to zero the forces acting on it resolved along two convenient perpendicular directions (generally horizontal and vertical).

(4) Also equate to zero the moments of the forces about any convenient point.

(5) Write down any geometrical relations connecting the lengths or angles involved in the figure.

63. *Resultant force and couple corresponding to any base point O of a system of coplanar forces.*

Through O take any pair of rectangular axes Ox and Oy.

At P_1, the point (x_1, y_1), let there act a force whose components parallel to the axes are X_1 and Y_1.

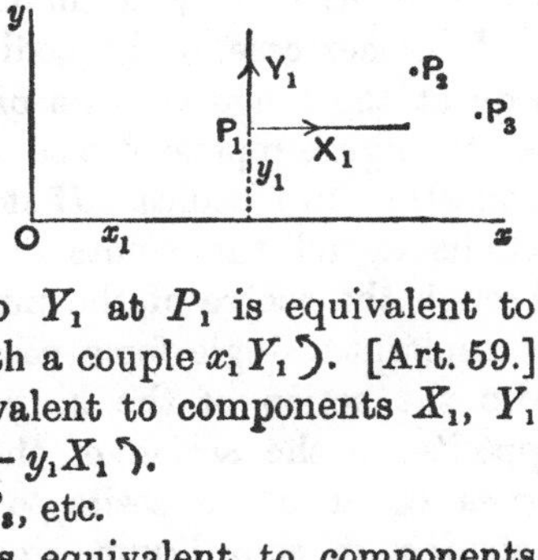

Then X_1 at P_1 is equivalent to a parallel force X_1 at O together with a couple y_1X_1↻. So Y_1 at P_1 is equivalent to a parallel force Y_1 at O together with a couple x_1Y_1↺. [Art. 59.]

Hence the force at P_1 is equivalent to components X_1, Y_1 along Ox, Oy and a couple $x_1Y_1 - y_1X_1$↺.

So for the other forces at P_2, P_3, etc.

Hence the system of forces is equivalent to components X, Y along Ox, Oy, and a couple G↺ about O, such that

$$X = X_1 + X_2 + X_3 + \ldots = \Sigma X_1,$$

$$Y = Y_1 + Y_2 + Y_3 + \ldots = \Sigma Y_1,$$

and $G = (x_1Y_1 - y_1X_1) + (x_2Y_2 - y_2X_2) + \ldots = \Sigma(x_1Y_1 - y_1X_1)$.

X and Y compound into a single force R acting at O.

64. *Equation to the resultant of a system of forces in one plane.*

As in the last article the system can be reduced to components X and Y along any two rectangular axes Ox and Oy, and a couple G about O↺.

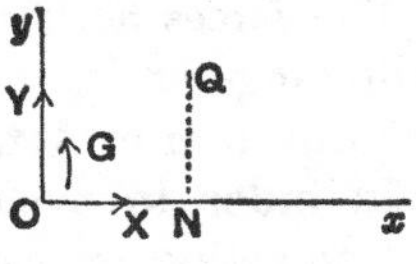

Let Q be any point (h, k) which lies on the resultant of the given system. By Art. 40 the moment of the system about it is equal to the moment of the resultant about it and is therefore zero.

Now the moment of the system about Q

$$= G + X . NQ - Y . ON = G - hY + kX,$$

so that $$G - hY + kX = 0.$$

Hence the locus of (h, k), *i.e.* the resultant, is the straight line

$$G - xY + yX = 0.$$

65. Bodies connected by smooth hinges. When two bodies are hinged together, it usually happens that, either a rounded end of one body fits loosely into a prepared hollow in the other body, as in the case of a ball-and-socket joint; or that a round pin, or other separate fastening, passes through a hole in each body, as in the case of the hinge of a door.

In either case, if the bodies be smooth, the action on each body at the hinge consists of a single force. Let the figure represent a section of the joint connecting two bodies. If it be smooth the actions at all the points of the joint pass through the centre of the pin and thus have as resultant a single force passing through O.

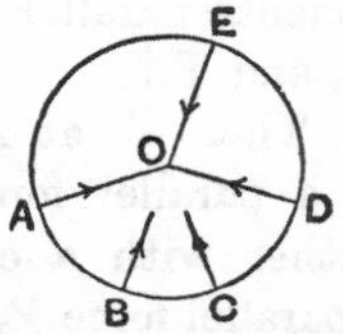

Also the action of the hinge on the one body is equal and opposite to the action of the hinge on the other body; for forces, equal and opposite to these actions, keep the pin, or fastening, in equilibrium, since its weight is negligible.

In solving questions concerning smooth hinges, the direction and magnitude of the action at the hinge are usually both unknown. Hence it is generally most convenient to assume the action of a smooth hinge on one body to consist of two unknown components at right angles to one another; the action of the hinge on the other body will then consist of components equal and opposite to these.

The forces acting on each body, together with the actions of the hinge on it, are in equilibrium, and the general conditions of equilibrium of Art. 60 will now apply.

In order to avoid mistakes as to the components of the reaction acting on each body, it is convenient, as in the second figure of the following example, not to produce the rods to meet but to leave a space between them.

66. Ex. *Three equal uniform rods, each of weight W, are smoothly jointed so as to form an equilateral triangle. If the system be supported at the middle point of one of the rods, shew that the action at the lowest angle is* $\frac{\sqrt{3}}{6}W$, *and that at each of the others is* $W\sqrt{\frac{13}{12}}$.

Let ABC be the triangle formed by the rods, and D the middle point of the side AB at which the system is supported.

Let the action of the hinge at A on the rod AB consist of two components, respectively equal to Y and X, acting in vertical and horizontal directions; hence the action of the hinge on AC consists of components equal and opposite to these. Since the whole system is symmetrical about the vertical line through D, the action at B will consist of components, also equal to Y and X, as in the figure.

Let the action of the hinge C on CB consist of Y_1 vertically upwards, and X_1 horizontally to the right, so that the action of the same hinge on CA consists of two components opposite to these, as in the figure.

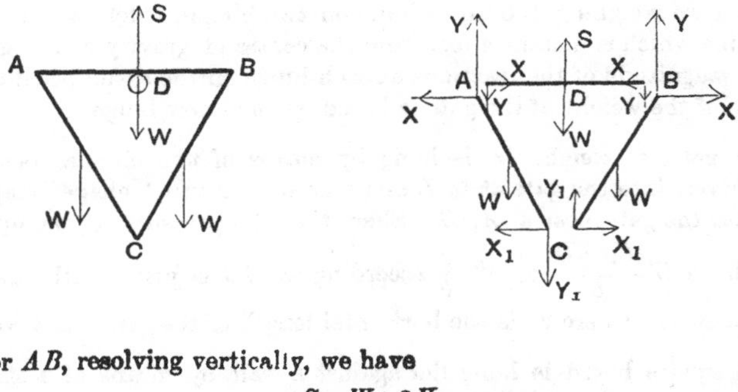

For AB, resolving vertically, we have

$$S = W + 2Y \quad \text{.................................(1)},$$

where S is the vertical reaction of the peg at D.

For CB, resolving horizontally and vertically, and taking moments about C, we have

$$X + X_1 = 0 \quad \text{.................................(2)},$$

$$W = Y + Y_1 \quad \text{.................................(3)},$$

and

$$W . a\cos 60° + X . 2a\sin 60° = Y . 2a\cos 60° \quad \text{...........(4)}.$$

For CA, by resolving vertically, we have

$$W = Y - Y_1 \quad \text{.................................(5)}.$$

Solving these equations, we have

$$X_1 = -\frac{\sqrt{3}}{6}W, \quad Y_1 = 0, \quad Y = W, \quad X = \frac{\sqrt{3}}{6}W \text{ and } S = 3W.$$

Hence the action of the hinge at B consists of a force $\sqrt{X^2+Y^2}$ $\left(i.e.\ W\sqrt{\frac{13}{12}}\right)$, acting at an angle $\tan^{-1}\frac{Y}{X}$ (*i.e.* $\tan^{-1}2\sqrt{3}$), to the horizon; also the action of the hinge at C consists of a horizontal force equal to $\frac{\sqrt{3}}{6}W$.

A priori reasoning would have shewn us that the action of the hinge at C must be horizontal; for the whole system is symmetrical about the line CD, and, unless the component Y_1 vanished, the reaction at C would not satisfy the condition of symmetry.

EXAMPLES

1. A pair of compasses, each of whose legs is a uniform bar of weight W, is supported, hinge downwards, by two smooth pegs placed at the middle points of the legs in the same horizontal line, the legs being kept apart at an angle $2a$ with one another by a weightless rod joining their extremities; shew that the thrust in this rod and the action at the hinge are each $\frac{1}{2}W\cot a$.

2. A gate weighing 100 lbs. is hung on two hinges, 3 feet apart, in a vertical line which is distant 4 feet from the centre of gravity of the gate. Find the magnitude of the reactions at each hinge on the assumption that the whole of the weight of the gate is borne by the lower hinge.

3. A gate, of weight W, is hung by means of two circular headed staples driven into the gate at C, D and placed over two L shaped staples driven into the gate post at A, B. Shew that the pressure at the upper hinge will be $W\frac{\sqrt{a^2+b^2}}{b}$ or $W.\frac{a}{b}$ according as CD is just a little less or greater than AB, where $2a$ is the horizontal length of the gate and $b=CD$

4. A square board is hung flat against a wall, by means of a string fastened to the two extremities of the upper edge and hung round a perfectly smooth rail; when the length of the string is less than the diagonal of the board, shew that there are three positions of equilibrium.

5. A square, of side $2a$, is placed with its plane vertical between two smooth pegs, which are in the same horizontal line and at a distance c; shew that it will be in equilibrium when the inclination of one of its edges to the horizon is either 45° or $\frac{1}{2}\sin^{-1}\frac{a^2-c^2}{c^2}$.

6. An isosceles triangular lamina rests, with its plane vertical and vertex downwards, between two smooth pegs in the same horizontal line; shew that there will be equilibrium if the base make an angle $\sin^{-1}(\cos^2 a)$ with the vertical, $2a$ being the vertical angle of the lamina and the length of the base being three times the distance between the pegs.

7. A prism, whose cross section is an equilateral triangle, rests with two edges horizontal on smooth planes inclined at angles α and β to the horizon. If θ be the angle that the plane through these edges makes with the vertical, shew that

$$\tan\theta = \frac{2\sqrt{3}\sin\alpha\sin\beta + \sin(\alpha+\beta)}{\sqrt{3}\sin(\alpha\sim\beta)}.$$

8. A triangle, formed of three rods, is fixed in a horizontal position and a homogeneous sphere rests on it; shew that the reaction on each rod is proportional to its length.

9. An elliptic lamina is acted on at the extremities of pairs of conjugate diameters by forces in its own plane tending outwards and normal to its edge; shew that there will be equilibrium if the force at the end of each diameter is proportional to the conjugate diameter.

10. A step-ladder in the form of the letter A, with each of its legs inclined at an angle α to the vertical, is placed on a horizontal floor, and is held up by a cord connecting the middle points of its legs, there being no friction anywhere; shew that, when a weight W is placed on one of the steps at a height from the floor equal to $\frac{1}{n}$ of the height of the ladder, the increase in the tension of the cord is $\frac{1}{n} W \tan\alpha$.

11. Three uniform beams AB, BC, and CD, of the same thickness, and of lengths l, $2l$, and l respectively, are connected by smooth hinges at B and C, and rest on a perfectly smooth sphere, whose radius is $2l$, so that the middle point of BC and the extremities, A and D, are in contact with the sphere; shew that the pressure at the middle point of BC is $\frac{91}{100}$ of the weight of the beams.

12. Three uniform rods AB, BC, and CD, whose weights are proportional to their lengths a, b, and c, are jointed at B and C and are in a horizontal position resting on two pegs P and Q; find the actions at the joints B and C, and shew that the distance between the pegs must be

$$\frac{a^2}{2a+b} + \frac{c^2}{2c+b} + b.$$

13. AB and AC are similar uniform rods, of length a, smoothly jointed at A. BD is a weightless bar, of length b, smoothly jointed at B, and fastened at D to a smooth ring sliding on AC. The system is hung on a small smooth pin at A. Shew that the rod AC makes with the vertical an angle

$$\tan^{-1}\frac{b}{a+\sqrt{a^2-b^2}}.$$

14. A square figure $ABCD$ is formed by four equal uniform rods jointed together, and the system is suspended from the joint A, and kept in the form of a square by a string connecting A and C; shew that the tension of the string is half the weight of the four rods, and find the direction and magnitude of the action at either of the joints B or D.

15. Three uniform rods AB, BC, CD, of lengths $2c$, $2b$, $2c$ respectively, rest symmetrically on a smooth parabolic arc whose axis is vertical and vertex upwards; the rods all touch the parabola and are hinged at B and C. If W be the weight of either of the slant rods, shew that the reaction of the parabola on it is $W\dfrac{a^2c}{(a^2+b^2)b}$, where $4a$ is the latus rectum of the parabola.

16. A light wire, in the shape of a quadrant of an ellipse cut off by the principal axes, has two equal weights fixed at its ends and rests on a smooth peg; shew that the eccentric angle of the point of contact with the peg lies between 45° and 60°.

17. Two equal smooth spheres, each of weight W and radius r, are placed inside a hollow cylinder open at both ends which rests on a horizontal plane; if a, $(<2r)$, be the radius of the cylinder, shew that the least weight it can have so as not to be upset is $2W\left(1-\dfrac{r}{a}\right)$.

18. A tipping basin, whose interior surface is spherical, is free to turn round an axis at a distance c below the centre of the sphere and at a distance a above the centre of gravity of the basin, and a heavy ball is laid at the bottom of the basin; shew that it will tip over if the weight of the ball exceed the fraction $\dfrac{a}{c}$ of the weight of the basin.

19. A circular disc, of weight W and radius a, is suspended horizontally by three equal vertical strings, of length b, attached symmetrically to its perimeter. Shew that the magnitude of the horizontal couple required to keep it twisted through an angle θ is

$$Wa^2\frac{\sin\theta}{\sqrt{b^2-4a^2\sin^2\dfrac{\theta}{2}}}.$$

67. Astatic equilibrium. When forces in one plane act at given points of a body, and keep it in equilibrium, it is not in general true that these forces keep the body in equilibrium when they are turned about their points of application about any angle (the same for each). When this is the case the equilibrium is said to be Astatic.

68. *If all the forces in a coplanar system are rotated about their points of application through the same angle in their own plane, their resultant passes through a fixed point in the body.*

Let P_1 be any force of the system acting at a point (x_1, y_1) in a direction inclined at θ_1 to the axis of x, and let X_1, Y_1 be its components parallel to the axes; so for the other forces.

Let X, Y be the components of the system along the axes Ox, Oy and G the moment of the forces about the origin, so that

$$X = \Sigma P_1 \cos\theta_1,$$

$$Y = \Sigma P_1 \sin\theta_1,$$

and $$G = \Sigma P_1 (x_1 \sin\theta_1 - y_1 \cos\theta_1).$$

As in Art. 64, the equation to the resultant of the system is

$$G + yX - xY = 0 \text{ ....(1)}.$$

Let all the forces be turned through an angle α about their points of application so that θ_1 becomes $\theta_1 + \alpha$, and hence

$$P_1 \cos\theta_1 \text{ becomes } P_1 \cos\theta_1 \cos\alpha - P_1 \sin\theta_1 \sin\alpha,$$

$$P_1 \sin\theta_1 \text{ becomes } P_1 \sin\theta_1 \cos\alpha + P_1 \cos\theta_1 \sin\alpha,$$

and $$P_1 (x_1 \sin\theta_1 - y_1 \cos\theta_1)$$

becomes $$P_1 \begin{Bmatrix} x_1 (\sin\theta_1 \cos\alpha + \cos\theta_1 \sin\alpha) \\ - y_1 (\cos\theta_1 \cos\alpha - \sin\theta_1 \sin\alpha) \end{Bmatrix},$$

i.e. $\cos\alpha \,.\, P_1 (x_1 \sin\theta_1 - y_1 \cos\theta_1) + \sin\alpha \,.\, P_1 (x_1 \cos\theta_1 + y_1 \sin\theta_1)$.

Hence $$X \text{ becomes } X\cos\alpha - Y\sin\alpha,$$

$$Y \text{ becomes } X\sin\alpha + Y\cos\alpha,$$

and $$G \text{ becomes } G\cos\alpha + V\sin\alpha,$$

where $V \equiv \Sigma (X_1 x_1 + Y_1 y_1)$ and is called the Virial of the system.

The equation to the new resultant of the system then becomes

$$G\cos\alpha + V\sin\alpha + y(X\cos\alpha - Y\sin\alpha) - x(X\sin\alpha + Y\cos\alpha) = 0,$$

i.e. $$\cos\alpha [G + yX - xY] + \sin\alpha [V - yY - xX] = 0 \text{......(2)}.$$

Whatever be the value of α, (2) always passes through the point whose coordinates are given by

$$G + yX - xY = 0,$$

$$V - yY - xX = 0,$$

i.e. through the point whose coordinates are

$$\frac{GY + VX}{X^2 + Y^2}, \quad \frac{VY - GX}{X^2 + Y^2}.$$

This point is called the Astatic Centre.

69. Suppose the forces to be in equilibrium before their displacement so that

$$X = 0, \quad Y = 0, \text{ and } G = 0.$$

Then they are in equilibrium after displacement if

$$X \cos \alpha - Y \sin \alpha = 0, \quad X \sin \alpha + Y \cos \alpha = 0,$$

and

$$G \cos \alpha + V \sin \alpha = 0.$$

They are thus in equilibrium after being rotated through any angle if $V = 0$, *i.e.* if $\Sigma (X_1x_1 + Y_1y_1) = 0$, which is thus the condition that the equilibrium is astatic.

It also follows that if the original forces are in equilibrium and each is turned through the same angle α, they are then equivalent to a couple whose moment is $V \sin \alpha$.

70. The quantity X_1x_1 is equal, as we shall see in Chapter V, to the work that would be done in a displacement in which the point of application of X_1 is moved from the origin to the point (x_1, y_1), and $X_1x_1 + Y_1y_1$ is the work done by the force P_1 as its point of application is moved from the origin to (x_1, y_1).

Hence the virial of the system is the work that would be done by all the forces of the system as their points of application are moved from the origin to their actual positions.

71. It may be easily seen geometrically that if each of a system of forces is turned through the same angle α, then their resultant always passes through a definite point whatever α may be.

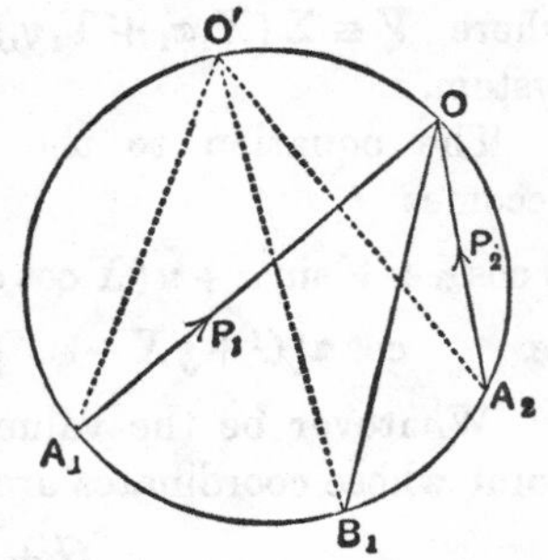

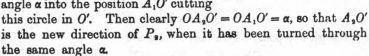

For let two of the forces, P_1 and P_2 acting at A_1 and A_2, meet in O and let OB_1 be the direction of their resultant meeting in B_1 the circle through O, A_1, and A_2.

Let P_1 be turned through any angle α into the position A_1O' cutting this circle in O'. Then clearly $OA_2O' = OA_1O' = \alpha$, so that A_2O' is the new direction of P_2, when it has been turned through the same angle α.

Also $\angle B_1O'A_2 = \angle B_1OA_2$, so that $O'B_1$ is the new position of the resultant of the new forces P_1 and P_2 acting along A_1O' and A_2O'.

Hence the point B_1 is a point through which the resultant of P_1 and P_2 acts, whatever be the angle through which the forces P_1 and P_2 are turned about A_1 and A_2. Also the angle $OB_1O' = \angle OA_1O' = \alpha$, so that the resultant has turned through the same angle as the component forces. Similarly, taking another of the given system P_3 acting at A_3, we find a point B_2 through which always acts the resultant of P_3 and the resultant of P_1 and P_2 acting at B_1, *i.e.* B_2 is the point through which always acts the resultant of P_1, P_2, and P_3.

By continuing in this manner until all the forces have been exhausted, we find a point through which the resultant of all the forces act whatever be the angle through which they have been turned.

If the forces P_1 and P_2 are parallel then O is at infinity and the circle becomes a straight line through A_1 and A_2. B_1 being the point where the resultant of the parallel forces P_1 and P_2 meets A_1A_2 is, by Art. 31, then such that

$$P_1 . A_1B_1 = P_2 . B_1A_2.$$

Ex. Shew that three coplanar forces P_1, P_2, P_3 acting at point A_1, A_2, A_3 are in astatic equilibrium if they meet in a point O situated on the circumcircle of A_1, A_2, A_3, and if

$$P_1 : P_2 : P_3 :: A_2A_3 : A_3A_1 : A_1A_2.$$

CHAPTER IV

FRICTION

72. In Art. 14 we defined smooth bodies to be bodies such that, if they be in contact, the only action between them is perpendicular to both surfaces at the point of contact. With smooth bodies, therefore, there is no force tending to prevent one body sliding over the other. If a perfectly smooth body be placed on a perfectly smooth inclined plane, there is no action between the plane and the body to prevent the latter from sliding down the plane, and hence the body will not remain at rest on the plane unless some external force be applied to it.

Practically, however, there are no bodies which are perfectly smooth; there is always *some* force between two bodies in contact to prevent one sliding upon the other. Such a force is called the force of friction.

Friction. Def. *If two bodies be in contact with one another, the property of the two bodies, by virtue of which a force is exerted between them at their point of contact to prevent one body sliding on the other, is called friction; also the force exerted is called the force of friction.*

73. Friction is a self-adjusting force; no more friction is called into play than is sufficient to prevent motion. Let a heavy slab of iron with a plane base be placed on a horizontal table. If we attach a piece of string to some point of the body, and pull in a horizontal direction passing through the centre of gravity of the slab, a resistance is felt which prevents our moving the body; this resistance is exactly equal to the force which we exert on the body. If we now stop pulling, the force of friction also ceases to act; for, if the force of friction did not cease to act, the body would move.

The amount of friction which can be exerted between two bodies is not, however, unlimited. If we continually increase the force which we exert on the slab, we find that finally the friction is not sufficient to overcome this force, and the body moves.

74. Friction plays an important part in the mechanical problems of ordinary life. If there were no friction between our boots and the ground, we should not be able to walk; if there were no friction between a ladder and the ground, the ladder would not rest, unless held, in any position inclined to the vertical; without friction nails and screws would not remain in wood, nor would a locomotive engine be able to draw a train.

75. The laws of statical friction are as follows:

LAW I. *When two bodies are in contact, the direction of the friction on one of them at its point of contact is opposite to the direction in which this point of contact would commence to move.*

LAW II. *The magnitude of the friction is, when there is equilibrium, just sufficient to prevent the body from moving.*

The above laws hold good, in general; but the amount of friction that can be exerted is limited, and equilibrium is sometimes on the point of being destroyed, and motion often ensues.

LIMITING FRICTION. DEF. *When one body is just on the point of sliding upon another body, the equilibrium is said to be limiting, and the friction then exerted is called limiting friction.*

The direction of the limiting friction is given by Law I. Its magnitude is given by the three following laws:

LAW III. *The magnitude of the limiting friction always bears a constant ratio to the normal reaction, and this ratio depends only on the substances of which the bodies are composed.*

LAW IV. *The limiting friction is independent of the extent and shape of the surfaces in contact, so long as the normal reaction is unaltered.*

LAW V. *When motion ensues, by one body sliding over the other, the direction of friction is opposite to the direction of motion; the magnitude of the friction is independent of the velocity, but the ratio of the friction to the normal reaction is slightly less than when the body is at rest and just on the point of motion.*

The above laws are experimental, and are by no means rigorously accurate, though they represent, however, to a fair degree of accuracy the facts under ordinary conditions.

For example, if one body be pressed so closely on another that the surfaces in contact are on the point of being crushed, Law III is no longer true; the friction then increases at a greater rate than the normal reaction.

76. Coefficient of Friction. The constant ratio of the limiting friction to the normal pressure is called the coefficient of friction, and is generally denoted by μ; hence, if F be the friction, and R the normal pressure, between two bodies when equilibrium is on the point of being destroyed, we have $\frac{F}{R} = \mu$, and hence $F = \mu R$.

The values of μ are widely different for different pairs of substances in contact; no pairs of substances are, however, known for which the coefficient of friction is so great as unity.

Angle of Friction. When the equilibrium is limiting, if the friction and the normal reaction be compounded into one single force, the angle which this force makes with the normal is called the angle of friction, and the single force is called the resultant reaction.

Let A be the point of contact of the two bodies, and let AB and AC be the directions of the normal force R and the friction μR. Let AD be the direction of the resultant reaction S, so that the angle of friction is BAD. Let this angle be λ.

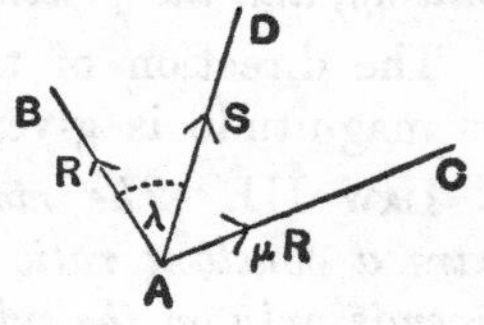

Then $S \cos \lambda = R$, and $S \sin \lambda = \mu R$.

Hence $S = R\sqrt{1 + \mu^2}$, and $\tan \lambda = \mu$.

Hence we see that *the coefficient of friction is equal to the tangent of the angle of friction.*

Since the greatest value of the friction is μR, it follows that the greatest angle which the direction of resultant reaction can make with the normal is λ, *i.e.* $\tan^{-1} \mu$.

Hence, if two bodies be in contact and if, with the common normal as axis, and the point of contact as vertex, we describe a cone whose semi-vertical angle is $\tan^{-1}\mu$, it is possible for the resultant reaction to have any direction lying within, or upon, this cone, but it cannot have any direction lying without the cone.

This cone is called the **Cone of friction.**

77. The following table, taken from Prof. Rankine's *Machinery and Millwork*, gives the coefficients and angles of friction for a few substances.

Substances	μ	λ
Wood on wood —Dry	·25 to ·5	14° to $26\frac{1}{2}$°
,, ,, ,, —Soaped	·04 to ·2	2° to $11\frac{1}{2}$°
Metals on metals —Dry	·15 to ·2	$8\frac{1}{2}$° to $11\frac{1}{2}$°
,, ,, ,, —Wet	·3	$16\frac{1}{2}$°
Leather on metals—Dry	·56	$29\frac{1}{2}$°
,, ,, ,, —Wet	·36	$20\frac{1}{2}$°
,, ,, ,, —Oily	·15	$8\frac{1}{2}$°

78. Equilibrium on a rough inclined plane. *A body is placed on a rough plane inclined to the horizon at an angle greater than the angle of friction, and is supported by a force, acting in a vertical plane through the line of greatest slope; to find the limits between which the force must lie.*

Let α be the inclination of the plane to the horizon, W the weight of the body, and R the normal reaction.

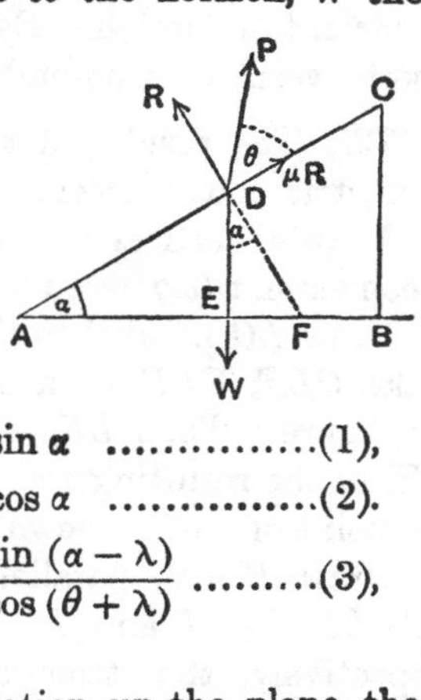

If the force P act at an angle θ with the inclined plane when the body is on the point of motion *down* the plane, and the friction therefore acts *up* the plane, the equations of equilibrium are

$$P\cos\theta + \mu R = W\sin\alpha \quad\text{..............(1)},$$

and

$$P\sin\theta + R = W\cos\alpha \quad\text{..............(2)}.$$

Hence
$$P = W\frac{\sin\alpha - \mu\cos\alpha}{\cos\theta - \mu\sin\theta} = W\frac{\sin(\alpha-\lambda)}{\cos(\theta+\lambda)} \quad\text{........(3)},$$

and R is easily found.

When the body is on the point of motion up the plane, the

friction μR acts down the plane. Hence, on changing the sign of μ, we have

$$P_1 = W\frac{\sin(\alpha+\lambda)}{\cos(\theta-\lambda)}\quad\text{.........................}(4).$$

For any value of the force, between P and P_1, there is equilibrium, but the body is not on the point of motion in either direction.

The force that will just be on the point of moving the body up the plane is least when (4) is least,

i.e. when $\cos(\theta-\lambda)$ is unity, *i.e.* when $\theta=\lambda$.

Hence the force required to move the body up the plane will be least when it is applied in a direction making with the inclined plane an angle equal to the angle of friction.

From (3) it follows that P is zero if $\lambda=\alpha$, *i.e.* a body will rest on an inclined plane and be on the point of motion if the inclination of the plane to the horizon is just equal to the angle of friction. On account of this property the angle of friction is sometimes called the angle of repose.

From this result also the coefficient of friction between two substances may be experimentally obtained. For let the inclined plane be made of one substance and let the body be a slab, with a plane face, of the other substance. If the angle of inclination of the plane be gradually increased until the slab just slides, the tangent of the angle of inclination is the coefficient of friction. By this method the laws of Art. 75 may also be verified; Coulomb used it in the year 1785.

79. The results of the previous article may be found by geometric construction.

Draw a vertical line KL to represent W on any scale that is convenient (*e.g.* one inch per lb. or one inch per 10 lbs.).

Draw LO parallel to the direction of the normal reaction R. Make OLF, OLF_1 each equal to the angle of friction λ, as in the figure. Then LF, LF_1 are parallel to the directions DH, DH_1 of the resulting reaction at D according as the body is on the point of motion down or up the plane.

Draw KMM_1 parallel to the supporting force P to meet LF, LF_1 in M and M_1. Then clearly KLM and KLM_1 are respectively the triangles of forces for the two extreme positions of equilibrium. Hence, on the same scale that KL

represents W, KM and KM_1 represent the P and P_1 of the previous article.

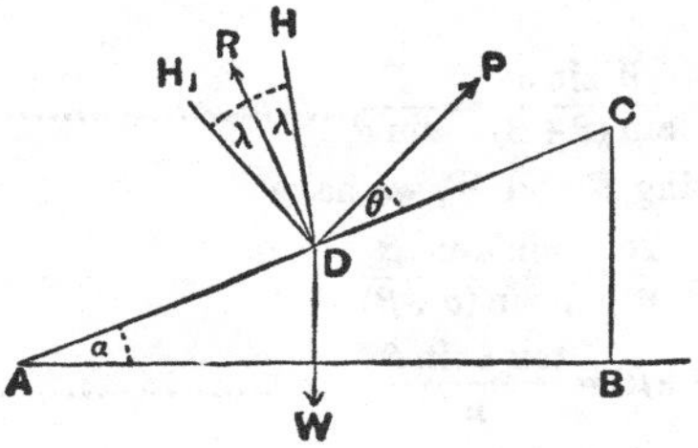

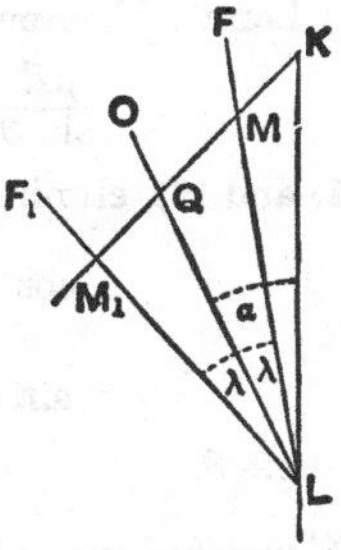

Clearly $OLK = \angle$ between R and the vertical $= \alpha$, so that

$$\angle MLK = \alpha - \lambda \text{ and } \angle M_1LK = \alpha + \lambda.$$

Similarly $\angle KQO = \angle$ between the directions of R and $P = 90° - \theta$, so that

$$\angle KQL = 90° + \theta, \quad \angle KM_1L = 90° + \theta - \lambda,$$

and $$\angle KML = 90° + \theta + \lambda.$$

Hence

$$\frac{P}{W} = \frac{KM}{KL} = \frac{\sin KLM}{\sin KML} = \frac{\sin(\alpha - \lambda)}{\sin(90° + \theta + \lambda)} = \frac{\sin(\alpha - \lambda)}{\cos(\theta + \lambda)},$$

and $$\frac{P_1}{W} = \frac{KM_1}{KL} = \frac{\sin KLM_1}{\sin KM_1L} = \frac{\sin(\alpha + \lambda)}{\sin(90° + \theta - \lambda)} = \frac{\sin(\alpha + \lambda)}{\cos(\theta - \lambda)}.$$

It is clear that KM_1 is least when it is drawn perpendicular to LF_1, *i.e.* when P_1 is inclined at a right angle to the direction, DH_1, of the resultant reaction, and therefore at an angle λ to the inclined plane.

80. *A particle is placed on a rough plane, whose inclination to the horizon is α, and is acted upon by a force P acting parallel to the plane and in a direction making an angle β with the line of greatest slope in the plane; if the coefficient of friction be μ and the equilibrium be limiting, find the direction in which the body will begin to move.*

Let W be the weight of the particle, and R the normal reaction.

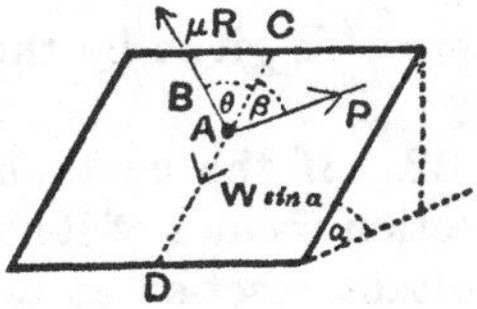

The forces perpendicular to the inclined plane must vanish.

$$\therefore R = W \cos \alpha \text{............(1).}$$

The other component of the weight will be $W \sin \alpha$, acting down the line of greatest slope

Let the friction, μR, act in the direction

AB, making an angle θ with the line of greatest slope, so that the particle would begin to move in the direction BA produced.

Since the forces acting along the surface of the plane are in equilibrium, we have, by Lami's Theorem,

$$\frac{\mu R}{\sin\beta}=\frac{W\sin\alpha}{\sin(\theta+\beta)}=\frac{P}{\sin\theta}\text{.........................(2)}.$$

From (1) and (2), eliminating R and W, we have

$$\cos\alpha=\frac{R}{W}=\frac{\sin\alpha\sin\beta}{\mu\sin(\theta+\beta)}.$$

Hence $$\sin(\theta+\beta)=\frac{\tan\alpha\sin\beta}{\mu}\text{ ....(3)},$$

giving the angle θ.

81. *Equilibrium of a particle constrained to rest on a rough curve under any given forces.*

Let the curve be plane, and X, Y the component forces parallel to the axes of coordinates.

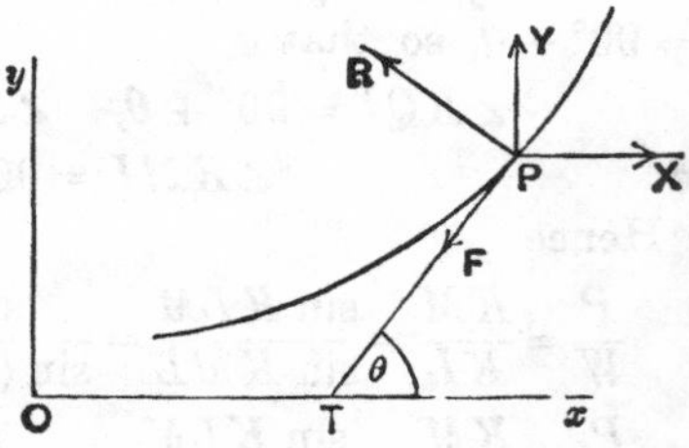

Let R be the normal reaction and F the friction along the tangent PT which makes an angle θ with the axis of x.

Resolving along and perpendicular to the tangent, we then have

$$X\cos\theta+Y\sin\theta=F\text{ ....(1)},$$

and $$X\sin\theta-Y\cos\theta=R\text{ ....(2)}.$$

If μ be the coefficient of friction, there will be equilibrium provided that F is not greater than μR or less than $-\mu R$, *i.e.* provided the numerical value, without regard to sign, of $X\cos\theta+Y\sin\theta$ is equal to or less than that of

$$\mu(X\sin\theta-Y\cos\theta),$$

i.e. if $$(X+Y\tan\theta)^2 \leqq \mu^2(X\tan\theta-Y)^2,$$

i.e. if $$\left(X+Y\frac{dy}{dx}\right)^2 \leqq \mu^2\left(X\frac{dy}{dx}-Y\right)^2,$$

where $\frac{dy}{dx}$ is given by the known equation of the curve.

82. If the curve is not a plane curve let (l, m, n) be the direction cosines of its tangent at any point P. Then since the resultant reaction at any point of a rough curve cannot make an angle greater than λ with the normal, and thus not an angle less than $\frac{\pi}{2} - \lambda$ with the tangent, there can be equilibrium if the angle that the resultant force makes with the tangent is equal to or greater than $\frac{\pi}{2} - \lambda$,

i.e. if $$\cos^{-1}\left(\frac{lX + mY + nZ}{\sqrt{X^2 + Y^2 + Z^2}}\right) \geqq \frac{\pi}{2} - \lambda,$$

i.e. if $$\frac{lX + mY + nZ}{\sqrt{X^2 + Y^2 + Z^2}} \leqq \sin \lambda,$$

i.e. if $$\frac{(lX + mY + nZ)^2}{X^2 + Y^2 + Z^2} \leqq \frac{\mu^2}{1 + \mu^2},$$

since $\tan \lambda = \mu$.

83. If the particle rest in contact with a rough surface, whose equation is $\phi(x, y, z) = 0$, under forces whose components are X, Y, Z, the conditions of equilibrium are easily found. For the direction cosines of the normal to the surface at the point (x, y, z) are proportional to ϕ_x, ϕ_y, and ϕ_z.

There will be equilibrium if the resultant force makes with the normal an angle not greater than λ,

i.e. if $$\cos^{-1}\left[\frac{X\phi_x + Y\phi_y + Z\phi_z}{\sqrt{X^2 + Y^2 + Z^2}\sqrt{\phi_x^2 + \phi_y^2 + \phi_z^2}}\right] \leqq \lambda,$$

i.e. if $$\frac{(X\phi_x + Y\phi_y + Z\phi_z)^2}{(X^2 + Y^2 + Z^2)(\phi_x^2 + \phi_y^2 + \phi_z^2)} \geqq \cos^2 \lambda,$$

i.e. $$\geqq \frac{1}{1 + \mu^2}.$$

EXAMPLES

1. Shew that the least force which will move a weight W along a rough horizontal plane is $W \sin \phi$, where ϕ is the angle of friction.

2. At what angle of inclination should the traces be attached to a sledge that it may be drawn up a given hill with the least exertion?

3. A weight W is laid upon a rough plane $\left(\mu = \frac{1}{\sqrt{3}}\right)$, inclined at 45° to the horizon, and is connected by a string passing through a smooth ring, A, at the top of the plane, with a weight P hanging vertically. If $W = 3P$, shew that, if θ be the greatest possible inclination of the string AW to the line of greatest slope in the plane, then

$$\cos \theta = \frac{2\sqrt{2}}{3}.$$

Find also the direction in which W would commence to move.

4. A weight W rests on a rough plane inclined at an angle a to the horizon, and the coefficient of friction is $2\tan a$. Shew that the least horizontal force along the plane which will move the body is $\sqrt{3}\,W\sin a$, and that the body will begin to move in a direction inclined at 60° to the line of greatest slope on the plane.

5. A heavy particle is placed on a rough plane inclined at an angle a to the horizon, and is connected by a stretched weightless string AP to a fixed point A in the plane. If AB be the line of greatest slope and θ the angle PAB when the particle is on the point of slipping, shew that

$$\sin\theta = \mu\cot a.$$

Interpret the result when $\mu\cot a$ is greater than unity.

6. Two weights, A and B, are connected by a string and placed on a horizontal table whose coefficient of friction is μ. A force P, $<\mu(A+B)$, is applied to A in the direction BA and its direction is gradually turned round an angle θ in the horizontal plane. If P be greater than $\mu\sqrt{A^2+B^2}$, shew that both B and A will slip when

$$\cos\theta = \frac{\mu^2(B^2-A^2)+P^2}{2\mu BP},$$

but if P be less than $\mu\sqrt{A^2+B^2}$ and greater than μA, then A alone will slip when $\sin\theta = \frac{\mu A}{P}$.

7. A cycloid is placed with its axis vertical and vertex downward. Shew that a particle can rest on it at any point which is not higher than $2a\sin^2\epsilon$ above its lowest point, where ϵ is the angle of friction and a is the radius of the generating circle of the cycloid.

8. A particle rests on the surface $xyz=c^3$ under the action of a constant force parallel to the axis of z; shew that the curve of intersection of the surface with the cone $\frac{1}{x^2}+\frac{1}{y^2}=\frac{\mu^2}{z^2}$ will separate the part of the surface on which equilibrium is possible from that on which it is not possible.

9. The ellipsoid $\frac{x^2}{a^2}+\frac{y^2}{b^2}+\frac{z^2}{c^2}=1$ is placed with the axis of x vertical and its surface is rough. Shew that a heavy particle will rest on it anywhere above its intersection with the cylinder

$$y^2c^4(\mu^2b^2+a^2)+z^2b^4(\mu^2c^2+a^2)=\mu^2b^4c^4,$$

where μ is the coefficient of friction.

10. The paraboloid $\frac{x^2}{a}+\frac{y^2}{b}=2z$ is placed with its axis vertical and its vertex uppermost; if μ be the coefficient of friction, shew that a particle will rest on it at any point above its curve of intersection with the cylinder

$$\frac{x^2}{a^2}+\frac{y^2}{b^2}=\mu^2.$$

11. A surface is formed by the revolution of a rectangular hyperbola about a vertical asymptote; shew that a particle will rest on it anywhere beyond its intersection with a certain circular cylinder.

12. A rough paraboloid of revolution, of latus rectum $4a$ and of coefficient of friction $\cot\beta$, revolves with uniform angular velocity ω about its axis which is vertical; if $\omega > \sqrt{\frac{g}{2a}\cot\frac{\beta}{2}}$ or $< \sqrt{\frac{g}{2a}\tan\frac{\beta}{2}}$, shew that a particle can rest anywhere except within a certain belt, but that for any angular velocity between these limiting values equilibrium is possible for all positions of the particle.

[The problem is the same as if the surface were at rest if we put on an additional "centrifugal" force, $m\omega^2 y$, outwards along a perpendicular from the particle upon the axis of the paraboloid.]

84. Rough joints or hinges. In the figure of Art. 65 the resultant reaction will not be normal to the joint at A if there be friction. In this case the reaction P at A is equivalent to a parallel force P through the centre O together with a couple whose moment is $P \times$ the perpendicular from O upon the line AP. Similarly for the reactions at any other points of contact. Thus the resultant actions consist of a set of forces acting through O together with a number of couples. These compound into a single force through O and a single couple. Hence, when there is friction at a joint and there is contact at more than one point, we must, in addition to the unknown force of Art. 65, or unknown component forces in known directions, assume also an unknown couple.

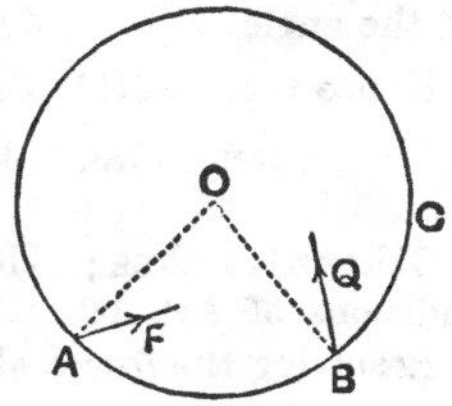

If the joint be rough but contact take place at only one point, such as A in the above figure, then, just as in the case of a smooth joint, the reaction may be assumed to consist of one force only, which passes through the point of contact.

85. We give some examples illustrative of the applications of the laws of friction.

Ex. 1. *A uniform rod rests in limiting equilibrium within a rough hollow sphere; if the rod subtend an angle $2a$ at the centre of the sphere, and if λ be the angle of friction, shew that the angle of inclination of the rod to the horizon is*

$$\tan^{-1}\left[\frac{\sin 2\lambda}{2\cos(a+\lambda)\cos(a-\lambda)}\right].$$

Let AB be the rod, G its middle point, and O the centre of the sphere, so that

$$\angle GOA = \angle GOB = \alpha.$$

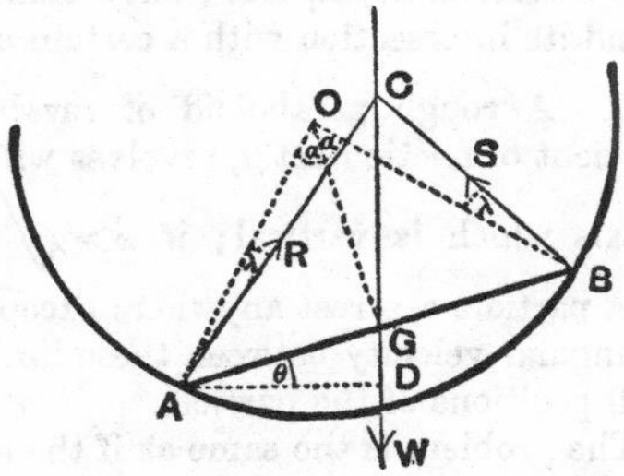

Through A and B draw lines AC and BC making an angle λ with the lines joining A and B to the centre. By Art. 76, these are the directions of the resultant reactions, R and S, at A and B respectively.

Since these reactions and the weight keep the rod in equilibrium, the vertical line through G must pass through C.

Let AD be the horizontal line drawn through A to meet CG in D so that the angle GAD is θ.

The angle $\quad CAG = \angle OAG - \lambda = 90° - \alpha - \lambda$,

and the angle $\quad CBG = \angle OBG + \lambda = 90° - \alpha + \lambda$.

Hence theorem (2) of Art. 55 gives

$$2\tan\theta = \tan(\alpha+\lambda) - \tan(\alpha-\lambda) = \frac{\sin 2\lambda}{\cos(\alpha+\lambda)\cos(\alpha-\lambda)} \quad \text{......(1).}$$

Otherwise thus; The solution may be also obtained by using the conditions of Art. 60.

Resolving the forces along the rod, we have

$$R\sin(\alpha+\lambda) - S\sin(\alpha-\lambda) = W\sin\theta \quad \text{..................(2).}$$

Resolving perpendicular to the rod, we have

$$R\cos(\alpha+\lambda) + S\cos(\alpha-\lambda) = W\cos\theta \quad \text{..................(3).}$$

By taking moments about A, we have

$$2S\cos(\alpha-\lambda) = W\cos\theta \quad \text{............................(4).}$$

From equations (3) and (4),

$$R\cos(\alpha+\lambda) = S\cos(\alpha-\lambda) = \tfrac{1}{2}W\cos\theta.$$

Substituting these values of R and S in (2), we have

$$\tan(\alpha+\lambda) - \tan(\alpha-\lambda) = 2\tan\theta.$$

Ex. 2. *A beam AB rests with one end A in contact with a rough horizontal floor, and the other end B in contact with a rough vertical wall, the vertical plane through AB being perpendicular to the wall; to discuss its equilibrium, the inclination of the beam to the horizontal being given.*

Let the normal reaction and friction at A be R and F, and those at B be S and F', as in the figure.

By resolving in two directions and taking moments about a point we obtain three, and only three equations, between the four unknown quantities R, S, F, and F'. They are thus indeterminate.

This may be also seen geometrically. For draw AL and BM inclined to the normals AC and BC at angles λ and λ', equal to the angles of friction at A and B, and let the vertical through G, the centre of gravity of the beam, meet them in U and V. Then, provided that V is within the space QAL and U within the space CBM as in the figure, *i.e.* provided that the vertical through G cuts the space $CQKN$, we may take any point P between U and V and the resultant reactions at A and B may have the directions AP, BP and still give equilibrium. If P were at U, the equilibrium would be limiting at A and not so at B; if P were at V, the equilibrium would be limiting at B and not so at A; and there could be any arrangement of forces between these two extreme cases.

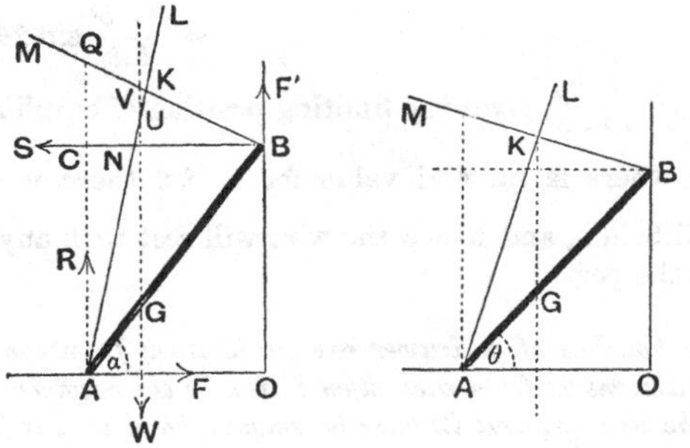

If the vertical through G were to the right of K, then there could be found no point P on it such that AP was within the cone of friction at A, and at the same time BP within the cone of friction at B. Hence equilibrium would be impossible.

If the ladder be in such a position that the equilibrium was limiting, and hence the ladder on the point of sliding down, the points U and V would coincide with K, the point of intersection of AL and BM, as in the second figure. If $AG=a$, $GB=b$, then, by the theorem of Art. 55, we have

$$(a+b)\cot KGB = a\cot AKG - b\cot GKB,$$

i.e.

$$(a+b)\tan\theta = a\cot\lambda - b\tan\lambda',$$

so that

$$\theta = \tan^{-1}\frac{a-b\mu\mu'}{\mu(a+b)}.$$

Ex. 3. *A uniform heavy elliptical wire, whose semiaxes are a and b, is hung over a small rough peg. Shew that, if the wire can be in equilibrium with any point of it in contact with the peg, the coefficient of friction must not be less than* $\dfrac{a^2-b^2}{2ab}$.

Suppose that the wire is in limiting equilibrium when the point of contact is P. Let PN be perpendicular to the major axis; let PG be the normal and C the centre. The resulting reaction at P, which makes an angle λ with the normal at P, must then just balance the weight. Hence CP must be vertical and $\angle CPG=\lambda$.

If θ be the eccentric angle of P, we have

$$\tan PCG = \frac{b}{a}\tan\theta,$$

and $$\tan PGN = \frac{PN}{GN} = b\sin\theta \div \left(\frac{b^2}{a^2}.CN\right) = \frac{a}{b}\tan\theta.$$

$$\therefore \tan\lambda = \tan CPG = \tan(PGN - PCN) = \frac{\left(\frac{a}{b}-\frac{b}{a}\right)\tan\theta}{1+\tan^2\theta}$$

$$= \frac{a^2-b^2}{2ab}\sin 2\theta.$$

Hence $\sin 2\theta = \frac{2ab\mu}{a^2-b^2}$ gives the limiting position of equilibrium.

If $\frac{2ab\mu}{a^2-b^2} > 1$, there is no real value for θ, *i.e.* there is no *limiting* position of equilibrium, and hence the wire will rest with any point of it in contact with the peg.

Ex. 4. *The handles of a drawer are equidistant from the sides of the drawer and are distant 2c from each other; if μ be the coefficient of friction at the sides of the drawer, and its base be smooth, shew that it is impossible to pull it out by pulling one handle straight outwards unless the length of the drawer from front to back exceeds $2\mu c$.*

Let $ABCD$ be the drawer and let its length and depth, AB and BC, be $2a$ and $2b$. If E be the handle nearest B, the effect of the pull P at E will be to jam the corners C and A against the fitment, so that the thrusts on the sides AD and BC will be R and S at A and C, as marked. The maximum resistance to the motion of the body will be when the frictions at A and C are μR and μS.

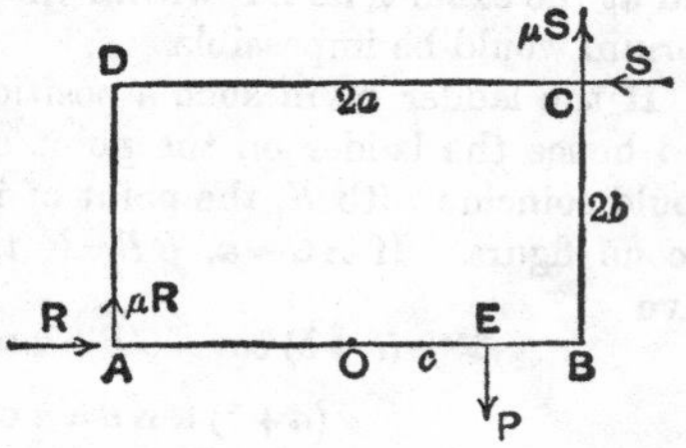

Resolving parallel to AB, $R = S$ ....(1).

Taking moments about A, we have

$$P(a+c) = \mu S.2a + S.2b \quad \text{.......................(2).}$$

Also, if there is to be motion in the direction DA, we must have

$$P > \mu(R+S), \text{ i.e. } > \mu.\frac{a+c}{\mu a+b}P,$$

so that $$(b-\mu c).P > 0.$$

This requires that $b > \mu c$, and then the magnitude of P is immaterial. Thus if $b > \mu c$, any pull P, however small, will move the drawer; whilst, if $b < \mu c$, no pull P, however great, will move it.

Ex. 5. *The length of the line joining the lowest points of the wheels of a bicycle is 2a, and the centre of gravity is at a height h above this line and at a distance x in front of its middle point. No account being taken of axle friction, find the slope of the greatest incline on which the bicycle can rest without slipping, according as the back or front wheel is braked.*

The hind wheel being braked, the friction at it is μS. If G be the centre of gravity then $O'D=a+x$; $DO=a-x$; $CG=h$.

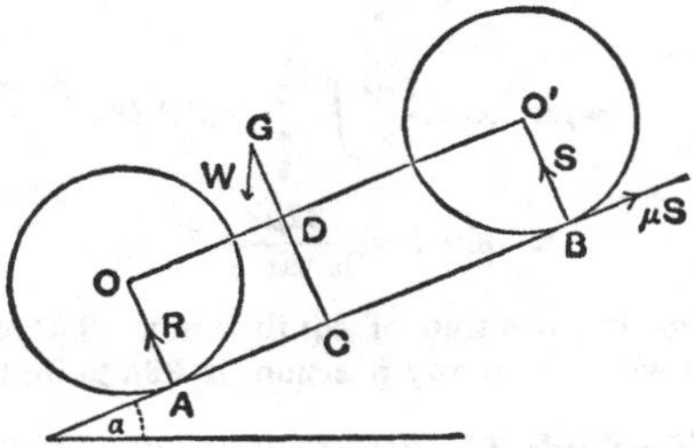

Resolving along and perpendicular to the inclined plane and taking moments about G, we have

$$\mu S = W \sin \alpha \quad \text{.........................(1)},$$

$$R+S = W \cos \alpha \quad \text{.........................(2)},$$

and
$$S[\mu h + a + x] = R(a-x) \quad \text{.........................(3)}.$$

(2) and (3) give
$$S[\mu h + 2a] = (a-x) W \cos \alpha.$$

Hence from (1)
$$\tan \alpha = \frac{\mu(a-x)}{\mu h + 2a} \quad \text{.........................(4)}.$$

If the front wheel be braked, the friction is μR acting at A, and, if β be the corresponding inclination of the plane, we have similarly

$$\mu R = W \sin \beta,$$

$$R+S = W \cos \beta,$$

and
$$R[a - x - \mu h] = S(a+x).$$

Hence, as before,
$$\tan \beta = \frac{\mu(a+x)}{2a - \mu h} \quad \text{.........................(5)}.$$

It is clear that $\beta > \alpha$, *i.e.* the bicycle can rest on an incline of greater slope when the front wheel is fixed than when the back wheel is fixed.

Ex. 6. *A flat heavy circular disc lies on a rough plane and can turn freely about a pin in its circumference; shew that it will rest in any position if the coefficient of friction is* $> \frac{9\pi \tan \alpha}{32}$, *where α is the inclination of the plane to the horizon, it being assumed that the weight of the disc is uniformly distributed over its area.*

Assume that the equilibrium of the disc is limiting when the diameter OA of the disc through the pin O is inclined at an angle ϕ to the line of greatest slope through O. Let w be the weight of the disc per unit of area, and a its radius, so that its weight is $\pi a^2 w$.

If P be any point of the disc, such that $OP=r$ and $\angle AOP=\theta$, the element of area at P is $r\delta\theta.\delta r$. The friction on this element is thus $\mu w.r\delta\theta\delta r.\cos a$ and acts through P at right angles to OP, since P would move, if it did move, at right angles to OP. Hence, by taking moments about O, we have

$$\pi a^2 w \sin a . a \sin\phi = \int_0^{2a\cos\theta}\int_{-\frac{\pi}{2}}^{+\frac{\pi}{2}} \mu w . r d\theta dr \cos a . r$$

$$= \mu w \cos a . \frac{8a^3}{3}\int_{-\frac{\pi}{2}}^{+\frac{\pi}{2}} \cos^3\theta d\theta = \frac{32a^3}{9}\mu w \cos a;$$

$$\therefore \sin\phi = \frac{32\mu}{9\pi\tan a}.$$

This gives the limiting position of equilibrium. There is no *limiting* position, *i.e.* the disc will rest in any position, if $32\mu > 9\pi\tan a$.

Ex. 7. *Two uniform rods AB, BC, of weights W and W′, are smoothly jointed at B, and are placed so as to be in a straight line on a rough horizontal table; the end A is acted on by a gradually increasing force P in a direction perpendicular to the rods. Find how the equilibrium is broken.*

Assume that, when equilibrium is on the point of being broken, the rod AB is on the point of turning about a point I of itself and the rod BC about I', where

$AI=x$, $BI'=y$, $AB=a$, $BC=b$.

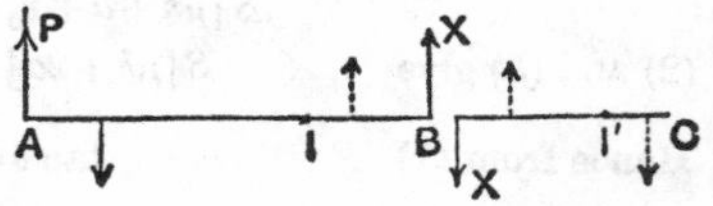

Then the frictions on AI and IB are in opposite directions as marked and similarly for BI', $I'C$.

Let X be the reaction at B which is clearly perpendicular to each rod.

The friction on AI is $\mu W.\frac{x}{a}$ and acts at a distance $\frac{x}{2}$ from I; so that on IB is $\mu W\frac{a-x}{a}$ and acts at a distance $\frac{a-x}{2}$ from I.

Resolving perpendicular to the rod AB and taking moments about I, we have

$$P+X=\mu W\frac{x}{a}-\mu W\frac{a-x}{a}=\frac{\mu W}{a}(2x-a)\quad\text{..............(1)},$$

and

$$P.x-X(a-x)=\mu W\frac{x}{a}.\frac{x}{2}+\mu W\frac{a-x}{a}.\frac{a-x}{2}=\frac{\mu W}{a}\left[x^2-ax+\frac{a^2}{2}\right]\quad\text{...(2)}.$$

So for BC we have

$$X=\frac{\mu W'}{b}(2y-b)\quad\text{..............................(3)},$$

and

$$X.y=\frac{\mu W'}{b}\left[y^2-yb+\frac{b^2}{2}\right]\quad\text{..........................(4)}.$$

(3) and (4) give $y=\frac{b}{\sqrt{2}}$ and $X=\mu W'(\sqrt{2}-1)$.

Also (1) and (2) give

$$\frac{\mu W}{a}\left[x^2-\frac{a^2}{2}\right]=Xa=\mu W'(\sqrt{2}-1)a.$$

$$\therefore \quad x^2=\frac{a^2}{2}\left[1+\frac{2W'}{W}(\sqrt{2}-1)\right],$$

and P is found from (1).

This is true so long as $x \leqq a$, *i.e.* so long as $\frac{2W'}{W}(\sqrt{2}-1) \leqq 1$.

If $\frac{2W'}{W}(\sqrt{2}-1)>1$, so that $x>a$, the above solution no longer holds and the frictions at the different points of AB are all in the same direction. We then easily have $P=X=\frac{\mu W}{2}$, so that in this case X is less than $\mu W'(\sqrt{2}-1)$, which we have found above to be the force requisite to move BC. The latter rod is therefore not on the point of motion, but only AB.

EXAMPLES

1. A uniform rod MN has its ends in two fixed straight rough grooves OA and OB, in the same vertical plane, which make angles α and β with the horizon; shew that, when the end M is on the point of slipping in the direction AO, the tangent of the angle of inclination of MN to the horizon is $\frac{\sin(\alpha-\beta-2\epsilon)}{2\sin(\beta+\epsilon)\sin(\alpha-\epsilon)}$, where ϵ is the angle of friction.

2. A vertical rectangular beam, of weight W, is constrained by guides to move only in its own direction, the lower end resting on a smooth floor. If a smooth inclined plane of given slope be pushed under it by a horizontal force acting at the back of the inclined plane, find the force required.

If there be friction between the floor and the inclined plane, but nowhere else, what must be the least value of μ so that the inclined plane may remain, when left in a given position under the beam, without being forced out?

3. A heavy rod, of length $2a$, lies over a rough peg with one extremity leaning against a rough vertical wall; if c be the distance of the peg from the wall and λ be the angle of friction both at the peg and the wall, shew that, when the point of contact of the rod with the wall is above the peg, then the rod is on the point of sliding downwards when $\sin^3\theta=\frac{c}{a}\cos^2\lambda$, where θ is the inclination of the rod to the wall.

If the point of contact of the rod and wall be below the peg, prove that the rod is on the point of slipping downwards when

$$\sin^2\theta\sin(\theta+2\lambda)=\frac{c}{a}\cos^2\lambda,$$

and on the point of slipping upwards when $\sin^2\theta\sin(\theta-2\lambda)=\frac{c}{a}\cos^2\lambda$.

4. If a uniform beam, of length $2h$, can rest with one end on a rough horizontal plane, and against the top of a wall of height h, in a vertical plane perpendicular to the wall and at any inclination to the wall which is geometrically possible, shew that the angle of friction between the beam and both wall and ground, supposed to be equally rough, must be not less than

$$\frac{1}{2}\sin^{-1}\frac{4}{3\sqrt{3}}.$$

5. Two equal uniform rods, of length $2a$, are smoothly jointed at one extremity by a hinge, and rest symmetrically upon a rough fixed sphere of radius c. Find the limiting position of equilibrium, and shew that, if the coefficient of friction be $c \div a$, the limiting inclination of each rod to the vertical is $\tan^{-1}\sqrt[3]{c \div a}$.

6. If a pair of compasses rest across a smooth horizontal cylinder of radius c, shew that the frictional couple at the joint to prevent the legs of the compasses from slipping must be

$$W(c \cot \alpha \operatorname{cosec} \alpha - a \sin \alpha),$$

where W is the weight of each leg, 2α the angle between the legs, and a the distance of the centre of gravity of a leg from the joint.

7. A rod, resting on a rough inclined plane, whose inclination α to the horizon is greater than the angle of friction λ, is free to turn about one of its ends, which is attached to the plane; shew that, for equilibrium, the greatest possible inclination of the rod to the line of greatest slope is $\sin^{-1}(\tan \lambda \cot \alpha)$.

8. Prove that an ordinary drawer cannot be pushed in by a force applied to one handle until it has been pushed in a distance $2\mu c$ by forces applied in some other manner, where $2c$ is the distance between the handles and μ is the coefficient of friction.

9. If one cord of a sash-window break, find the least coefficient of friction between the sash and the window-frame in order that the other weight may still support the window.

10. A hemispherical shell rests on a rough plane, whose angle of friction is λ; shew that the inclination of the plane base of the rim to the horizon cannot be greater than $\sin^{-1}(2 \sin \lambda)$.

[The centre of gravity bisects the radius perpendicular to the base of the shell.]

11. A solid homogeneous hemisphere rests on a rough horizontal plane and against a smooth vertical wall. Shew that, if the coefficient of friction be greater than $\frac{3}{8}$, the hemisphere can rest in any position and, if it be less, the least angle that the base of the hemisphere can make with the vertical is $\cos^{-1}\frac{8\mu}{3}$.

If the wall be rough (coefficient of friction μ'), shew that this angle is $\cos^{-1}\left(\frac{8\mu}{3}\cdot\frac{1+\mu'}{1+\mu\mu'}\right)$.

[The centre of gravity divides the radius perpendicular to the base of the hemisphere in the ratio 3 : 5.]

12. If a hemisphere rest in equilibrium with its curved surface in contact with a rough plane inclined to the horizon at an angle a, shew that the inclination of the plane base of the hemisphere to the horizontal is $\sin^{-1}(\frac{8}{3}\sin a)$, provided that a is less than $\sin^{-1}\frac{3}{8}$ and also is less than the angle of friction.

13. A uniform hemisphere, of radius a and weight W, rests with its spherical surface on a horizontal plane, and a rough particle, of weight W', rests on the plane surface; shew that the distance of the particle from the centre of the plane face is not greater than $\frac{3W\mu a}{8W'}$, where μ is the coefficient of friction.

14. A sphere, whose radius is a and whose centre of gravity is at a distance c from the centre, rests in limiting equilibrium on a rough plane inclined at an angle a to the horizon; shew that it may be turned through an angle

$$2\cos^{-1}\left(\frac{a\sin a}{c}\right),$$

and still be in limiting equilibrium.

15. A uniform rectangular board, whose sides are $2a$ and $2b$, rests in limiting equilibrium in contact with two rough pegs in the same horizontal line at a distance d apart. Shew that the inclination θ of the side $2a$ to the horizontal is given by the equation

$$d\cos\lambda\cos(\lambda+2\theta)=a\cos\theta-b\sin\theta,$$

where λ is the angle of friction.

16. A rigid framework in the form of a rhombus, of side a and acute angle a, rests on a rough peg whose coefficient of friction is μ. Shew that the distance between the two extreme positions which the point of contact with the peg can have is $a\mu\sin a$.

17. A boy, of weight w, stands on a sheet of ice and pushes with his hands against the smooth vertical side of a chair of weight nw. Shew that he can incline his body to the horizon at any angle greater than $\cot^{-1}2\mu$ or $\cot^{-1}2\mu n$, according as the chair or the boy is the heavier, the coefficient of friction between the boy and the ice, or the chair and the ice, being μ

18. A uniform heavy rod lies on a rough horizontal table and is pulled perpendicularly to its length by a string attached to any point. About what point will it commence to turn?

Shew also that the ratio of the forces, required to move the rod, when applied at the centre and through the end of the rod perpendicular to the rod, is $\sqrt{2}+1:1$.

19. A uniform rough beam AB lies horizontally upon two others at points A and C; shew that the least horizontal force applied at B in a direction perpendicular to BA, which is able to move the beam, is the lesser of the two forces $\frac{1}{2}\mu W$ and $\mu W \frac{b-a}{2a-b}$, where AB is $2a$, AC is b, W is the weight of the beam, and μ the coefficient of friction.

20. A uniform rough beam AB, of length $2a$, is placed horizontally on two equal and equally rough balls, the distance between whose centres is b, touching them in C and D; shew that, if b be not greater than $\frac{4a}{3}$, a position of the beam can be found in which a force P exerted at B perpendicular to the beam will cause it to be on the point of motion both at C and D at the same time.

21. A uniform plank, of length $2a$ and weight W, rests with its middle point upon a rough horizontal cylinder, whose axis is perpendicular to the plank; shew that the greatest weight that can be attached to one end of the plank, without its sliding off the cylinder, is $\frac{b\lambda}{a-b\lambda} . W$, where b is the radius of the cylinder and λ the angle of friction.

22. A surface is generated by the revolution of an ellipse, of eccentricity e and foci S and H, about its major axis. On the surface is placed a rough particle P which is attracted towards the foci by forces which vary directly as the distances PS and PH respectively; shew that the particle will rest anywhere on the surface if the coefficient of friction be

$$> \frac{e^2}{2\sqrt{1-e^2}}.$$

23. A circular disc, of weight W, rests in a vertical plane with a point A in contact with a rough table, and is pressed at a point B by a man's finger. If the line AB make an angle α with the vertical, and if λ and λ' be the angles of friction at A and B, prove that if $\alpha > \lambda'$ the ring rolls along the table immediately, however small the pressure that is applied at B, that if $\lambda' > \alpha > \lambda$ the ring will slip at A when a normal pressure $W \cos\alpha \sin\lambda \operatorname{cosec}(\alpha - \lambda)$ is applied by the man, and that if α is less than both λ and λ' no force applied by the finger will make it move.

24. Two uniform beams, AC and BC, are connected by a smooth hinge at C and placed in a vertical plane with their lower ends resting on a rough horizontal plane. If equilibrium is broken, shew that the end of the longer beam will slide and that the other beam will rotate.

25. Shew that the least force which when applied to the surface of a heavy uniform sphere will just maintain it in equilibrium against a rough vertical wall is

$$W \cos\epsilon \text{ or } W \tan\epsilon \left[\tan\epsilon - \sqrt{\tan^2\epsilon - 1}\right],$$

according as $\epsilon <$ or $> \cos^{-1}\frac{\sqrt{5}-1}{2}$, where W is the weight of the sphere and ϵ is the angle of friction.

26. Shew that two cylindrical logs, of equal radii but unequal weights W and W', where $W' > W$, can rest in contact on an inclined plane with their axes horizontal and the heavier log uppermost, if the coefficient of friction μ (supposed the same at each of the lines of contact) exceeds $\frac{W'+W}{W'-W}$, and the inclination of the plane is less than

$$\tan^{-1}\frac{2\mu W'}{(\mu+1)(W'+W)}.$$

If the inclination of the plane is gradually increased beyond this value, how is the equilibrium broken?

27. An easel has its front legs and its back legs inclined at 30° to the vertical. Each set of legs is uniform and of weight W. A blackboard of weight W just covers the top half of the front legs. A lecturer presses normally on the central point of the blackboard with a force equal to $\frac{3}{2}$ times the weight of the board. If μ, the coefficient of friction between the legs and the floor, is $\frac{1}{3}$, prove that equilibrium will be broken by the back legs slipping.

28. Three uniform rods, of lengths a, b, and c, are rigidly connected to form a triangle ABC which is hung over a rough peg so that the side BC may rest in contact with it; find the length of the portion of the rod over which the peg may range, shewing that if

$$\mu > \frac{a(a+b+c)}{b(b+c)}\operatorname{cosec} C + \tan\frac{C-B}{2},$$

where $C > B$, the triangle will rest in any position.

29. A perfectly rough plane is inclined at an angle α to the horizon; shew that the least eccentricity of the ellipse which can rest on the plane is

$$\sqrt{\frac{2\sin\alpha}{1+\sin\alpha}}.$$

30. An elliptic cylinder rests in limiting equilibrium between a rough vertical and an equally rough horizontal plane, the axis of the cylinder being horizontal and the major axis of the ellipse inclined to the horizon at an angle of 45°; shew that the coefficient of friction is $\frac{\sqrt{1+2e^2-e^4}-1}{2-e^2}$, where e is the eccentricity of the cross section of the cylinder.

31. Three equal cylindrical rods are placed symmetrically round a fourth of the same radius, and the bundle is then surrounded by two equal elastic strings symmetrically placed with respect to the ends; if the unstretched length of each string be equal to the circumference of each rod, shew that the force necessary to pull out the middle rod is $\frac{54\mu\lambda}{\pi}$, where μ is the coefficient of friction and λ is the modulus of elasticity.

32. Three equal spheres, of radius a, are placed in contact on a rough horizontal plane and a fourth sphere, of radius b, is placed on them; if there is equilibrium, shew that the least angle of friction between the upper and lower spheres is $\frac{1}{2}\sin^{-1}\left[\frac{2a}{\sqrt{3}(a+b)}\right]$.

33. The length of the line joining the lowest points of the wheels of a bicycle is a, and the centre of gravity is at a height h above this line and at a distance x in front of its middle point. No account being taken of axle friction or of road resistance to rolling, shew that when the brake is hard on the back wheel the slope of the greatest incline on which the bicycle can be held up without slipping back is α, where $\tan\alpha = \frac{\mu a - 2x}{2a - \mu h}$ and μ is the coefficient of friction between the tyres and the ground.

34. Two wheels, A and B, of equal radius a and of weights W and W_1, are connected by a light bar of length b which is attached to their centres; the wheels are placed on a rough plane with their common plane vertical, A being the higher, and the inclination of the plane is gradually increased; shew that, if slipping commenced at the same inclination, whichever wheel be locked, then $\mu\frac{a}{b} = \frac{W - W_1}{W + W_1}$.

35. A reel, consisting of a spindle of radius c with two circular ends of radius a, is placed on a rough inclined plane and has a thread wound on it which unwinds when the reel rolls downwards. If μ be the coefficient of friction and α be the inclination of the plane to the horizontal, shew that the reel can be drawn up the plane by means of the thread if μ be not less than $\frac{c\sin\alpha}{a - c\cos\alpha}$. If μ be just equal to this value, shew that the corresponding direction of the thread is horizontal.

36. The cylindrical axle of a wheel is supported on two parallel rails which constitute an inclined plane; a thread is wound round the circumference of the wheel; under what circumstances will pulling the thread downwards parallel to the plane cause the wheel to roll up the plane?

37. A reel of thread, whose rim and spindle have radii a and b respectively, rests on a rough horizontal table and the loose end of the thread passes under the spindle and lies along the table. The whole system is symmetrical about a plane perpendicular to the axis of the reel. Shew that, if the loose thread be raised to an angle θ with the horizontal, the slightest tension in it will in general cause the reel to roll, and the motion will be towards or from the hand of the experimenter according as θ is less or greater than a certain value. When θ has this critical value, shew that there will be no motion unless the tension exceeds a certain finite limit

38. When the shafts of a dog-cart are horizontal its centre of gravity is just over the axle of the wheel. The wheel is of radius a, and it turns freely on a rough axle, of radius b, and the ground is rough enough to prevent any slipping. Shew that the least force which, when applied at the end of the shaft of length l, will just move the dog-cart passes through the point of contact of the wheel with the ground and is equal to

$$\frac{wk\sqrt{a^2+l^2}}{ak+l\sqrt{1-k^2}},$$

where $k=\frac{b}{a}\sin\epsilon$, ϵ is the angle of friction between the wheel and its axle, and w is the weight of the dog-cart.

39. A heavy carriage wheel is to be dragged over an obstacle, which touches the wheel at C, by means of a rope which is tied to a spoke of the wheel and pulled horizontally. Shew that the wheel will roll round C, and not slip both at C and the ground, if the height of the rope above C is less than $a\sin\alpha\cot(\alpha-\epsilon)$, where a is the radius of the wheel, ϵ the angle of friction at C, and α is the angle made with the vertical by the radius of the wheel through C.

40. A solid circular cylinder is placed with its base on a rough horizontal plane and is capable of free motion about its axis; if W be its weight, and a the radius of its cross-section, shew that the moment of the least couple that will move it is $\frac{2}{3}\mu W.a$, assuming that its weight is borne uniformly by the plane.

41. A heavy elliptic disc, placed on an imperfectly rough table and acted on by a horizontal force F, is on the point of motion. Prove that, if its weight be equally distributed over the area and if it begin to turn about a focus, the force F must act along an ordinate at a distance $\frac{2a}{3}\cdot\frac{1-e^2}{e}$ from the centre.

42. A uniform disc in the shape of a cardioid lies on a rough plane inclined at α to the horizon, and is capable of turning freely round a pin at its pole. When just about to slip, its axis makes an angle β with the line of greatest slope; shew that $\sin\beta=\frac{8}{5}\mu\cot\alpha$, where μ is the coefficient of friction.

Prove also that the direction of the action at the pin makes an angle $\tan^{-1}(\frac{1}{2}\tan\beta)$ with the axis of the cardioid.

43. A hoop is placed upon a rough horizontal plane and a string fastened to it at any point P is pulled in the direction of the tangent at P. Shew that the hoop will begin to turn about the other end of the diameter through P.

CHAPTER V

WORK. VIRTUAL WORK

86. Work. A force is said to do work when its point of application moves in the direction of the force.

The force exerted by a horse, in dragging a waggon, does work.

The pressure of the steam, in moving the piston of an engine, does work.

When a man winds up a watch or a clock he does work.

The measure of the work done by a force is the product of the force and the distance through which it moves its point of application in the direction of the force.

Suppose that a force acting at a point A of a body moves the point A to D, then the work done by P is measured by the product of P and AD. If the point D be on the side of A toward which the force acts, this work is positive; if D lie on the opposite side, the work is negative.

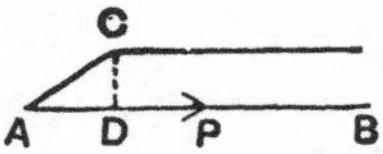

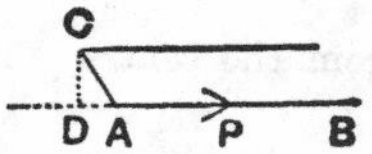

Next, suppose that the point of application of the force is moved to a point C, which does not lie on the line AB. Draw CD perpendicular to AB, or AB produced. Then AD is the distance through which the point of application is moved in the direction of the force. Hence in the first figure the work done is $P \times AD$; in the second figure the work done is $-P \times AD$. When the work done by the force is negative, this is sometimes expressed by saying that the force has work done against it.

In the case when AC is at right angles to AB, the points A and D coincide, and the work done by the force P vanishes. Thus if a body be moved about on a horizontal table the work done by its weight is zero. So, again, if a body be moved on an inclined plane, no work is done by the normal reaction of the plane.

87. The unit of work, used in Statics, is called a Foot-Pound, and is the work done by a force, equal to the weight of a pound, when it moves its point of application through one foot in its own direction. A better, though more clumsy, term than "Foot-Pound" would be Foot-Pound-weight.

88. It will be noticed that the definition of work, given in Art. 86, necessarily implies motion. A man may use great exertion in *attempting* to move a body, and yet do no work on the body. For example, suppose a man pulls at the shafts of a heavily-loaded van, which he cannot move. He may pull to the utmost of his power, but, since the force which he exerts does not move its point of application, he does no work (in the technical sense of the word).

89. *To shew that the work done in raising a number of particles from one position to another is Wh, where W is the total weight of the particles, and h is the distance through which the centre of gravity of the particles has been raised.*

Let $w_1, w_2, w_3, \dots w_n$ be the weights of the particles; in the initial position let $x_1, x_2, x_3, \dots x_n$ be their heights above a horizontal plane, and $\bar{x}$ that of their centre of gravity, so that, as in Art. 34, we have

$$W.\bar{x} = w_1x_1 + w_2x_2 + \dots + w_nx_n \quad \dots\dots\dots\dots(1).$$

In the final position let $x_1', x_2', \dots x_n'$ be the heights of the different particles, and $\bar{x}'$ the height of the new centre of gravity, so that

$$W.\bar{x}' = w_1x_1' + w_2x_2' + \dots + w_nx_n' \quad \dots\dots\dots\dots(2).$$

By subtraction we have

$$w_1(x_1' - x_1) + w_2(x_2' - x_2) + \dots = W(\bar{x}' - \bar{x}).$$

But the left-hand member of this equation gives the total work done in raising the different particles of the system from their initial position to their final position; also the right-hand side

$= W \times$ height through which the centre of gravity has been raised

$= W.h.$

90. It will be noted that the result of the last article does not in any way depend on the initial or final arrangement of the particles amongst themselves, except in so far as the initial and final positions of the centre of gravity depend on these arrangements.

For example, a hole may be dug in the ground, the soil lifted out, and spread on the surface of the earth at the top of the hole. We only want the positions of the C.G. of the soil initially and finally, and then the work done is known. This work is quite independent of the path by which the soil went from its initial to its final position.

91. Power. Def. *The power of an agent is the amount of work that would be done by the agent if it worked uniformly for the unit of time.*

The unit of power used by engineers is called a **Horse-Power.** An agent is said to be working with one horse-power when it performs 33,000 foot-pounds in a minute, *i.e.* when it would raise 33,000 lbs. through a foot in a minute, or when it would raise 330 lbs. through 100 feet in a minute, or 33 lbs. through 1000 feet in a minute.

This estimate of the power of a horse was made by Watt, but is above the capacity of ordinary horses. The word Horse-Power is usually abbreviated into H.P.

92. *Graphic representation of the work done by a force.*

It is sometimes difficult to calculate directly the work done by a varying force, but it may be quite possible to obtain the result to a near degree of approximation.

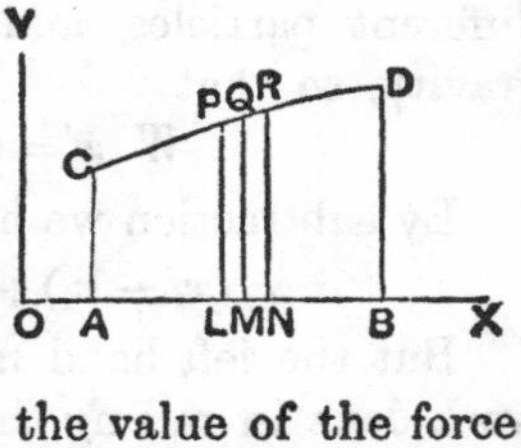

Suppose the force to always act in the straight line OX, and let us find the work done as its point of application moves from A to B. At A and B erect ordinates AC and BD to represent the value of the force for these two points of application. For each intermediate point of application L erect the ordinate LP to represent the corresponding value of the acting force; then the tops of these ordinates will clearly lie on some such curve as CPD.

Take M a very near point to L, so near that the force may be considered to have remained constant as its point of application moved through the small distance LM.

Then the work done by the force

= its magnitude × distance through which its point of application has moved

$= LP \times LM =$ area PM very nearly.

Similarly whilst the point of application moves from M to N the work done

= area QN very nearly, and so on.

Hence it follows that the work done as the point of application moves from A to B is, when the lengths LM, MN, ... are taken indefinitely small, equal more and more nearly to the area $ACDB$.

93. As an example of the above construction let us find the work done in stretching an elastic string from length b $(=OB)$ to length c $(=OC)$, the unstretched length of the string being $a\,(=OA)$.

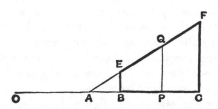

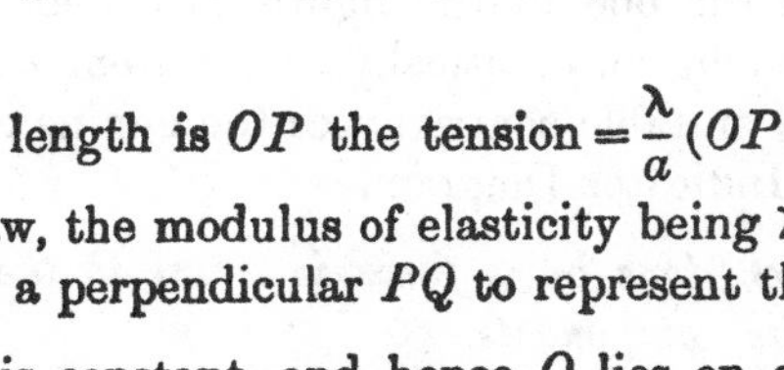

When the length is OP the tension $=\frac{\lambda}{a}(OP-a)=\frac{\lambda}{a}.PA$, by Hooke's Law, the modulus of elasticity being λ.

At P erect a perpendicular PQ to represent this tension.

Then $\frac{PQ}{PA}$ is constant, and hence Q lies on a straight line AEF passing through A. If this straight line meet the perpendiculars through B and C in E and F, the required work is, as in the last article, represented by the area $BEFC$, and hence is $\frac{1}{2}BC \times (BE + CF)$, *i.e.* it

= *Extension produced multiplied by the mean of the initial and final tensions.*

Or, by using Integral Calculus, the work done

$$=\int_b^c T\,.\,dx=\frac{\lambda}{a}\int_b^c (x-a)\,dx=\frac{\lambda}{2a}\Big[(x-a)^2\Big]_b^c$$

$$=\frac{\lambda}{2a}(c-b)(c+b-2a)=\frac{c-b}{2}[T_B+T_C].$$

94. As another example let us take the case of the Indicator Diagram of a steam-engine.

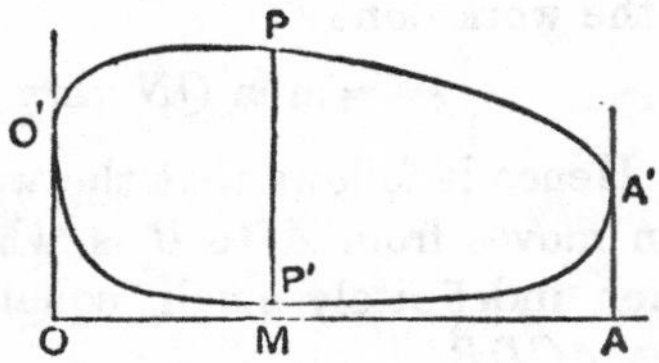

Suppose OA to represent the distance traversed by the piston of an engine. When it is at the position M in its forward motion, erect a perpendicular MP to denote the pressure of the steam on it, so that the curve $O'PA'$ represents the steam pressure during the forward motion. Similarly during the backward motion of the piston, when the steam has been cut off, let the curve $A'P'O'$ represent the pressure on the same face of the piston as before. Then during the forward motion the area of the curve $OO'PA'A$ gives the work done by the steam on the piston. So during the return motion the area of the curve $A'P'O'OA$ represents the work done by the steam *against* the piston.

Hence, during the complete stroke, the net work done by the steam on one face of the piston is given by the area $O'PA'P'O'$, and hence can be found.

A curve, like the one in the figure, is called an Indicator Diagram and can be automatically found from the motion of the piston, *i.e.* by suitable contrivances the engine can be made to draw its own Indicator Diagram.

95. *The work done by a force is equal to the sum of the works done by its components.*

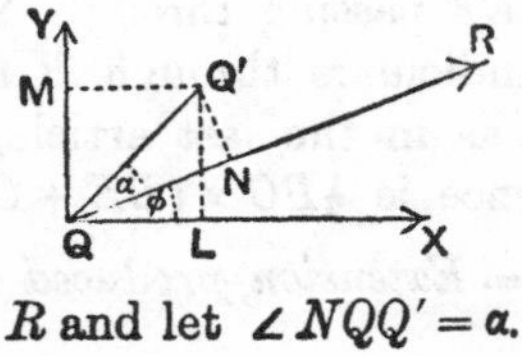

Let the components of R in two directions at right angles be X and Y, R being inclined at an angle ϕ to the direction of X, so that

$$X=R\cos\phi \text{ and } Y=R\sin\phi.$$

Let the point of application Q of R be removed to a point Q' in the plane of the paper, and draw $Q'N$ perpendicular to R and let $\angle NQQ'=\alpha$.

The sum of the works done by X and Y

$$= X . QL + Y . QM$$
$$= R\cos\phi . QQ'\cos(\phi+\alpha) + R\sin\phi . QQ'\sin(\phi+\alpha)$$
$$= R . QQ'\cos\alpha = R . QN = \text{the work done by } R.$$

96. If the forces and displacement be not in one plane the same result easily follows. For let the direction cosines of QR referred to any three rectangular axes Qx, Qy, Qz be (l, m, n) so that $X = lR$, $Y = mR$, and $Z = nR$. Let the displacement QQ' be through a distance δs along a line whose direction cosines are (l_1, m_1, n_1) so that $\delta x = l_1 . \delta s$, $\delta y = m_1 . \delta s$, and $\delta z = n_1 . \delta s$. The work done by the component forces

$$= X\delta x + Y\delta y + Z\delta z = R\delta s\,(ll_1 + mm_1 + nn_1)$$
$$= R . \delta s \times \cos Q'QR = R \times \text{projection of } QQ' \text{ on the direction of } R$$
$$= \text{the work done by } R.$$

If a particle move along a smooth curve in space, and if the force acting on it at any point (x, y, z) have as components X, Y, Z, it follows that the total work done on the particle as it moves from a point A to a point $B = \int_A^B (X\,dx + Y\,dy + Z\,dz)$.

97. *Work done by a couple.* Let the forces of the couple be each P and let its arm be AB of length a.

Suppose the couple to be moved into another position so that AB goes to $A'B'$, the angle between AB and $A'B'$ being the small angle $\delta\theta$.

First, move the forces parallel to themselves so that the arm AB takes up the parallel position $A'C$. The work done by the equal and opposite forces P during this displacement is zero.

Now turn the forces through the angle $\delta\theta$ about A'. The force P acting at A' has no displacement and thus does no work. The displacement of the point of application of the other force P is $a . \delta\theta$, and the total work done is thus $P . a . \delta\theta$, *i.e.* the moment of the couple multiplied by the elementary angle turned through. If the total angle turned through by the couple be α, the corresponding work $= \int_0^\alpha P . a . \delta\theta = P . a . \alpha$, so that in all cases the work done by a couple, when it is rotated about an axis perpendicular to its own plane, is equal to the moment of the couple multiplied by the angle of rotation.

98. *Potential Energy.* The potential energy of a body due to a given system of forces is the work the system can do on the body as it passes from its present configuration to some standard configuration usually called the zero position.

Thus the potential energy of a particle of weight W at a height h above the ground is Wh.

If however we take into consideration the variation of gravity and assume that the attraction of the Earth, supposed to be a sphere of radius a, is $\frac{\mu}{x^2}$ at a distance x from its centre, the potential energy at a height h, when the Earth's surface is taken to be the zero position,

$$=\int_{a+h}^{a}\left(-\frac{\mu}{x^2}\right)dx=\mu\left[\frac{1}{a}-\frac{1}{a+h}\right]=\frac{\mu h}{a(a+h)}=\frac{ah}{a+h}W,$$

since $W=\frac{\mu}{a^2}$, the attraction of the Earth at a point on its surface.

Again, if one end of an elastic string of natural length a be tied to a fixed point, the potential energy of a particle tied to its other end is, by Art. 93, $\frac{\lambda x^2}{2a}$ when the stretched length is $a+x$.

99. **Ex. 1.** *A spherical shot, of weight W lbs. and radius a feet, lies at the bottom of a cylindrical bucket, of radius b feet, which is filled up to a depth h feet, $(h>2a)$, with water. Shew that the work done in lifting the shot just clear of the water must exceed* $W\left(h-\frac{4a^3}{3b^2}\right)-W'\left(h-a-\frac{2a^3}{3b^2}\right)$ *foot-pounds, the weight of the water displaced by the shot being W′ lbs.*

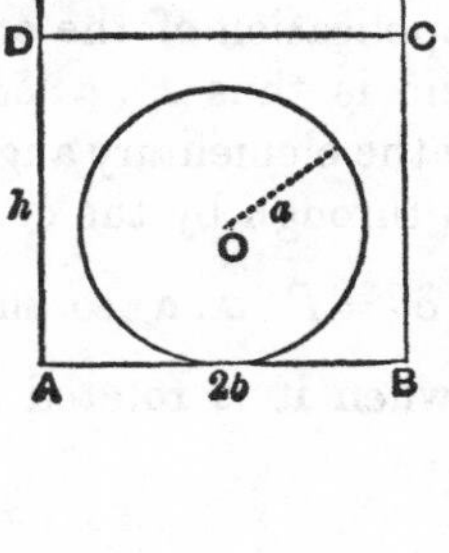

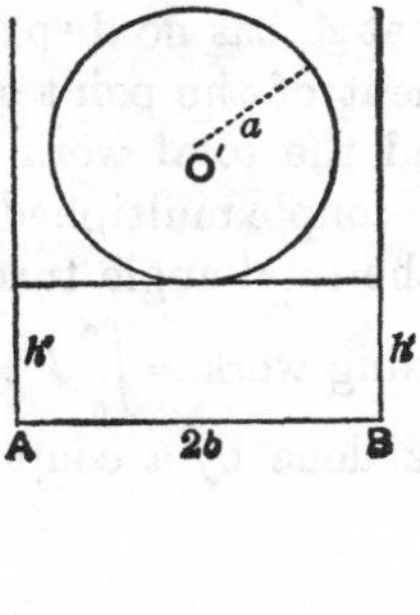

If w be the weight of a unit volume of the water, and σ the specific gravity of the shot, we have

$$W=\frac{4\pi}{3}a^3\sigma w \text{ and } W'=\frac{4}{3}\pi a^3 w.$$

The work done must at least be equal to the increase in the potential energy of the system.

In the first case the potential energy

$=$ that of (a cylinder $ABCD-$sphere) of water

$+\sigma$ that of the sphere of water equal to the given sphere

$=$ potential energy of a cylinder $ABCD$ of water

$+(\sigma-1)$ that of the sphere of water

$$=\pi b^2 hw\,.\frac{h}{2}+\frac{4}{3}\pi a^3(\sigma-1)\,.\,aw=\frac{3}{8}\frac{b^2h^2}{a^3}W'+(W-W')a \quad\text{.................(1).}$$

When the sphere has been lifted out let h' be the depth of the water, so that $\pi b^2\,.\,h'=$ volume of the water $=\pi b^2 h-\frac{4}{3}\pi a^3$,

i.e.
$$h'=h-\frac{4a^3}{3b^2}\quad\text{...............................(2).}$$

In the second case the potential energy

$$=\pi b^2 h'w\,.\frac{h'}{2}+W(h'+a)=\frac{3}{8}\frac{b^2h'^2}{a^3}W'+W(h'+a)\quad\text{........(3).}$$

The increase in the potential energy

$$=\frac{3}{8}\frac{b^2W'}{a^3}(h'^2-h^2)+Wh'+W'a$$

$$=W\left(h-\frac{4a^3}{3b^2}\right)-W'\left(h-a-\frac{2a^3}{3b^2}\right).$$

Ex. 2. *A quantity V of gas at a pressure Π is contained in a cylindrical vessel; if it be allowed to expand so that the length alters from x_0 to x_1, the temperature remaining constant, shew that the work done is $\Pi V \log_e \frac{x_1}{x_0}$.*

If it expand adiabatically, i.e. so that no heat passes into or out of the gas and the relation between the pressure p and the volume v is therefore $pv^k=$ constant, shew that the corresponding work is $\frac{\Pi V}{k-1}\left[1-\left(\frac{x_0}{x_1}\right)^{k-1}\right]$.

Let p be the pressure when the length occupied by the gas is x, so that by Boyle's Law $px=\Pi x_0$. Whilst the length changes from x to $x+\delta x$, the work done

$=p\,.\,A\delta x$, where A is the section of the cylinder,

$$=\frac{\Pi x_0}{x}\,.\,\frac{V}{x_0}\,\delta x.$$

Hence the work required $=\int_{x_0}^{x_1}\frac{\Pi V}{x}dx=\Pi V\log_e\frac{x_1}{x_0}$.

In the second case we have $px^k = \Pi x_0^k$, and the work done

$$= \int_{x_0}^{x_1} p \cdot \frac{V}{x_0} dx = \Pi V x_0^{k-1} \int_{x_0}^{x_1} \frac{dx}{x^k}$$

$$= -\Pi V \frac{x_0^{k-1}}{k-1} \left[\frac{1}{x_1^{k-1}} - \frac{1}{x_0^{k-1}} \right] = \frac{\Pi V}{k-1} \left[1 - \left(\frac{x_0}{x_1} \right)^{k-1} \right].$$

Ex. 3. If in the previous question the gas be contained in a vessel of any shape, and be allowed to expand from a volume V at pressure Π to any volume V_1, shew that the work done is

$$\Pi V \log \frac{V_1}{V} \text{ or } \frac{\Pi V}{k-1} \left[1 - \left(\frac{V}{V_1} \right)^{k-1} \right],$$

according to the condition under which the expansion takes place.

EXAMPLES

1. A steamer is going at the rate of 15 miles per hour; if the effective H.P. of her engines be 10,000, what is the resistance to her motion?

[$111\frac{17}{28}$ tons wt.]

2. A man is cycling at the rate of 6 miles per hour up a hill whose slope is 1 in 20; if the weight of the man and the machine be 200 lbs., prove that he must at the least be working at the rate of ·16 H.P.

3. A man rowing 40 strokes per minute propels a boat at the rate of 10 miles an hour, and the resistance to his motion is equal to 8 lbs. wt.; find the work he does in each stroke and the H.P. at which he is working.

[176 ft.-lbs.; $\cdot 21\dot{3}$ H.P.]

4. An elastic cord, whose natural length is 10 inches, can be kept stretched to a length of 15 inches by a force of 5 lbs. wt.; find the amount of work done in stretching it from a length of 12 inches to a length of 15 inches.

[$\frac{7}{8}$ ft.-lb.]

5. A spiral spring requires a force of one pound weight to stretch it one inch. How much work is done in stretching it three inches more?

[$\frac{5}{8}$ ft.-lb.]

6. A force acts on a particle, its initial value being 20 lbs. wt. and its values being 25, 29, 32, 31, 27, and 24 lbs. wt. in the direction of the particle's motion when the latter has moved through 1, 2, 3, 4, 5, and 6 feet respectively; find, by means of a graph, the work done by the force, assuming that it varies uniformly during each foot of the motion.

[166 ft.-lbs.]

7. If the axis of a screw be vertical and the distance between the threads 2 inches, and a door, of weight 100 lbs., be attached to the screw as to a hinge, find the work done in turning the door through a right angle.

[$4\frac{1}{6}$ ft.-lbs.]

8. Prove that the tension of a stay is equal to 9 tons' weight if it be set up by a force of 49 lbs. at a leverage of 2 feet acting on a double screw having a right-handed screw of 5 threads to the inch and a left-handed one of 6 threads to the inch.

[For one complete turn of the screw its ends are brought nearer by a distance of $(\frac{1}{5}+\frac{1}{6})$ inch. Hence the principle of work gives

$$T\times(\tfrac{1}{5}+\tfrac{1}{6})\times\tfrac{1}{12}=49\times 2\pi\,.\,2,$$

where T is the tension of the stay in lbs. wt.]

9. A Venetian blind consists of n thin bars, besides the top fixed bar, and the weight of the movable part is W. When let down the length of the blind is a, and when pulled up it is b; shew that the work done against gravity in drawing up the blind is

$$W\,.\,\frac{n+1}{2n}(a-b).$$

10. A solid hemisphere, of weight 12 lbs. and radius 1 foot, rests with its flat face on a table. How many foot-lbs. of work are required to turn it over so that it may rest with its curved surface in contact with the table? [Use the results of Arts. 89 and 148.] [3 ft.-lbs.]

11. A uniform log weighing half a ton is in the form of a triangular prism, the sides of whose cross section are $1\frac{1}{2}$ ft., 2 ft., and $2\frac{1}{2}$ ft. respectively, and the log is resting on the ground on its narrowest face. Prove that the work which must be done to raise it on its edge so that it may fall over on to its broadest face is approximately ·27 ft.-tons. [Use Art. 137.]

12. A cyclist always working at the rate of $\frac{2}{25}$ of a horse-power rides at 10 miles an hour on level ground and 8 miles an hour up an incline of 1 in 150. Suppose the man and his machine to weigh 180 lbs., and the resistance on a level road to consist of two parts, one constant and the other proportional to the square of the velocity, shew that when the velocity is v miles per hour the resistance is

$$\left(1\cdot 75+\frac{v^2}{80}\right)\text{ lbs. wt.}$$

13. A cylindrical cork, of length l and radius r, is slowly extracted from the neck of a bottle. If the normal pressure per unit of area between the bottle and the unextracted part of the cork at any instant be constant and equal to P, shew that the work done in extracting it is $\pi\mu r l^2 P$, where μ is the coefficient of friction.

14. A weight W is drawn up along the surface of a rough cone, whose height is h and whose vertical angle is $2a$, and the path cuts all the lines of greatest slope at the same angle β. If the coefficient of friction is μ, shew that the work done when the weight arrives at the vertex of the cone is $Wh(1+\mu\tan a\sec\beta)$.

$$\left[\text{The work}=W.\,h+\int_0^h \mu R\,.\,\frac{dx}{\cos\beta\cos a}\text{, where } R=W\sin a.\right]$$

15. A particle, of weight W, is at the bottom of a rough hemispherical bowl which is fixed with its vertex at its lowest point. The particle is fastened to a string which passes over the rim of the bowl and the particle is slowly drawn by the string up the bowl in a vertical plane through the axis of the bowl; prove that the work done in drawing the particle up to the rim is

$$Wa\left[1+\frac{\pi}{2}\sin^2\epsilon - \sin 2\epsilon \log\frac{1+\tan\epsilon}{\sqrt{2}}\right],$$

where a is the radius of the bowl and ϵ is the angle of friction.

[If R be the reaction of the bowl when the radius to the particle is inclined at θ to the horizontal, we have, by resolving perpendicular to the string,

$$R = W\cos\epsilon \sin\frac{\theta}{2}\sec\left(\frac{\theta}{2}-\epsilon\right).$$

Hence the work done against friction in dragging the particle up

$$=\int_{\frac{\pi}{2}}^{0}\mu R(-a\,.\,d\theta) = Wa\sin\epsilon\int_0^{\frac{\pi}{2}}\sin\frac{\theta}{2}\sec\left(\frac{\theta}{2}-\epsilon\right)d\theta$$

$$=2\,Wa\sin\epsilon\int_{-\epsilon}^{\frac{\pi}{4}-\epsilon}\sin(\phi+\epsilon)\sec\phi\,.\,d\phi = 2\,Wa\sin\epsilon\int_{-\epsilon}^{\frac{\pi}{4}-\epsilon}[\sin\epsilon+\cos\epsilon\tan\phi]\,d\phi,$$

etc.

Also the work done against the weight $= W\,.\,a$.]

16. A solid homogeneous cone of height h, radius r, and specific gravity s is placed inside a vertical cylinder, of radius r, their bases being in contact, and water is poured into the cylinder to the height h so that the cone is just immersed. Shew that to raise the cone vertically so as to be just clear of the water work must be done equal to

$$\tfrac{3}{4}Wh\left(1-\frac{7}{8s}\right),$$

where W is the weight of the cone, s being greater than unity.

100. Virtual Work. When we have a system of forces acting on a body in equilibrium and we suppose that the body undergoes a slight displacement, which is consistent with the geometrical conditions under which the system exists, and if a point Q of the body, with this imagined displacement, goes to Q', then QQ' is called the Virtual Velocity, or Displacement, of the point Q. The word Virtual is used to imply that the displacement is an imagined, and not an actual, displacement.

If a force R act at the point Q and if $Q'N$ be the perpendicular from Q' on the direction of R, then the product $R\,.\,QN$ is called the Virtual Work or Virtual Moment of the force R.

As in Art. 86 this work is positive, or negative, according as QN is in the same direction as R, or in the opposite direction.

101. The principle of virtual work states that *If a system of forces acting on a body be in equilibrium and the body undergo a slight displacement consistent with the geometrical conditions of the system, the algebraic sum of the virtual works is zero; and conversely, if this algebraic sum be zero, the forces are in equilibrium. In other words, if each force P have a virtual displacement δp in the direction of its line of action, then, to the first order of small quantities, $\Sigma(P.\delta p) = 0$; also conversely, if $\Sigma(P.\delta p)$ be zero, the forces are in equilibrium.*

If the body be a single particle then, by Art. 96, it follows that, if the sum of the virtual works of all the forces which act on a particle is zero, the virtual work of the resultant is zero, and hence that the resultant vanishes and the particle is at rest.

In the next article we give a proof of this theorem for coplanar forces. In Art. 175 will be found a proof for forces in three dimensions.

102. *Proof of the principle of virtual work for any system of forces in one plane.*

Take any two straight lines at right angles to one another in the plane of the forces and let the body undergo a slight displacement. This can clearly be done by turning the body through a suitable small angle α radians about O and then moving it through suitable distances a and b parallel to the axes.

[The student may illustrate this by moving a book from any position on a table into any other position, the book throughout the motion being kept in contact with the table.]

Let Q be the point of application of any force R, whose coordinates referred to O are x and y and whose polar coordinates are r and θ, so that $OQ = r$ and $XOQ = \theta$.

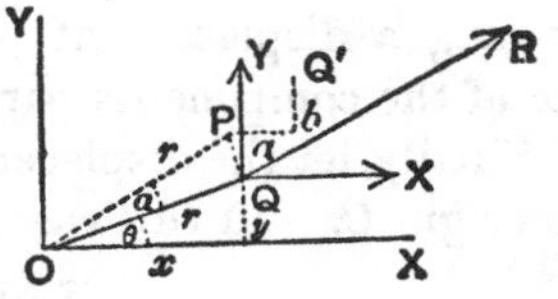

When the small displacement has been made the coordinates of the new position Q' of Q are

$$r\cos(\theta+\alpha)+a \quad \text{and} \quad r\sin(\theta+\alpha)+b,$$

i.e. $\quad r\cos\theta - \alpha . r\sin\theta + a \quad \text{and} \quad r\sin\theta + \alpha . r\cos\theta + b,$

if squares of the small angle α be neglected.

The changes in the coordinates of Q are therefore

$$a-\alpha . r\sin\theta \text{ and } b+\alpha . r\cos\theta,$$

i.e. $$a-\alpha y \text{ and } b+\alpha x.$$

If then X and Y be the components of R, the virtual work of R, which is equal to the sum of the virtual works of X and Y,

$$= X(a-\alpha y)+Y(b+\alpha x)=a.X+b.Y+\alpha(Yx-Xy).$$

Similarly we have the virtual work of any other force of the system, a, b, and α being the same for each force.

The sum of the virtual works will therefore be zero if

$$a\Sigma(X)+b\Sigma(Y)+\alpha\Sigma(Yx-Xy) \text{ be zero.}$$

If the forces be in equilibrium then $\Sigma(X)$ and $\Sigma(Y)$ are, by Art. 60, separately equal to zero.

Also $\Sigma(Yx-Xy)=$ sum of the moments of all the forces about O, and this sum is zero, by Art. 60.

It follows that if the forces be in equilibrium the sum of their virtual works is zero.

103. Conversely, if the sum of the virtual works be zero for any displacement whatever, the forces are in equilibrium.

With the same notation as in the last article, the sum of the virtual works is

$$a\Sigma(X)+b\Sigma(Y)+\alpha\Sigma(Yx-Xy)............(1),$$

and this is given to be zero for all displacements.

Choose a displacement such that the body is moved only through a distance a parallel to the axis of x. For this displacement b and α vanish, and (1) then gives

$$a\Sigma(X)=0,$$

i.e. the sum of the components parallel to OX is zero. Similarly, choosing a displacement parallel to the axis of y, we have the sum of the components parallel to OY zero also.

Finally, let the displacement be one of simple rotation round the origin O. In this case a and b vanish and (1) gives

$$\Sigma(Yx-Xy)=0,$$

so that the sum of the moments of the forces about O vanish.

The three conditions of equilibrium given in Art. 60 therefore hold and the system of forces is in equilibrium.

104. ***Forces which may be omitted in forming the equation of Virtual Work.***

(1) *The tension of an inextensible string.*

For let OA be such a string whose tension is T. In the displaced position let $O'A'$ be the string, and draw perpendiculars $O'M$ and $A'N$ on OA. It is easy to shew that, to the first order of small quantities, $OM = AN$.

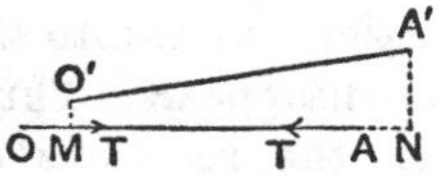

For, taking O as origin and OA as axis of x, let O' be the point (x_1, y_1, z_1) and A' the point $(a + x_2, y_2, z_2)$, where $OA = a$, and $x_1, y_1, z_1, x_2, y_2, z_2$ are all small quantities.

Since $O'A' = OA$, the string being inextensible,

$$\therefore\ (a + x_2 - x_1)^2 + (y_2 - y_1)^2 + (z_2 - z_1)^2 = a^2.$$

$$\therefore\ 2a(x_2 - x_1) + \text{sqs. of small quantities} = 0.$$

$$\therefore\ x_2 = x_1, \text{ to the first order of small quantities,}$$

i.e. $$OM = AN.$$

Hence the virtual work of the tension

$$= T.OM + T(-AN) = 0.$$

Similarly for any other force along the line joining two particles, P and Q, of the system, the distance between which remains invariable.

(2) *The reaction R of any smooth surface with which the body is in contact.*

For if the surface be smooth the reaction R is normal to the surface at the point of contact P, so that if P move to a neighbouring near point P', PP' is at right angles to the force; its virtual work is thus zero.

If the surface be rough the work done by the friction F, viz. $F.(-PP')$, must come into the equation, since it is not in general zero.

(3) *The reaction at any point of contact P with a fixed surface on which the body rolls without sliding.*

For the point of contact P of the body is for the moment at rest, and so its displacement is zero. The normal reaction at P and the friction at P have then zero displacements.

(4) ***The reactions between any two bodies of the material system considered.***

For these reactions are equal and opposite on the two bodies. Hence provided we write down the equation of virtual work for the two bodies taken together the virtual work of any such reaction comes into the equation twice, with opposite signs, and thus disappears. Thus if we are considering a system of jointed rods, the reactions at the joints can be omitted from the equation, as in the examples of Art. 106.

105. We may, if we please, choose a displacement which does not satisfy the geometrical conditions of the system, and it is often convenient to choose such a displacement. But if we do make such a choice we must bring into the equation the corresponding force.

Thus if we assume such a displacement as will make a string vary in length, as in Ex. 2 of the next article, we must bring into the equation the term tension × increase in length of the string.

106. Ex. 1. *Six equal rods AB, BC, CD, DE, EF, and FA are each of weight W and are freely jointed at their extremities so as to form a hexagon; the rod AB is fixed in a horizontal position and the middle points of AB and DE are joined by a string; prove that its tension is $3W$.*

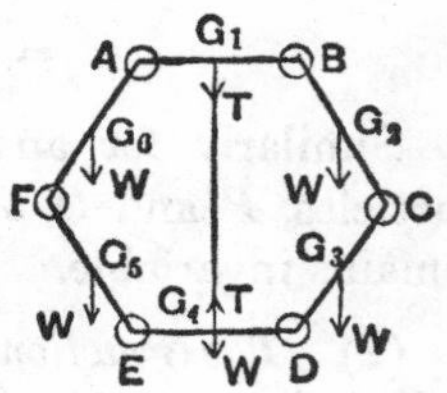

Let G_1, G_2, G_3, G_4, G_5, and G_6 be the middle points of the rods. Since, by symmetry, BC and CD are equally inclined to the vertical the depths of the points C, G_3 and D below AB are respectively 2, 3, and 4 times as great as that of G_2.

Let the system undergo a displacement in the vertical plane of such a character that D and E are always in the vertical lines through B and A and DE is always horizontal. If G_2 descend a vertical distance x, then G_3 will descend $3x$, G_4 will descend $4x$, whilst G_5 and G_6 will descend $3x$ and x respectively.

The sum of the virtual works done by the weights

$$=W.x+W.3x+W.4x+W.3x+W.x=12W.x.$$

If T be the tension of the string, the virtual work done by it will be

$$T\times(-4x).$$

For the displacement of G_4 is in a direction opposite to that in which T acts and hence the virtual work done by it is negative.

The principle of virtual work then gives

$$12W.x+T(-4x)=0, \quad i.e. \ T=3W.$$

Ex. 2. *Four equal uniform rods are jointed to form a rhombus ABCD, which is placed in a vertical plane with AC vertical and A resting on a horizontal plane. The rhombus is kept in the position in which $\angle BAC=\theta$ by a light string joining B and D. Shew that its tension is $2W\tan\theta$, where W is the weight of a rod.*

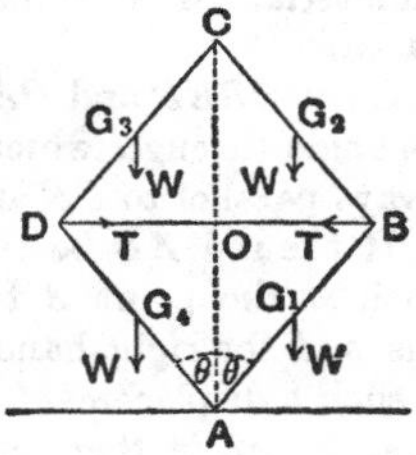

Let x be the height above A of the middle points of AB and AD, so that $3x$ is clearly the height of the middle points of BC and CD.

Let $BO=y=OD$, where O is the centre of the rhombus.

Choose as our displacement one in which θ becomes $\theta+\delta\theta$, and hence x becomes $x+\delta x$ and y becomes $y+\delta y$.

Then, T being the tension of BD, the equation of virtual work is

$$2T(-\delta y)+W(-\delta x)+W(-\delta x)+W\{-\delta(3x)\}+W\{-\delta(3x)\}=0.$$

$$\therefore\ T=-4W\frac{\delta x}{\delta y}.$$

Now, if $AB=2a$, we have $x=a\cos\theta$ and $y=2a\sin\theta$.

$$\therefore\ \frac{\delta x}{\delta y}=\frac{-a\sin\theta\,.\,\delta\theta}{2a\cos\theta\,.\,\delta\theta}=-\tfrac{1}{2}\tan\theta.$$

$$\therefore\ T=2W\tan\theta.$$

[The reaction at A is omitted because it has no displacement; the reaction at B is omitted because it comes in twice, for the rod AB and for the rod BC, with opposite signs in the two cases.]

Ex. 3. Roberval's Balance. This balance, which is a common form of letter-weigher, consists of four rods AB, BE, ED, and DA freely jointed at the corners A, B, E, and D so as to form a parallelogram, whilst the middle points, C and F, of AB and ED are attached to fixed points C and F which are in a vertical straight line. The rods AB and DE can freely turn about C and F.

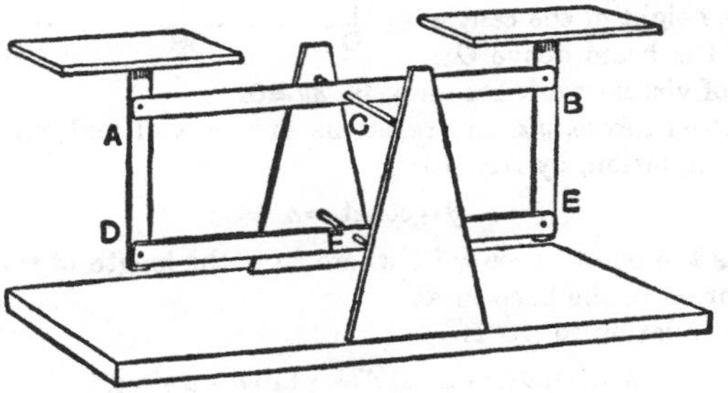

To the rods AD and BE are attached equal scale-pans. In one of these is placed the substance W which is to be weighed and in the other the counterbalancing weight P.

We shall apply the Principle of Virtual Work to prove that it is immaterial on what part of the scale-pans the weights P and W are placed.

Since $CBEF$ and $CADF$ are parallelograms it follows that, whatever be the angle through which the balance is turned, the rods BE and AD are always parallel to CF and therefore are always vertical.

If the rod AB be turned through a small angle the point B rises as much as the point A falls. The rod BE therefore rises as much as AD falls, and the right-hand scale-pan rises as much as the left-hand one falls. In such a displacement the virtual work of the weights of the rod BE and its scale-pan is therefore equal and opposite to the virtual work of the weights of AD and its scale-pan. These virtual works therefore cancel one another in the equation of virtual work.

Also if the displacement of the right-hand scale-pan be p upwards, that of the left-hand one is p downwards. The equation of virtual work therefore gives

$$P.p+W(-p)=0, \text{ i.e. } P=W.$$

Hence, if the machine balance in any position whatever, the weights P and W are equal, and this condition is independent of the position of the weights in the scale-pans. The weights therefore may have any position on the scale-pans. It follows that the scale-pans need not have the same shape, nor be similarly attached to the machine, provided only that their weights are the same.

Ex. 4. *A uniform beam rests tangentially upon a smooth curve in a vertical plane and one end of the beam rests against a smooth vertical wall; if the beam is in equilibrium in any position, find the equation to the curve.*

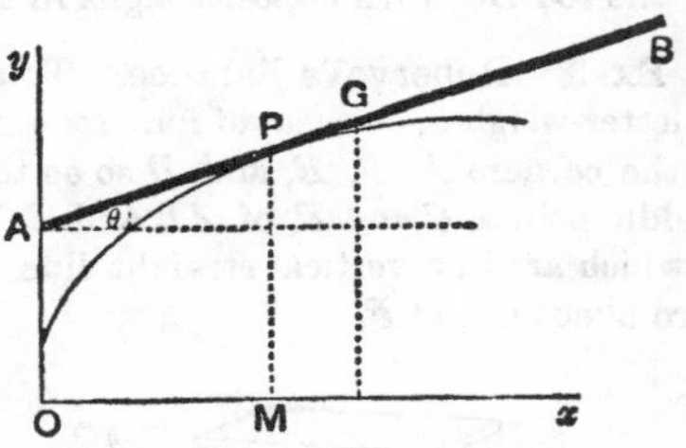

Take the wall as the axis of y and any point O on it as the origin.

If $\bar{y}$ be the height of the centre of gravity of the beam above Ox, the equation of virtual work becomes $W.\delta\bar{y}=0$.

[For the other forces, *viz.* the reactions of the wall and curve, do not enter into the equation, by Art. 104.]

$$\therefore\ \bar{y}=\text{const.}=h.$$

Hence G is the point $(a\cos\theta, h)$, where $2a$ is the length of the rod, and θ is its inclination to the horizontal.

Hence the equation to AG is

$$y-h=\tan\theta\,(x-a\cos\theta)=x\tan\theta-a\sin\theta.$$

For its envelope, differentiating with respect to θ, we have $x=a\cos^3\theta$ and $y-h=-a\sin^3\theta$.

$\therefore\ x^{\frac{2}{3}}+(y-h)^{\frac{2}{3}}=a^{\frac{2}{3}}$, so that the required curve is a portion of a four-cusped hypocycloid.

EXAMPLES

1. Four equal heavy uniform rods are freely jointed so as to form a rhombus which is freely suspended by one angular point, and the middle points of the two upper rods are connected by a light rod so that the rhombus cannot collapse. Prove that the tension of this light rod is $4W\tan\alpha$, where W is the weight of each rod and 2α is the angle of the rhombus at the point of suspension.

2. A string, of length a, forms the shorter diagonal of a rhombus formed of four uniform rods, each of length b and weight W, which are hinged together. If one of the rods be supported in a horizontal position, prove that the tension of the string is

$$\frac{2W(2b^2-a^2)}{b\sqrt{4b^2-a^2}}.$$

3. A regular hexagon $ABCDEF$ consists of six equal rods which are each of weight W and are freely jointed together. The hexagon rests in a vertical plane and AB is in contact with a horizontal table; if C and F be connected by a light string, prove that its tension is $W\sqrt{3}$.

4. A square framework, formed of uniform heavy rods of equal weights W jointed together, is hung up by one corner. A weight W is suspended from each of the three lower corners and the shape of the square is preserved by a light rod along the horizontal diagonal. Prove that its tension is $4W$.

5. Four equal jointed rods, each of length a, are hung from an angular point, which is connected by an elastic string with the opposite point. If the rods hang in the form of a square, and if the modulus of elasticity of the string be equal to the weight of a rod, shew that the unstretched length of the string is $\frac{a\sqrt{2}}{3}$.

6. Four rods are jointed together to form a parallelogram, the opposite joints are joined by strings forming the diagonals, and the whole system is placed on a smooth horizontal table. Shew that their tensions are in the same ratio as their lengths.

7. Six equal heavy beams are freely jointed at their ends to form a hexagon, and are placed in a vertical plane with one beam resting on a horizontal plane; the middle points of the two upper slant beams, which are inclined at an angle θ to the horizon, are connected by a light cord. Shew that its tension is $6W\cot\theta$, where W is the weight of each beam.

8. A regular hexagon is composed of six equal heavy rods freely jointed together, and two opposite angles are connected by a string, which is horizontal, one rod being in contact with a horizontal plane; at the middle point of the opposite rod is placed a weight W_1; if W be the weight of each rod, shew that the tension of the string is $\frac{3W+W_1}{\sqrt{3}}$.

9. Six equal heavy rods, freely hinged at their ends, form a regular hexagon *ABCDEF* which when hung up by the point *A* is kept from altering its shape by two light rods *BF* and *CE*. Prove that the thrusts of these rods are $\frac{5\sqrt{3}}{2}W$ and $\frac{\sqrt{3}}{2}W$, where *W* is the weight of either rod.

[First give the system a virtual displacement in which *AB* and *AF* remain fixed and *BC* and *FE* become equally inclined to the vertical, and so obtain the tension of *CE*; then give the system a displacement in which *BC* and *FE* both remain vertical, and the rest of the rods are still equally inclined to the vertical.]

10. A flat semi-circular board with its plane vertical and curved edge upwards rests on a smooth horizontal plane and is pressed at two given points of its circumference by two beams which slide in smooth vertical tubes. If the board is in equilibrium, find the ratio of the weights of the beams.

11. Two equal uniform rods *AB* and *AC*, each of length $2b$, are freely jointed at *A* and rest on a smooth vertical circle of radius a. Shew that, if 2θ be the angle between them, then

$$b\sin^3\theta = a\cos\theta.$$

[The weights, *W*, are the only forces that come into the equation of virtual work, and the height of the centre of gravity of each rod above the centre of the circle is $\frac{a}{\sin\theta} - b\cos\theta$.

$$\therefore\ 2W\delta\left[\frac{a}{\sin\theta} - b\cos\theta\right]=0,\quad i.e.\ -\frac{a}{\sin^2\theta}\cos\theta\,.\,\delta\theta + b\sin\theta\,.\,\delta\theta = 0, \text{ etc.}]$$

12. A prism whose cross section is an equilateral triangle rests with two edges on smooth planes inclined at angles α, β to the horizon. If θ be the angle which the plane containing these edges makes with the vertical, shew that

$$\tan\theta = \frac{2\sqrt{3}\sin\alpha\sin\beta + \sin(\alpha+\beta)}{\sqrt{3}\sin(\alpha-\beta)}.$$

13. Two small smooth rings of equal weight slide on a fixed elliptical wire, whose major axis is vertical, and they are connected by a string which passes over a small smooth peg at the upper focus; shew that the weights will be in equilibrium wherever they are placed.

14. Four equal uniform rigid rods, each of weight *W*, jointed together at their ends so as to form a rhombus are hung from one corner and kept approximately in the form of a square by means of weightless rods which form the diagonals. Assuming that the very small extensions or compressions, whichever the diagonals undergo, are proportional to the tensions or thrusts they exert, prove that these forces are each equal to *W*.

15. Two small rings, of equal weight, slide on a smooth wire in the shape of a parabola, whose axis is vertical and vertex upwards, and attract one another with a force which varies as the distance; if they can rest in any symmetrical position on the wire, shew that they will rest in all symmetrical positions.

16. A smooth rod passes through a smooth ring at the focus of an ellipse whose major axis is horizontal, and rests with its lower end on the quadrant of the curve which is furthest removed from the focus. Find its position of equilibrium, and shew that its length must at least be $\frac{3a}{4}+\frac{a}{4}\sqrt{1+8e^2}$, where $2a$ is the major axis and e is the eccentricity.

17. One end of a beam rests against a smooth vertical wall and the other end on a smooth curve in a vertical plane perpendicular to the wall; if the beam rests in all positions, shew that the curve is an ellipse whose major axis lies along the horizontal line described by the centre of gravity of the beam.

18. A heavy rod AB, of length $2l$, rests upon a fixed smooth peg at C and with its end B upon a smooth curve. If it rests in all positions, shew that the curve is a conchoid whose polar equation, with C as origin, is $r=l+\frac{a}{\sin\theta}$.

19. A small heavy ring P slides on a smooth wire whose plane is vertical, and is connected by a string passed over a small pulley O in the plane of the curve with another weight W which hangs freely. If the ring is in equilibrium in any position on the wire, shew that the form of the latter must be that of a conic section whose focus is at the pulley.

[If in the position of equilibrium OP is r and is inclined at θ to the vertical, the equation of virtual work gives

$$P\delta(r\cos\theta)+W\delta(l-r)=0.$$

$$\therefore\ Pr\cos\theta+W(l-r)=\text{const. etc.}]$$

20. AB is a heavy beam which can turn about a horizontal axis at A; a cord fastened to B passes over a smooth pulley C, vertically above A, and is tied at the other end to a given weight P which moves on a given smooth curve; find the form of the curve if there is equilibrium in all positions.

[If x be the depth of the middle point of the beam below A, and W be its weight, then

$$P\delta(r\cos\theta)+W\delta x=0,$$

so that $$Pr\cos\theta+Wx=\text{const.}$$

Also $(l-r)^2=c^2+4a^2+4cx$, where $AB=2a$, $AC=c$, and l is the length of the cord. Eliminate x.]

CHAPTER VI

GRAPHIC SOLUTIONS

107. THE resultant of a system of forces acting at a point may be obtained graphically by means of the Polygon of Forces. For, (Fig. Art. 24,) forces acting at a point O and represented in magnitude and direction by the sides of the polygon $ABCDEF$ are in equilibrium. Hence the resultant of forces represented by AB, BC, CD, DE, and EF must be equal and opposite to the remaining force FA, *i.e.* the resultant must be represented by AF.

It follows that the resultant of forces P, Q, R, S, and T acting on a particle may be obtained thus; take a point A and draw AB parallel and proportional to P, and in succession BC, CD, DE, and EF parallel and proportional respectively to Q, R, S, and T; the required resultant will be represented in magnitude and direction by the line AF. The same construction would clearly apply for any number of forces.

Many problems which would be difficult or, at any rate, very laborious to solve by analytical methods are comparatively easy to solve graphically. These questions are of common occurrence in engineering and other practical work. There is generally little else involved besides the use of the Triangle of Forces and Polygon of Forces.

108. Ex. 1. *$ACDB$ is a string whose ends are attached to two points, A and B, which are in a horizontal line and are seven feet apart. The lengths of AC, CD, and DB are $3\frac{1}{2}$, 3, and 4 feet respectively, and at C is attached a one-pound weight. An unknown weight is attached to D of such a magnitude that, in the position of equilibrium, CDB is a right angle. Find the magnitude of this weight and the tensions of the strings.*

Let T_1, T_2, and T_3 be the required tensions and let x lbs. be the weight at D. Take a vertical line OL, one inch in length, to represent the weight, one pound, at C. Through O draw OM parallel to AC, and through L draw LM parallel to CD.

By the triangle of forces OM represents T_1, and LM represents T_2.

Produce OL vertically downwards and through M draw MN parallel to BD. Then, since LM represents T_2, it follows that T_3 is represented by MN, and x by LN. By actual measurement, we have

$OM = 3{\cdot}05$ ins., $LM = 2{\cdot}49$ ins., $MN = 5{\cdot}1$ ins., and $NL = 5{\cdot}63$ ins.

Hence the weight at D is 5·63 lbs. and the tensions are respectively 3·05, 2·49, and 5·1 lbs. wt.

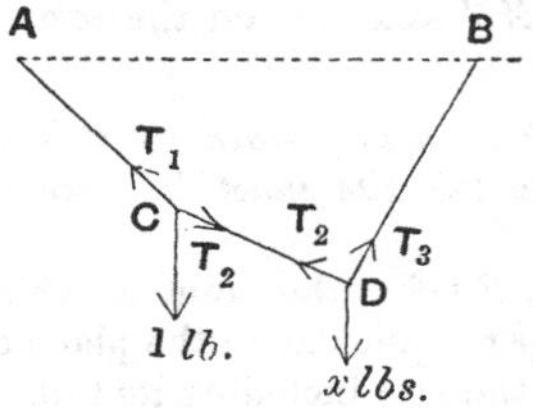

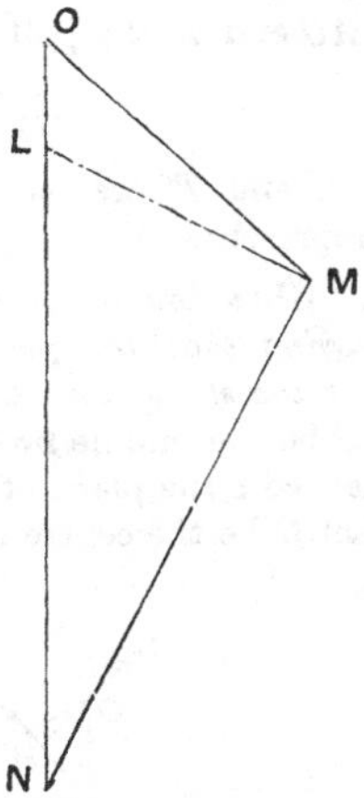

Ex. 2. The Crane. The essentials of a Crane are represented in the annexed figure. AB is a vertical post; AC a beam, called the jib, capable of turning about its end A; it is supported by a wooden bar, or chain, CD, called the tie, which is attached to a point D of the post AB. At C

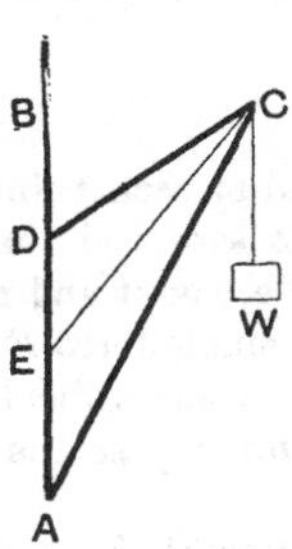

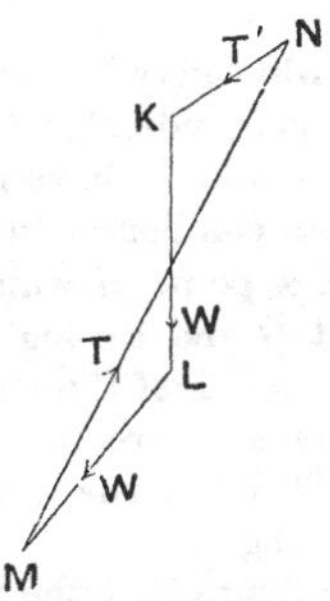

is a pulley, over which passes a chain one end of which is attached to a weight to be lifted and to the other end of which, E, is applied the force which raises W. This end is usually wound round a drum or cylinder. The tie CD is sometimes horizontal, and often the direction of the chain

CE coincides with it. In the above crane the actions in the jib and tie may be determined graphically as follows.

Draw *KL* vertically to represent *W* on any scale, and then draw *LM* equal to *KL* and parallel to *CE*; through *M* draw *MN* parallel to *AC* and through *K* draw *KN* parallel to *DC*.

Then *KLMN* is a polygon of forces for the equilibrium of *C*; for we assume the tension of the chain to be unaltered in passing over the pulley *C*, and hence that the tension of *CE* is equal to *W*. Hence, if *T* be the thrust of *AC* and *T'* the pull of *CD*, we have

$$\frac{T}{MN}=\frac{T'}{NK}=\frac{W}{KL}.$$

Hence *T* and *T'* are represented by *MN* and *NK* on the same scale that *KL* represents *W*.

Ex. 3. *Shew how the forces which act on a kite maintain it in equilibrium, proving that the perpendicular to the kite must lie between the direction of the string and the vertical.*

Let *AB* be the middle line of the kite, *B* being the point at which the tail is attached; the plane of the kite is perpendicular to the plane of the paper. Let *G* be the centre of gravity of the kite including its tail.

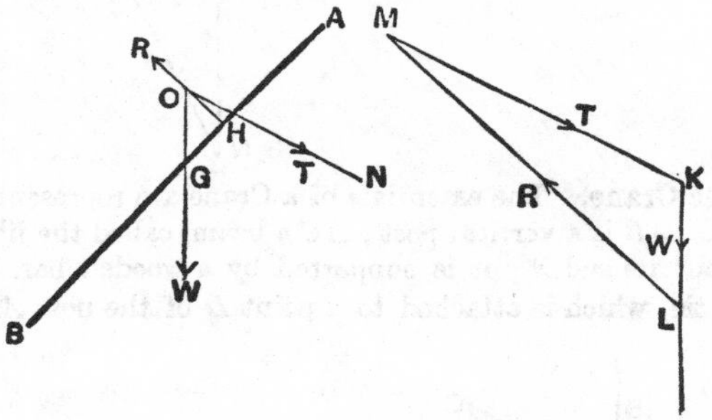

The action of the wind may be resolved at each point of the kite into two components, one perpendicular to the kite and the other along its surface. The latter components have no effect on it and may be neglected. The former components compound into a single force *R* perpendicular to the kite which acts at a point *H* which is a short distance above *G*. *R* and *W* meet at a point *O* and through it must pass the direction of the third force, *viz.* the tension *T* of the string.

Draw *KL* vertically to represent the weight *W*, and *LM* parallel to *HO* to represent *R*. Then, by the triangle of forces, *MK* must represent the tension *T* of the string.

It is clear from the figure that the line *MK* must make a greater angle with the vertical *LK* than the line *LM*, *i.e.* the perpendicular to the kite must lie between the vertical and the direction of the string.

From the triangle of forces it is clear that both *T* and *W* must be smaller than the force *R* exerted by the wind.

EXAMPLES

[The following examples are to be solved graphically.]

1. A heavy beam, AB, 10 feet long is supported, A uppermost, by two ropes attached to it at A and B which are respectively inclined at 55° and 50° to the horizontal; if AB be inclined at 20° to the horizontal, find at what distance from A its centre of gravity is. Also, if its weight be 200 lbs., find the tensions of the two ropes.

[3·16 ft.; 133 and 118·8 lbs. wt.]

2. AB is a uniform beam turning on a pivot at C and kept in equilibrium by a light string AD attached to the highest point A and to a point D vertically below C. If $AB=3$ ft., $AC=1$ ft., $CD=2$ ft., and $DA=2{\cdot}7$ ft., and the weight of the beam be 10 lbs., find the tension of the string and the reaction of the pivot. [6·75 and 16·6 lbs. wt.]

3. A cantilever consists of a horizontal rod AB hinged to a fixed support at A, and a rod DC hinged at a point C of AB and also hinged to a fixed point D vertically below A. A weight of 1 cwt. is attached at B; find the actions at A and C, given that $AB=6$ ft., $AC=2$ ft., and $AD=3$ ft., the weights of the rods being neglected. [2·83 and 3·61 cwt.]

4. The plane of a kite is inclined at 50° to the horizon, and its weight is 10 lbs. The resultant thrust of the air on it acts at a point 8 inches above its centre of gravity, and the string is tied at a point 10 inches above it. Find the tension of the string and the thrust of the air.

[26·8 and 32·1 lbs. wt.]

109. Funicular, *i.e.* Rope, Polygon. If a light cord have its ends attached to two fixed points, and if at different points of the cord there be attached weights, the figure formed by the cord is called a funicular polygon.

Let O and O_1 be the two fixed points at which the ends of the cord are tied, and let $A_1, A_2, \dots A_n$ be the points of the cord at which are attached bodies whose weights are $w_1, w_2, \dots w_n$ respectively.

Let the lengths of the portions $OA_1, A_1A_2, A_2A_3, \dots A_nO_1$ be $a_1, a_2, a_3, \dots a_{n+1}$ respectively, and let their inclinations to the horizon be

$$\alpha_1, \alpha_2, \dots \alpha_{n+1}.$$

Let h and k be respectively the horizontal and vertical distances between the points O and O_1, so that

$$a_1 \cos\alpha_1 + a_2\cos\alpha_2 + \dots + a_{n+1}\cos\alpha_{n+1} = h \quad \dots\dots(1),$$

and $$a_1 \sin\alpha_1 + a_2\sin\alpha_2 + \dots + a_{n+1}\sin\alpha_{n+1} = k \quad \dots\dots(2).$$

Let $T_1, T_2, \dots T_{n+1}$ be respectively the tensions of the portions of the cord.

Resolving vertically and horizontally for the equilibrium of the different weights in succession, we have

$$T_2 \sin \alpha_2 \quad - T_1 \sin \alpha_1 = w_1, \text{ and } T_2 \cos \alpha_2 \quad - T_1 \cos \alpha_1 = 0;$$

$$T_3 \sin \alpha_3 \quad - T_2 \sin \alpha_2 = w_2, \text{ and } T_3 \cos \alpha_3 \quad - T_2 \cos \alpha_2 = 0;$$

....

$$T_{n+1} \sin \alpha_{n+1} - T_n \sin \alpha_n = w_n, \text{ and } T_{n+1} \cos \alpha_{n+1} - T_n \cos \alpha_n = 0.$$

These $2n$ equations, together with the equations (1) and (2), are theoretically sufficient to determine the $(n+1)$ unknown tensions, and the $(n+1)$ unknown inclinations

$$\alpha_1, \alpha_2, \ldots \alpha_{n+1}.$$

From the right-hand column of equations, we have

$$T_1 \cos \alpha_1 = T_2 \cos \alpha_2 = T_3 \cos \alpha_3 = \ldots = T_{n+1} \cos \alpha_{n+1} = K \text{ (say)} \ldots(3),$$

so that the horizontal component of the tension of the cord is constant throughout and equal to K.

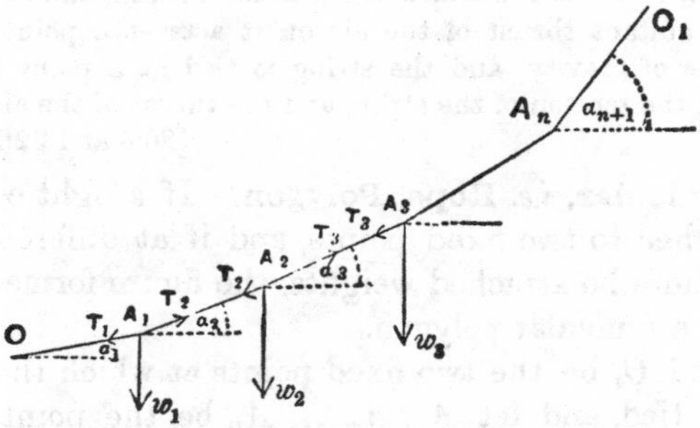

From (3), substituting for $T_1, T_2, \ldots T_{n+1}$ in the left-hand column of equations, we have

$$\tan \alpha_2 \quad - \tan \alpha_1 = \frac{w_1}{K},$$

$$\tan \alpha_3 \quad - \tan \alpha_2 = \frac{w_2}{K},$$

....

$$\tan \alpha_{n+1} - \tan \alpha_n = \frac{w_n}{K}.$$

If the weights be all equal, then $\tan \alpha_1, \tan \alpha_2, \ldots \tan \alpha_{n+1}$ are in arithmetical progression.

Hence when a set of equal weights are attached to different points of a cord, as above, the tangents of inclination to the horizon of successive portions of the cord form an arithmetical progression whose constant difference is the weight of any attached particle divided by the constant horizontal tension of the cords.

110. Graphic construction. If, in the Funicular Polygon, the inclinations of the different portions of cord be given we can easily, by geometric construction, obtain the ratios of $w_1, w_2, \dots w_n$.

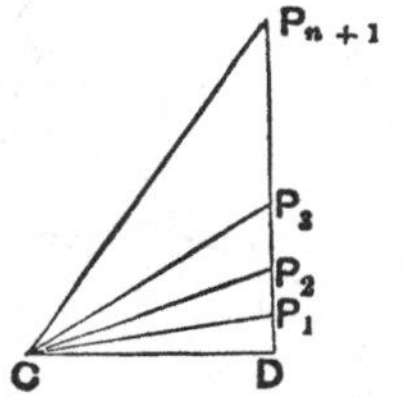

For let C be any point and CD the horizontal line through C. Draw CP_1, $CP_2, \dots CP_{n+1}$ parallel to the cords OA_1, $A_1A_2, \dots A_nO_1$, so that the angles P_1CD, $P_2CD, \dots$ are respectively $\alpha_1, \alpha_2, \dots$.

Draw any vertical line cutting these lines in $D, P_1, P_2, \dots$.

Then, by the previous article,

$$\frac{w_1}{K} = \tan\alpha_2 - \tan\alpha_1 = \frac{DP_2}{CD} - \frac{DP_1}{CD} = \frac{P_1P_2}{CD},$$

$$\frac{w_2}{K} = \tan\alpha_3 - \tan\alpha_2 = \frac{DP_3}{CD} - \frac{DP_2}{CD} = \frac{P_2P_3}{CD},$$

and so on.

Hence the quantities $K, w_1, w_2, \dots w_n$ are respectively proportional to the lines CD, P_1P_2, $P_2P_3, \dots P_nP_{n+1}$, and hence their ratios are determined.

This result also follows from the fact that CP_2P_1 is a triangle of forces for the weight at A_1, CP_3P_2 similarly for the weight at A_2, and so on.

Similarly, if the weights hung on at the joints be given and the directions of any two of the cords be also known, we can determine the directions of the others. We draw a vertical line and on it mark off P_1P_2, $P_2P_3, \dots$ proportional to the weights $W_1, W_2, \dots$. If the directions of the cords OA_1, A_1A_2 are given, we draw P_1C, P_2C parallel to them, and thus determine the point C. Join C to $P_3, P_4, \dots$ etc., and we have the directions of the rest of the cords.

111. ***To find, by a graphic construction, the resultant of any number of coplanar forces.***

Let the forces be P, Q, R, and S whose lines of action are as in the left-hand figure.

Draw the figure $ABCDE$ having its sides AB, BC, CD, and DE respectively parallel and proportional to P, Q, R, and S. Join AE, so that by the Polygon of Forces AE represents the required resultant in magnitude and direction.

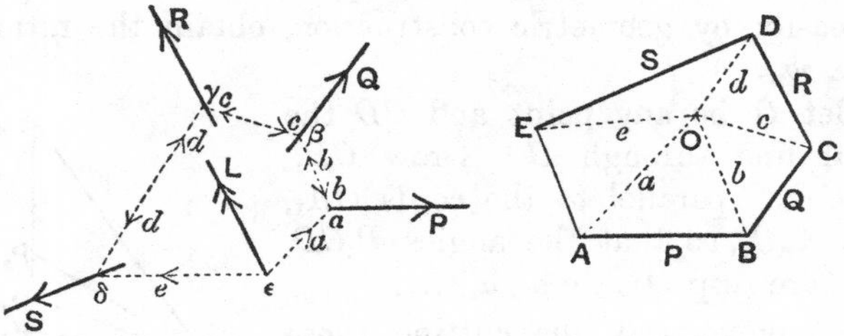

Take any point O and join it to A, B, C, D, and E; let the lengths of these joining lines be a, b, c, d, and e respectively.

Take any point α on the line of action of P; draw $\alpha\beta$ parallel to BO to meet Q in β, $\beta\gamma$ parallel to CO to meet R in γ, and $\gamma\delta$ parallel to DO to meet S in δ. Through δ and α draw lines parallel respectively to EO and OA to meet in ϵ.

Through ϵ draw ϵL parallel and equal to AE. Then ϵL shall represent the required resultant in magnitude and line of action, on the same scale that AB represents P.

For P, being represented by AB, is equivalent to forces represented by AO and OB and therefore may be replaced by forces equal to a and b in the directions $\epsilon\alpha$ and $\beta\alpha$. So Q may be replaced by b and c in directions $\alpha\beta$ and $\gamma\beta$, R by c and d in directions $\beta\gamma$ and $\delta\gamma$, and S by forces d and e in directions $\gamma\delta$ and $\epsilon\delta$.

The forces P, Q, R, and S have therefore been replaced by forces acting along the sides of the figure $\alpha\beta\gamma\delta\epsilon$, of which the forces along $\alpha\beta$, $\beta\gamma$, and $\gamma\delta$ balance. Hence we have left forces at ϵ which are parallel and equal to AO and OE, whose resultant is AE.

Since ϵL is drawn parallel and equal to AE, it therefore represents the required resultant in magnitude and line of action.

Such a figure as $ABCDE$ is called a Force Polygon and one such as $\alpha\beta\gamma\delta\epsilon$ is called a Link or Funicular Polygon, because it represents a set of links or cords in equilibrium.

112. If the point E of the Force Polygon coincides with the point A it is said to close, and then the resultant force vanishes.

If the Force Polygon closed, but the Funicular Polygon did not close, *i.e.* if $\delta\epsilon\alpha$ was not a straight line, we should have left forces acting at δ and α parallel to OE and AO, *i.e.* we should in this case have two equal, opposite, and parallel forces forming a couple.

If however the Funicular Polygon also closed, then $\delta\epsilon\alpha$ would be a straight line and these two equal, opposite, and parallel forces would now be in the same straight line and would balance.

Hence, if the forces P, Q, R, S are in equilibrium, both their Force and Funicular Polygons must close.

113. If the forces be parallel the construction is the same as in the previous article. The annexed figure is drawn for the case in which the forces are parallel and two of the five forces are in the opposite direction to that of the other three.

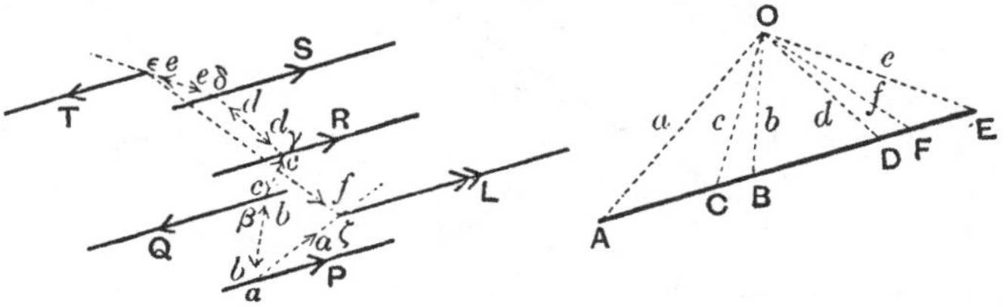

Since P, R, and S are in the same direction we have AB, CD, and DE in one direction, whilst BC and EF which represent Q and T are in the opposite direction.

The proof of the construction is the same as in the last article. The line ζL, equal and parallel to AF, represents the required resultant both in magnitude and line of action.

This construction clearly applies to finding the resultant weight of a number of weights.

Ex. *A uniform beam HK, of length* 12 *feet and weight* 5 *cwt., is supported at H and K so as to be horizontal; at points L and M on it, such that $HL = 2$ ft. and $HM = 8$ ft., weights of* 4 *cwt. and* 3 *cwt. are placed; find graphically the reactions at H and K.*

Measure AB, BC, CD vertically on a scale in which 1 cwt. is represented by half an inch, so that $AB=2$ ins., $BC=2\frac{1}{2}$ ins., and $CD=1\frac{1}{2}$ ins.

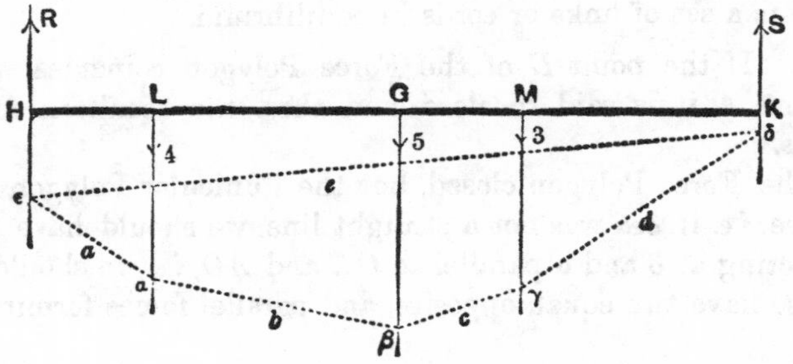

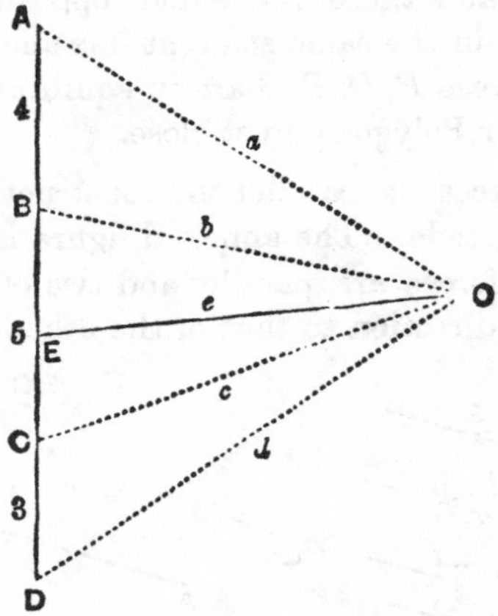

Take any convenient pole O.

From any point α on the vertical through L draw $\alpha\epsilon$, $\alpha\beta$ parallel to OA, OB to meet the verticals through H and G in ϵ and β; draw $\beta\gamma$ parallel to CO, $\gamma\delta$ parallel to DO, and join $\delta\epsilon$. Draw OE parallel to $\delta\epsilon$ to meet AB in E. Then DE, EA clearly represent S and R. On measurement, we have $R=6{\cdot}83$ cwt. and $S=5{\cdot}17$ cwt.

114. There is another system of lettering (known as Bow's or Henrici's system) that may be used conveniently in the work of Art. 111.

Let the space between the forces P and Q be called B; that between Q and R be called C and so on.

Then the line of action of P is the boundary between the spaces A and B, and hence in the second figure the line that represents it may conveniently be called *ab*. The Force Polygon therefore is named *abcde*.

When the pole *o* has been taken and the Funicular Polygon $\alpha\beta\gamma\delta\epsilon$ has been drawn, the space within the latter is then conveniently called *O*.

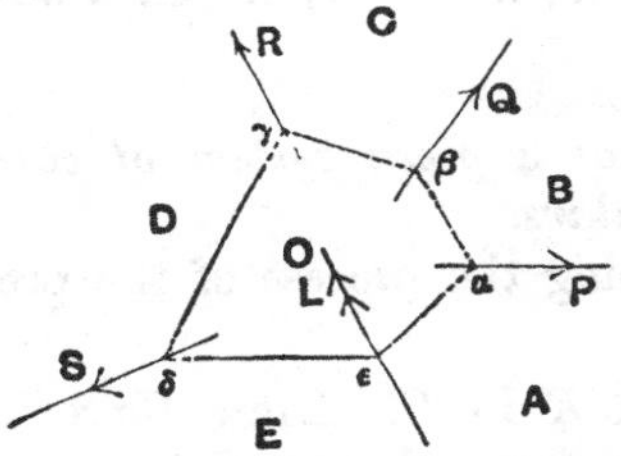

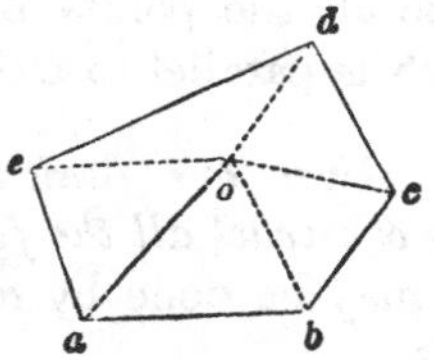

Thus a small letter attached to an angle of the Force Polygon corresponds to a big letter attached to a space of the Funicular Polygon.

The point of application, α, of the force P is the meeting point of the spaces A, B and O and hence may be called the point ABO; the corresponding triangle of forces for this point is *abo*. Similarly for the other forces.

115. *If any two funiculars of a given system of forces corresponding to two poles O and O′ be drawn, the locus of the intersection of their corresponding sides is a straight line which is parallel to OO′.*

Let $\alpha'\beta'\gamma'$... be the funicular, constructed as in Art. 111, corresponding to the second pole O'. Reverse all the forces $P, Q, R, \ldots$ acting at $\alpha', \beta', \gamma', \ldots$

Resolve Q at β into forces, equal to BO and OC, along $\alpha\beta$ and $\gamma\beta$; and the reversed Q at β' into forces, equal to $O'B$ and CO', along $\beta'\alpha'$ and $\beta'\gamma'$; and take moments about U, the intersection of $\alpha\beta$ and $\alpha'\beta'$.

Then since these four components are in equilibrium the sum of their moments about U is zero Also two of them pass through U. Hence the moments of the forces along $\gamma\beta$ and $\beta'\gamma'$ (equal to OC and CO' respectively) about U is zero.

Hence their resultant passes through U. But it clearly also passes through V, the intersection of $\beta\gamma$ and $\beta'\gamma'$. Hence their resultant is in the line UV.

But by the right-hand figure the resultant of forces represented by OC, CO' is parallel to OO', so that UV is parallel to OO'.

Similarly the line VW, joining V to the intersection, W, of the lines $\gamma\delta$ and $\gamma'\delta'$, is parallel to OO'.

Hence all the points U, V, W, ... clearly lie on a straight line which is parallel to OO'.

116. *Given one funicular of a given system of coplanar forces, to construct all the funiculars.*

This may be done by reversing the process of the previous article.

For let $\alpha\beta$ meet any line HK in U. Draw $U\alpha'\beta'$ in any arbitrary direction to meet the forces P and Q in α' and β'. Let $\beta\gamma$, $\gamma\delta$, $\delta\epsilon$, ... meet HK in $V, W, X, \ldots$.

Draw $V\beta'\gamma'$, $W\gamma'\delta'$, $X\delta'\epsilon'$, ... to meet $R, S, T, \ldots$ in $\gamma', \delta', \epsilon', \ldots$. Then, by the previous article, $\alpha'\beta'\gamma'\ldots$ is another funicular.

Also, since HK and $U\alpha'\beta'$ are both arbitrary, it is clear that an infinite number of funiculars can be thus obtained.

117. *Graphic representation of the moment of the resultant of given forces.*

Let the notation be as in Art. 111.

If we want the moment about a given point M, through it draw a line MUV parallel to the resultant L to meet in U and V the two sides of the funicular polygon which meet on this resultant. From O draw a perpendicular OH upon AE, and from M a perpendicular MN on the line of action of the resultant L.

Since the sides of the triangles $U\epsilon V$ and AOE are respectively parallel,

$$\therefore \frac{UV}{AE} = \frac{U\epsilon}{AO} = \frac{\text{perp}^{\text{r}} \text{ from } \epsilon \text{ on } UV}{\text{perp}^{\text{r}} \text{ from } O \text{ on } AE} = \frac{MN}{OH},$$

so that $AE.MN = UV.OH$.

Hence the sum of the moments of the four component forces P, Q, R, S about M = the moment of their resultant L about M

$$= L.MN = AE.MN$$

$$= UV.OH,$$

i.e. the moment about M is equal to the intercept, on the line through M parallel to the resultant, of the sides of the funicular polygon which meet on the resultant multiplied by the perpendicular from the pole O on the side of the force polygon which represents the resultant.

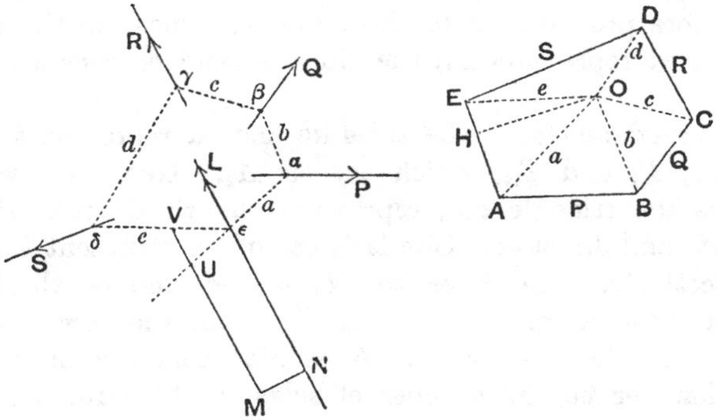

Similarly, the moment about M of any of the component forces P is equal to the intercept on the line through M parallel to P of the two sides through α of the funicular polygon multiplied by the perpendicular from the pole O on the side AB of the force polygon which represents P.

118. *A closed polygon of light rods freely jointed at their extremities is acted upon by a given system of forces acting at the joints which are in equilibrium; find the actions along the rods.*

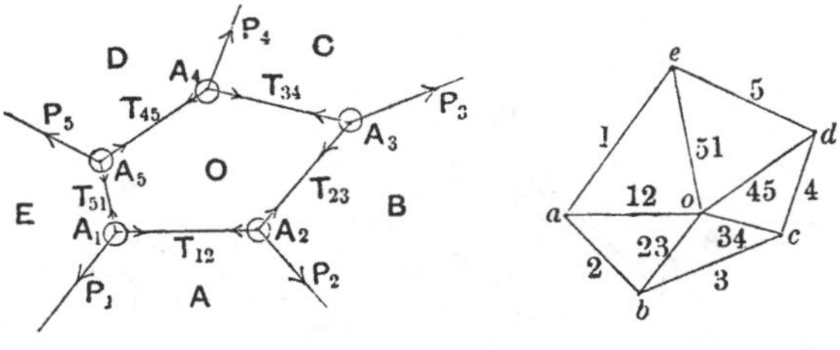

Let A_1A_2, A_2A_3, ... A_5A_1 be a system of five rods freely jointed at their ends, and at the joints let given forces P_1, P_2, P_3, P_4, and P_5 act as in the figure. Let the consequent actions along the rods be T_{12}, T_{23}, T_{34}, T_{45}, T_{51} as marked.

Draw the pentagon *abcde* having its sides parallel and proportional to the forces $P_1, P_2, \ldots P_5$. Since the forces are in equilibrium this force polygon is a closed figure.

Through *a* draw *ao* parallel to A_1A_2 and through *e* draw *eo* parallel to A_5A_1. Now the triangle *eoa* has its sides parallel to the forces P_1, T_{12}, and T_{51} which act on the joint A_1. Its sides are therefore proportional to these forces; hence, on the same scale that *ea* represents P_1, the sides *ao* and *oe* represent T_{12} and T_{51}.

Join *oc, od,* and *oe.* The sides *ab* and *oa* represent two of the forces, P_2 and T_{12}, which act on A_2. Hence *bo,* which completes the triangle *aob,* represents the third force T_{23} in magnitude and direction. Similarly *oc* and *od* represent T_{34} and T_{45} respectively. The lines *oa, ob, oc, od,* and *oe* therefore represent, both in magnitude and direction, the forces along the sides of the framework. A similar construction would apply whatever be the number of sides in the framework.

119. It is clear that the figure and construction of the preceding article are really the same as those of Art. 111.

If the right-hand figure represents a framework of rods *ab, bc, cd,* ... acted on at the joints by forces along *ao, bo,* ..., then the polygon $A_1A_2A_3A_4A_5$ of the left-hand figure is clearly its force polygon, since A_1A_2, A_2A_3, ... are respectively parallel to *ao, bo,* ....

Hence either of these two polygons may be taken as the Framework, or Funicular Polygon, and then the other is the Force Polygon. For this reason such figures are called Reciprocal.

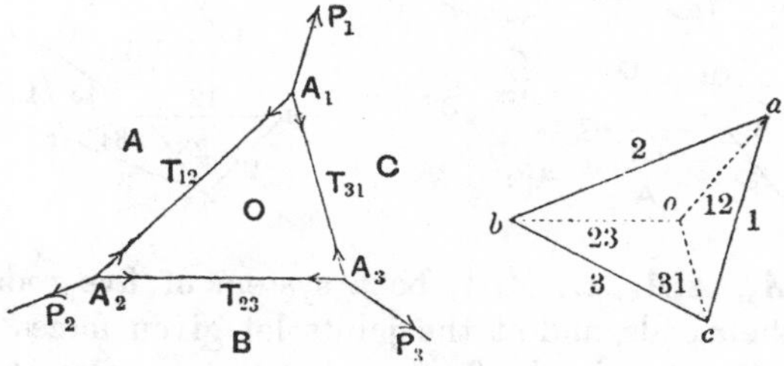

As another example we give a triangular framework acted

on at its joints by three forces P_1, P_2, P_3 in equilibrium whose force polygon is abc; conversely, $A_2A_3A_1$ is the force polygon for the triangle abc acted on by forces T_{12}, T_{23}, and T_{31}.

120. Ex. 1. *A framework, $A_1A_2A_3A_4$, consisting of light rods stiffened by a brace A_1A_3, is kept in a vertical plane by supports at A_1 and A_4, so that A_1A_4 is horizontal; the lengths of A_1A_2, A_2A_3, A_3A_4 and A_4A_1 are* 3, 2, 3, *and* 4 *feet respectively; also A_1A_4 and A_2A_3 are parallel, and A_1A_2 and A_3A_4 are equally inclined to A_1A_4. If weights of* 10 *and* 5 *cwt. respectively be placed at A_2 and A_3, find the reactions of the supports at A_1 and A_4, and the forces exerted by the different portions of the framework.*

Let the forces in the sides be as marked in the figure and let P_1 and P_4 be the reactions at A_1 and A_4.

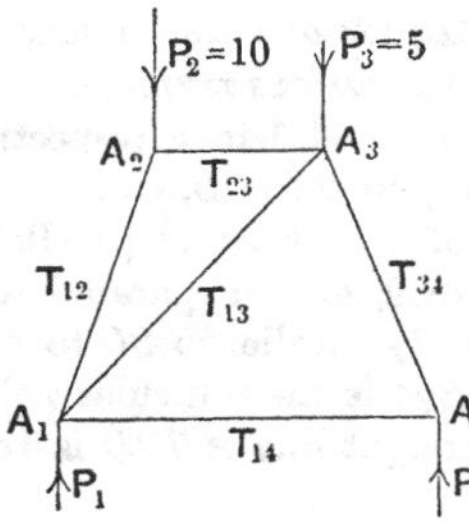

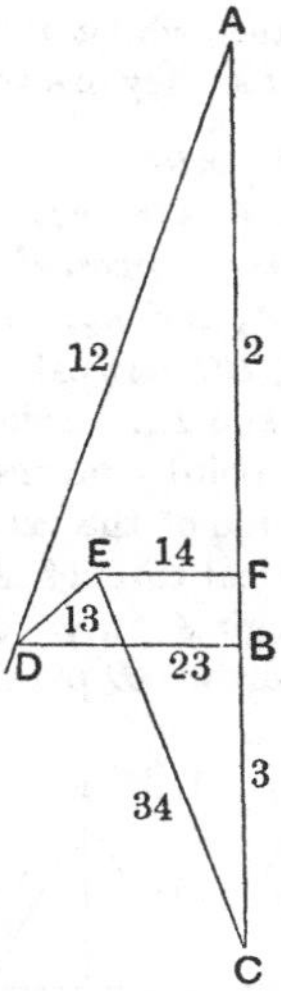

Draw a vertical line AB, 5 inches in length, to represent the weight 10 cwt. at A_2; also draw AD parallel to A_1A_2 and BD parallel to A_2A_3. Then ABD is the Triangle of Forces for the joint A_2.

Note that the force at A_2 in the bar A_2A_3 must be along A_2A_3 or A_3A_2, and that at A_3 along A_3A_2 or A_2A_3.

[Any bar, which undergoes stress, as in this case, is either resisting a tendency to compress it, or a tendency to stretch it. In the first case, the action at each end is from its centre towards its ends, in which case the bar is called a Strut; in the second case the action at each end is towards its centre, when the bar is called a Tie. In *either* case the actions at the two ends of the bar are equal and opposite.]

Draw BC vertical and equal to $2\frac{1}{2}$ ins. to represent the weight at A_3. Draw CE parallel to A_3A_4 and DE parallel to A_1A_3. Then $DBCE$ is the Polygon of Forces for the joint A_3.

Draw EF horizontal to meet AC in F. Then ECF is the Triangle of Forces for A_4, so that the reaction P_4 is represented by CF and T_{14} by FE.

Finally, for the joint A_1, we have the polygon $DEFA$, so that P_1 is represented by FA.

On measuring, we have, in inches,

$$EF=1{\cdot}10,\quad CE=3{\cdot}31,\quad DB=1{\cdot}77,\quad DA=5{\cdot}30,\quad DE={\cdot}91,$$
$$CF=3{\cdot}125, \text{ and } FA=4{\cdot}375.$$

Hence, since one inch represents 2 cwt., we have, in cwts.,

$$T_{14}=2{\cdot}20,\quad T_{34}=6{\cdot}62,\quad T_{23}=3{\cdot}54,\quad T_{12}=10{\cdot}60,\quad T_{13}=1{\cdot}82,$$
$$P_4=6{\cdot}25, \text{ and } P_1=8{\cdot}75.$$

Also from the order of the forces in the triangles and polygons of forces it is clear that the bars A_1A_3 and A_1A_4 are in a state of tension, *i.e.* they are ties, whilst the other bars of the framework are in a state of compression, *i.e.* they are struts.

Ex. 2. *A portion of a Warren girder consists of a light frame composed of three equilateral triangles $A_1A_2A_5$, $A_5A_2A_3$, $A_5A_3A_4$ and rests with $A_1A_5A_4$ horizontal, being supported at A_1 and A_4. Loads of 2 and 1 tons are suspended from A_2 and A_3; find the stresses in the various members.*

Draw AB, BC vertically, equal to 2 inches and 1 inch respectively, to represent P_2 and P_3. Take any pole O and join OA, OB, OC.

Take any point a on the line of action of P_2; draw $a\delta$ parallel to OA to meet the line of the reaction P_1 at A_1 in δ, and $a\beta$ parallel to OB to meet the vertical through A_3 in β, and then $\beta\gamma$ parallel to CO to meet the vertical through A_4 in γ. Join $\gamma\delta$; then $a\beta\gamma\delta$ is the funicular polygon of which (if we draw OD parallel to $\gamma\delta$) the straight line $ABCD$ is the force polygon.

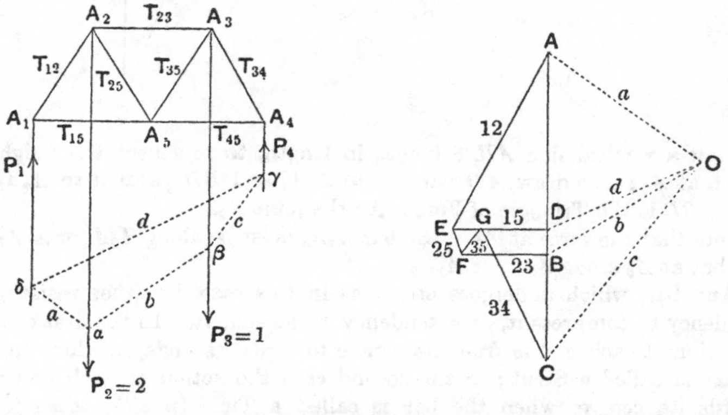

Hence P_1 and P_4 are represented by DA and CD. Let the forces exerted by the rods, whether thrusts or tensions, be T_{12}, T_{23}, ... as marked.

Draw DE, AE parallel to A_1A_5 and A_1A_2; then AED is a triangle of forces for A_1, so that AE and ED represent T_{12} and T_{15}.

Draw EF, BF parallel to A_2A_6 and A_2A_3, so that $EABF$ is the polygon of forces for the joint A_2, and thus EF and FB represent T_{25} and T_{23}.

Draw CG, FG parallel to A_3A_4 and A_3A_5 (which meet on DE), so that $FBCG$ is the polygon of forces for the joint A_3, and thus FG and GC represent T_{35} and T_{34}.

Then CDG is the triangle of forces for A_4, so that DG represents T_{45}.

Finally, $EDGF$ is the polygon of forces for the joint A_5.

On measuring off the various lengths in inches, we have

$$P_1 = 1{\cdot}75, \quad P_4 = 1{\cdot}25, \quad T_{12} = 2{\cdot}02, \quad T_{23} = {\cdot}87, \quad T_{34} = 1{\cdot}44, \quad T_{45} = {\cdot}72,$$
$$T_{15} = 1{\cdot}01, \quad T_{25} = T_{35} = {\cdot}29 \text{ ton's wt. respectively.}$$

Also, from the order of the forces in the triangles and polygons of forces, it is clear that A_1A_5, A_5A_4 and A_5A_3 are ties, and the others are struts.

[A girder, consisting of a number of portions like the part in the figure, is called a Warren girder after the name of its inventor, Capt. Warren, who introduced it about the year 1850.]

Ex. 3. *$A_1A_2A_3A_4A_5A_6$ is a roof-truss as in the figure; at the points A_2, A_3, A_4, A_6 act forces P_2, P_3, P_4, P_6 in the directions marked; equilibrium is kept by means of a reaction P_1 at A_1 which is unknown in magnitude and direction, and an unknown reaction P_5 at A_5 in the given direction A_5X; find these reactions and also the tensions or thrusts of the rods forming the truss.*

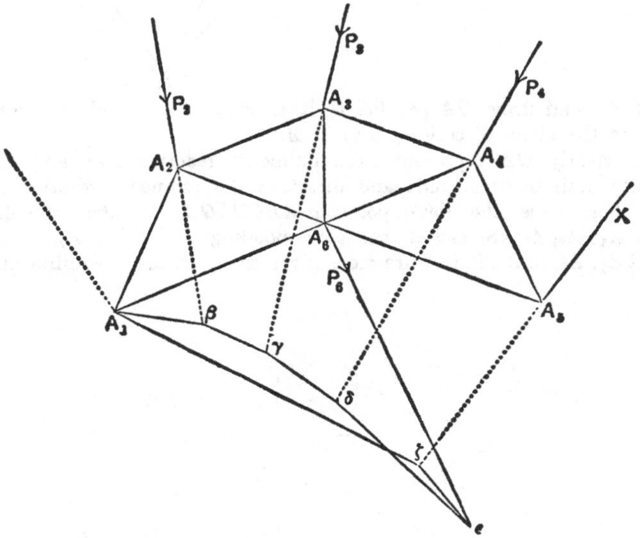

We must first find the magnitudes of P_1 and P_5.

On any convenient scale draw BC, CD, DE, EG to represent P_2, P_3, P_4, P_6 in magnitude and direction, and take any pole O.

Starting with A_1 (which is the only point we know on the line of action of the unknown force P_1) draw the funicular polygon $A_1\beta\gamma\delta$... in which $A_1\beta$, $\beta\gamma$, $\gamma\delta$, $\delta\epsilon$, $\epsilon\zeta$ are parallel respectively to OB, OC, OD, OE, and OG.

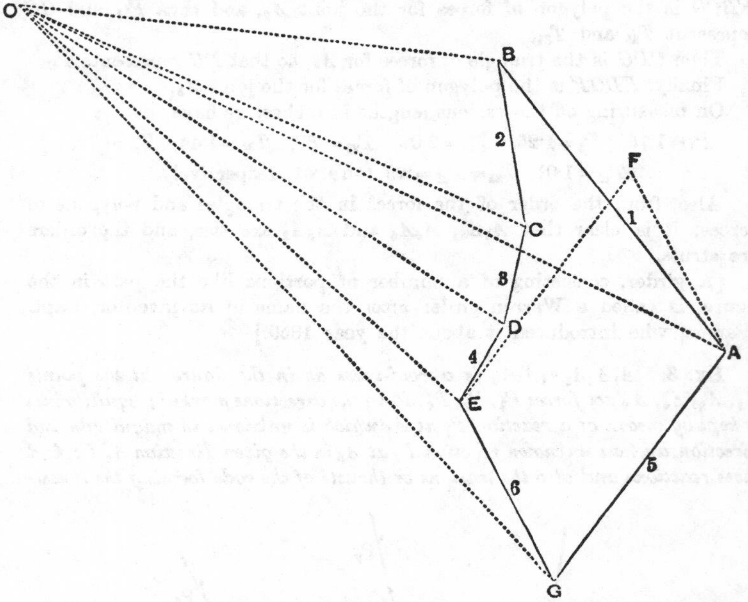

Join $A_1\zeta$ and draw OA parallel to it to meet GA (which is drawn parallel to the given direction A_5X) in A.

Then clearly GA represents the unknown reaction P_5; and AB represents both in magnitude and direction the unknown reaction P_1 at A_1. For, since the force polygon $ABCDEGA$ and the funicular polygon $A_1\beta\gamma\delta\epsilon\zeta A_1$ are closed, the corresponding forces P_1, P_2, ... P_6 acting at A_1, A_2, ... A_6, in the directions given or found, are in equilibrium.

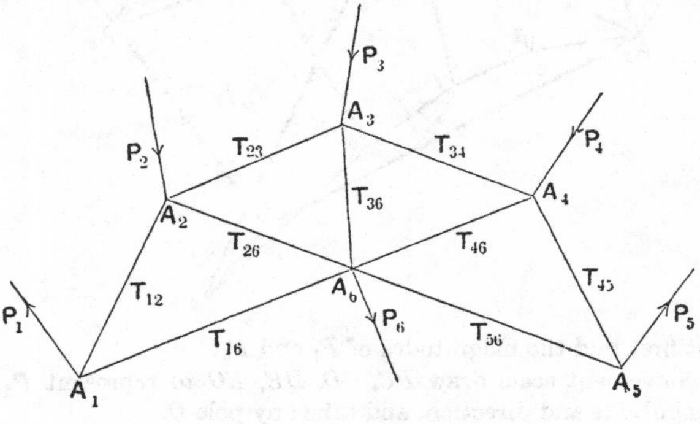

To draw the force polygon as neatly as possible we should have the forces in it in the order in which they are in the frame. Draw then *EF*, *FA* parallel to *GA*, *GE* and we have as the force polygon the figure *ABCDEFA*.

Draw *BK*, *AK* parallel to A_1A_2 and A_1A_6 so that *ABK* is a triangle of forces for A_1. Through *K*, *C* draw *KL*, *CL* parallel to A_2A_6, A_2A_3 so that *BCLK* is a polygon of forces for A_2.

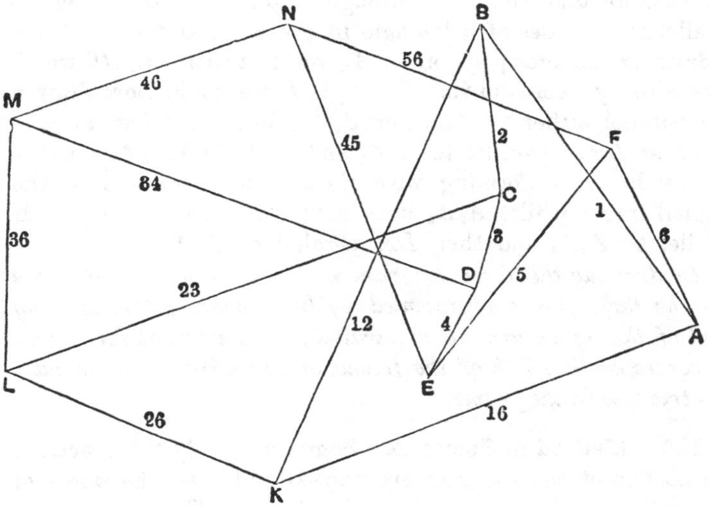

Through *L*, *D* draw *LM*, *DM* parallel to A_3A_6, A_3A_4 so that *CDML* is a polygon of forces for A_3.

Similarly, *DENM* is a polygon of forces for A_4. It is then found that *NF* is parallel to A_5A_6, so that *EFN* is a triangle of forces for A_5, and *FAKLMN* is a polygon of forces for A_6.

It is easily seen that the rods A_1A_2, A_2A_3, A_3A_4, A_4A_5 are in thrust, and that the other five are in tension.

121. There is one point to be noted which is of some importance in drawing the figure and which may be explained by the last example of the foregoing article. In drawing the forces for the joint A_2 we might have drawn a line through *C* parallel to A_2A_6 and one through *K* parallel to A_2A_3; this would not have been wholly wrong, but the figure resulting would have been more complicated than the preceding. The proper rule to observe is that the lines meeting at any point *C* of the figure should, if possible, be parallel to forces acting at two joints, and the link joining these joints, of the frame. Thus since at C there meet the forces acting at two joints A_2, A_3 of the frame, the third line through it should be a line parallel to A_2A_3; hence we draw CL parallel to A_2A_3.

But sometimes, as at the point L, we do not have lines parallel to any of the original external forces of the frame; in this case we take the lines through L to be such that they are parallel to the sides of a triangle in the original frame. Thus in drawing the force polygon for A_3, we start with LC, CD which have already been drawn. Through L we could now draw a line parallel either to A_3A_4, or A_3A_6; but the forces already drawn at L are parallel to A_2A_3 and A_2A_6, and A_3A_6 satisfies the condition of forming with these two a triangle in the original frame whilst A_3A_4 does not; we therefore draw LM parallel to A_3A_6 and then DM parallel to A_3A_4.

In drawing the force diagram we therefore should take care that the three forces represented by lines passing through any point of the figure are either parallel to two external forces and the corresponding link of the frame, or are parallel to the sides of a triangle in the frame.

122. Method of Sections. Sometimes only the reactions in a portion of the diagram are wanted and then the whole of the reciprocal diagram need not be drawn. Thus suppose in Ex. 3 of Art. 120 we only want T_{16}, T_{26}, T_{23}. Imagine a line QRS drawn to cut them so that Q, R, S are respectively on A_2A_3, A_2A_6, and A_1A_6. The portion to the left of QRS is then in equilibrium under the action of the forces on it, provided we include T_{23}, T_{26}, T_{16} acting along A_2A_3, A_2A_6, A_1A_6, *i.e.* it is in equilibrium under the action of the external forces P_1, P_2, T_{23}, T_{26}, T_{16}. These are equivalent to two forces P_1 and P_2 given in magnitude and direction, T_{16} whose line of action is given, and a force X [the resultant of the two unknown tensions T_{23} and T_{26}] of which we only know its point of action A_2. The solution may thus be completed by drawing successively triangles of forces for P_1, T_{16}, and T_{12}, then for T_{12}, P_2, and X, and finally for X, T_{23} and T_{26}.

Or we may take moments about A_2, and we have

$$T_{16} \times \text{perpendicular from } A_2 \text{ on } A_1A_6$$
$$= P_1 \times \text{perpendicular from } A_2 \text{ on } A_1P_1.$$

Similarly, moments about the intersection of A_2A_3 and A_1A_6, and moments about A_1 will give T_{23} and T_{26}.

In applying this method care must be taken that the section chosen does not cut more than three of the members of the frame.

123. A number of bars jointed together at their ends is known as a framework or, more simply, as a frame. When the forces acting on such a bar are such that it is in tension, it is called a tie; when it is in compression it is called a strut.

The simplest frame is a triangle of three rods AB, BC, CA connected by joints at A, B, and C. Since the shape of a triangle is fixed when its sides are given, such a framework is unalterable in shape and is said to be *stiff*, or to be a perfect frame, whatever be the loads applied to its joints.

A frame consisting of four rods AB, BC, CD, DA hinged at A, B, C, and D is clearly not necessarily of a constant shape; for no geometrical figure, except a triangle, is given in shape when the lengths of its sides are given. It is said to be imperfect, because it does not preserve the same shape if the forces applied to its joints vary. It may however be made to become stiff by adding a diagonal bar AC hinged to the others at A and C. The forces acting along the five sides of the frame are now determinate for any given system of forces applied to its joints.

Suppose that in addition to the diagonal bar AC we now add another diagonal bar BD. The frame is now said to be redundant, because it now contains one more link or member than is necessary to determine its form; the forces acting along its members due to any given system of loading would not now be determinate.

In general, a frame is stiff if it can be dissected into a number of triangles, but it may then be redundant.

124. Non-redundant stiff frames. In two dimensions, if n be the number of joints, the number of bars must be $2n-3$, if the frame is to be both stiff and non-redundant. For, if we have three joints A, B, C, the number of bars necessary to determine them is three, viz. BC, CA, AB. Any other joint D is given in position if we are given the bars (say AD, BD) joining it to any two of the previous joints. So for any other joint E. Hence after the first three joints, two extra bars are required for each joint. The total number for n joints thus

$$=3+2(n-3)=2n-3.$$

In three dimensions, the number is easily seen to be $3n-6$. For after the first three joints A, B, C are given, the position of a fourth joint D is given if the bars AD, BD, CD are given. The position of a fifth joint E is known if we are given the bars joining it to any three of the previous four, *e.g.* if AE, BE, DE are given. Hence for any joint after the first three, three additional bars must be given. The total number required therefore

$$=3+3(n-3)=3n-6.$$

125. In the examples of Art. 120 are several non-redundant stiff frames. Sometimes it is convenient to have a redundant frame. Thus in Ex. 1 of Art. 120, A_1A_3 is a tie. If the 10 cwt. at A_2 were removed, then A_1A_3 would clearly become a strut. But, if A_1A_3 were very flexible, it would not act well as a strut; in this case it would be better to have an additional tie A_2A_4. When the weight is all at A_3 then A_2A_4 would be in action and A_1A_3 of no importance; if all the weight were at A_2 then A_2A_4 would go out of use, and A_1A_3 only would be in action.

Such members as A_1A_3 or A_2A_4, capable of being used as a strut or tie only, are called semi-members.

EXAMPLES

1. Loads of 2, 4, 3 cwt. are placed on a beam 10 ft. long at distances of 1 ft., 3 ft., 7 ft. from one end. Find by an accurate drawing the line of action of the resultant. [3·9 ft. from the end.]

2. A horizontal beam 20 feet long is supported at its ends and carries loads of 3, 2, 5, and 4 cwt. at distances of 3, 7, 12, and 15 feet respectively from one end. Find by means of a funicular polygon the thrusts on the two ends. [7·15 and 6·85 cwt.]

3. Weights of 5, 10, 12, 8, and 6 lbs. rest upon a beam at distances 1, 4, 7, 9, and 12 feet from one end. The beam is supported at distances 5 and 15 feet from the same end. Find graphically the supporting forces. [34·2 and 6·8 lbs. wt.]

4. *ABCDEF* is a regular hexagon. Shew that the forces which must act along AC, AF, and DE to produce equilibrium with a force of 40 lbs. weight acting along EC are respectively 10, 17·32, and 34·64 lbs. weight.

5. Fig. 1 consists of a symmetrical system of light rods freely jointed and supported vertically at the extremities; vertical loads of 10 and 5 cwt. are placed at the points indicated; find the thrusts or tensions of the rods, if the side rods are inclined at 50° to the horizon.

[$T_1=13{\cdot}05$, $T_2=9{\cdot}79$, $T_3=3{\cdot}26$, $T_4=8{\cdot}39$, and $T_5=5$ cwt. T_4 and T_5 are ties; the others are struts.]

6. Fig. 2 consists of a symmetrical system of light rods freely jointed and supported by vertical reactions at A and B; if a weight of 10 cwt. be placed at D, find the thrusts or tensions in the rods, given that $\angle DAB=55°$ and $\angle CAB=35°$.

[$T_1=8{\cdot}39$, $T_2=11{\cdot}98$, and $T_3=9{\cdot}62$ cwt. T_2 is a strut and T_1 and T_3 are ties.]

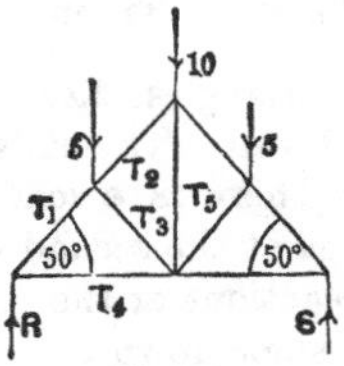

Fig. 1.

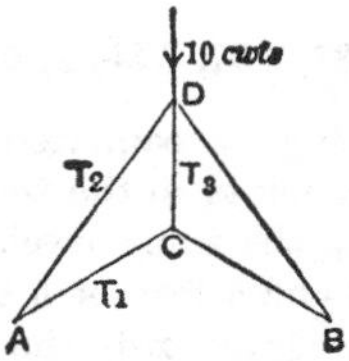

Fig. 2.

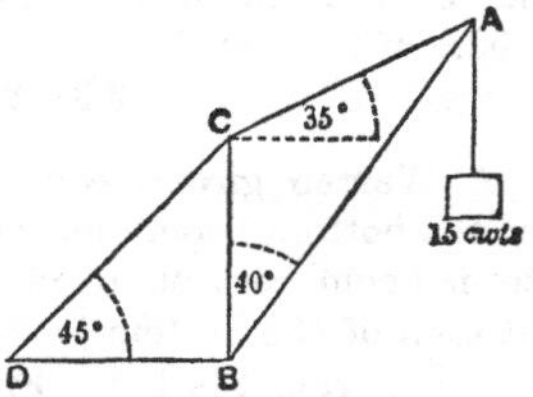

Fig. 3.

7. A crane is constructed as in Fig. 3, and 15 cwt. is hung on at A; find the forces along the parts AC and AB. If the post BC be free to move, and BD be rigidly fixed, find the pull in the tie CD.

[37·2, 47·5, and 43·1 cwt.]

8. A portion of a Warren girder consists of three equilateral triangles ABC, ADC, BCE, the lines AB, DCE being horizontal and the latter the uppermost. It rests on vertical supports at A and B and carries 5 tons at D and 3 tons at E. Find the reactions at the supports and the stresses in the four inclined members.

[6 tons and 2 tons; 5·77, 1·155, 1·155 and 3·464 tons; of the last four the first, third, and fourth are struts and the second is a tie.]

9. $ABCD$ consists of a quadrilateral consisting of four light rods loosely jointed, which is stiffened by a rod BD; at A and C act forces equal to 40 lbs. weight. Given that $AB=2$ ft., $BC=3$ ft., $CD=4$ ft., $DA=4\frac{1}{2}$ ft., and $DB=5$ ft., find the tensions or thrusts of the rods.

[The tensions of AB, BC, CD, and DA are 32·4, 36·4, 16·8, and 25·5 lbs. wt.; the thrust of BD is 36·7 lbs. wt.]

10. Two uniform equal rods, AB, BC, are freely jointed at B, and are suspended freely from a peg at A, the rods being maintained at right angles to one another by a massless string AC fastened to the peg at A; having given the weight W of each rod, shew graphically that the tension of the string is $\cdot 67\,W$, and that the stress across the joint B is equal to $\frac{1}{2}W$.

11. ABC is a horizontal line such that $AB=5$ ft., and $BC=15$ ft. D is a point vertically over B such that $BD=10$ feet and E bisects DC. AC, CD, DA, BD, BE are rods forming a framework; loads of 10 cwt. each are applied at D and E, and the system is supported at A and C. Draw the reciprocal figure, and determine the stresses in the different members of the framework.

[5·63, 13·13, 15·77, 6·76, 12·58, 5, and 9·01 cwt.]

12. A framework consisting of five bars AB, BC, CA, CD, DA, freely jointed at their extremities, is placed in a vertical plane. ABC is a right-angled triangle with AB horizontal and AC vertical and $AB=AC=10$ feet. The angle BAD is 135° and ACD is 120°. The framework supports a vertical load of 1 ton at D, equilibrium being maintained by vertical forces at A and B. Find the magnitude of these forces and the reactions in the various bars of the frame.

[3·37, 2·37, 2·37, 3·35, 1, 2·73, and 3·35 tons wt.]

13. A Warren girder, consisting of equilateral triangles, has five joints in the bottom boom and four joints in the top boom. The ends of the bottom boom rest on piers at the same level. There is a load of 3 tons at each of the bottom joints and a load of 5 tons at the second top joint, counting from the left. Find graphically the reactions at the piers and the stresses in the four inclined members which slope down towards the left. [10·625, 9·375, 8·80, 5·34, 3·90, and 7·36 tons wt.]

14. A Warren girder, of 48 feet span, has the lower boom divided into four segments and the lengths of the inclined members are 12 feet. It is loaded with a uniformly distributed load of 2 tons per foot run and a concentrated load of 50 tons at the centre. Find the magnitude and sense of the stresses in the members.

[Replace the load on each segment by a weight, equal to half the load, acting at each of the ends of the segment.]

15. A roof is in section half of a regular octagon $ABCDE$; the points A and D, and also the points B and E, are connected by tie beams and the whole of it is to be regarded as freely jointed at the points A, B, C, D, E and supported on walls of equal height at A and E. The roof is covered with a uniform covering of tiles. By a graphical method, or otherwise, obtain the magnitudes of the stresses in the different members of the roof in terms of the total weight of the roof, the weight of the framework being regarded as negligible in comparison. The weight of each section of the tiling covering the beams AB, BC, CD, DE may be assumed to act at the middle point of each beam.

16. Shew that the line of action of the resultant of any system of forces is the locus of the points of intersection of the extreme sides of all funiculars of the system.

CHAPTER VII

SHEARING STRESSES. BENDING MOMENTS

126. In this chapter we shall consider some examples of the internal actions upon a section of a beam.

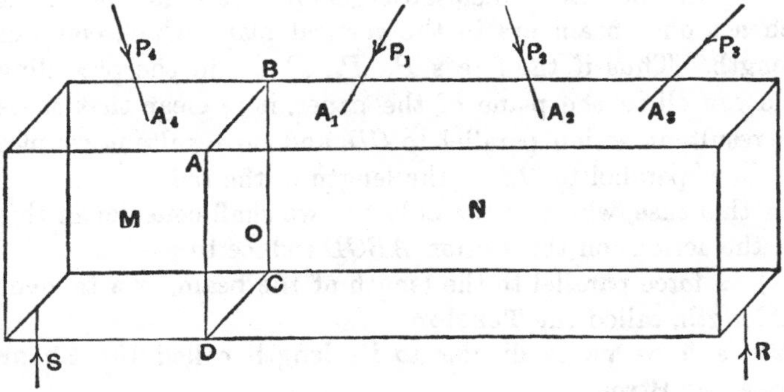

Take the case of a beam, as in the figure, supported at its ends and acted upon by forces P_1, P_2, P_3,... acting at points A_1, A_2, A_3,.... Take any section $ABCD$ of the beam. Call the part to the left of this section M and that to the right N. The action of M on N consists of a multitude of forces exerted by the fibres which cross the section $ABCD$. But, whatever these forces are, they must together with the forces P_1, P_2, P_3,... R keep the part N in equilibrium.

Hence the resultant of these forces across the section $ABCD$ must be equal and opposite to the resultant of the forces P_1, P_2, P_3,... R.

Now if we take any point O on the section $ABCD$ as origin, and lines parallel to the edges of the beam as axes, we shall find in Chap. X, that all the forces acting on N are equivalent to

three component forces at O parallel to the axes,

and three component couples at O about the axes.

These compound into a single force and a single couple.

Hence the resultant at the section $ABCD$ of the actions along the fibres must be equivalent to a single force and a single couple.

The actions of the part M on the part N are equal and opposite to those of N on M. Hence, whatever forces and couples we assume as the resultant of the actions on one side of the section $ABCD$, we must assume equal and opposite forces and couples on the other side of this section.

127. In the case which most generally occurs, the forces which act on a beam are in the vertical plane which contains its length. Thus if the forces $P_1, P_2, P_3, \ldots$ in the preceding article are all in the plane of the paper, it is clear that there is no resultant action parallel to CD, and no resultant couples about lines parallel to DA or the length of the rod.

In this case, which is the only one we shall consider in this book, the actions on the section $ABCD$ reduce to

(1) a force parallel to the length of the beam, or a tangent to its length, called the **Tension**;

(2) a force perpendicular to its length called the **Shear** or **Shearing Stress**;

(3) a resultant couple about a line perpendicular to the length of the beam called the **Bending Moment or Stress Couple.**

128. It is a matter of common experience that, in the case of a body like a lead pencil, it is the Bending Moment, and not the Tension, that causes it to break; in the case of a string, however, the Bending Moment is of no consequence, and it is the Tension which causes it to snap.

Since it is clear that the tendency of the rod to break is greater, the greater is the bending moment, the latter is always taken as the measure of the tendency to break.

The Shearing Stress and Bending Moment may both be

exhibited graphically by erecting at each point of the rod an ordinate proportional to either of them.

129. *If a horizontal beam be subjected to any vertical loading, to shew that the vertical shear S is equal to* $\frac{dM}{dx}$, *where M is the bending moment at the point considered.*

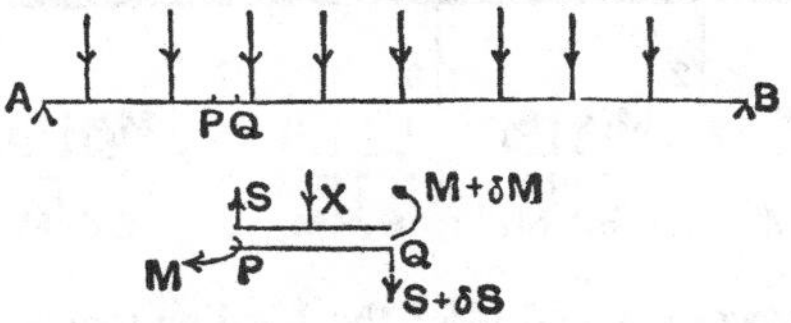

Consider any small element PQ ($= \delta x$) of the beam at a distance x from A. Let S be the upward shear at the face P and $S + \delta S$ the corresponding downward shear at the face Q. Let M be the bending moment ↻ at P and thus $M + \delta M$ the bending moment ↺ at Q.

[In the lower figure PQ is shewn magnified.]

Let X be the load per unit of length at P, so that the load on PQ may be taken to be $X\delta x$ acting at its middle point. Then, taking moments about P for the element PQ, we have

$$M = M + \delta M - (S + \delta S)\,\delta x - X\delta x \,.\, \frac{\delta x}{2},$$

$$\therefore \; S + \delta S = \frac{\delta M}{\delta x} - \tfrac{1}{2} X \,.\, \delta x.$$

Proceeding to the limit when δx is zero, we have, for any finite loading,

$$S = \frac{dM}{dx}.$$

It follows that, if we draw the curve of shearing stress and also the curve of bending moment, the ordinate to the former curve measures the slope of the latter curve.

Again, if we resolve vertically the forces acting on the element, we have

$$S = S + \delta S + X\delta x,$$

i.e. $$\frac{dS}{dx} = -X, \text{ in the limit.}$$

130. Ex. 1. ***A beam, 12 ft. long and of negligible weight, is supported at its ends and carries 2 tons at the quarter-length from one end, 3 tons at the middle, and 4 tons at the quarter-length from the other end; draw the curves of bending moment and shearing stress for the whole beam.***

Let R and R' be the reactions of the supports at the ends, so that, by taking moments about them, we easily have $R=4$ and $R'=5$ tons.

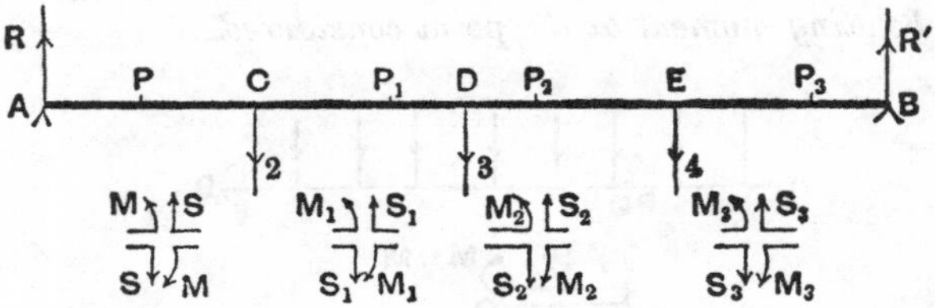

At a point P between A and C let the bending moment and shearing stress be M and S as marked; similarly for a point P_1 between C and D let these be M_1 and S_1, and so on.

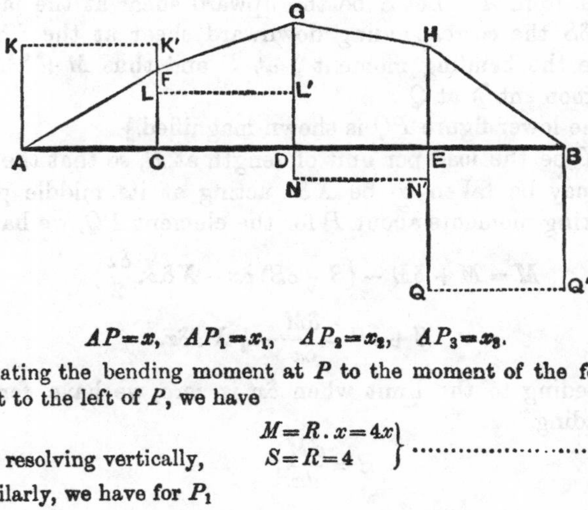

Let $\qquad AP=x, \quad AP_1=x_1, \quad AP_2=x_2, \quad AP_3=x_3.$

Equating the bending moment at P to the moment of the forces on the part to the left of P, we have

and, on resolving vertically,
$$\left.\begin{array}{l} M=R\,.\,x=4x \\ S=R=4 \end{array}\right\} \quad \text{.........................(1).}$$

Similarly, we have for P_1

and
$$\left.\begin{array}{l} M_1=R\,.\,x_1-2\,(x_1-AC)=2x_1+6 \\ S_1=R-2=2 \end{array}\right\} \quad \text{..................(2).}$$

For P_2 we have, considering the part to the right of P_2,

and
$$\left.\begin{array}{l} M_2=R'\,(12-x_2)-4\,(AE-x_2)=24-x_2 \\ S_2=4-R'=-1 \end{array}\right\} \quad \text{............(3).}$$

Finally for P_3, taking the part to the right of P_3,

and
$$\left.\begin{array}{l} M_3=R'\,(AB-x_3)=60-5x_3 \\ S_3=-R'=-5 \end{array}\right\} \quad \text{....................(4).}$$

If we erect ordinates at each point of the beam to represent the bending moment at the point, the locus of the ends will clearly be straight lines AF, FG, GH, and HB, where CF, DG, and EH represent 12, 18, and 15 foot-tons respectively.

Also the curves representing the shearing stresses will be horizontal straight lines KK', LL', NN', and QQ', where AK, CL, DN, and EQ represent 4, 2, -1, and -5 tons respectively.

In all cases where we have concentrated loads it will be found that the bending-moment curves are straight lines, so that all we need determine are the values of the bending moments at the points of loading.

There is discontinuity in the shearing stress at a point of loading such as C; this is due to the assumption that the load is concentrated at C and applied there at a single mathematical point; in practice a load cannot be applied at a single point, so that there is no such abrupt discontinuity.

Ex. 2. *AB is a stiff uniform beam, of weight W and length $2a$, supported at A and B so as to be horizontal; a weight $\frac{2W}{3}$ is placed at a point C of the beam such that $AC=\frac{a}{2}$; draw the curves of bending moment and shearing stress.*

If w is the weight of the beam per unit of length, then $W=2wa$.

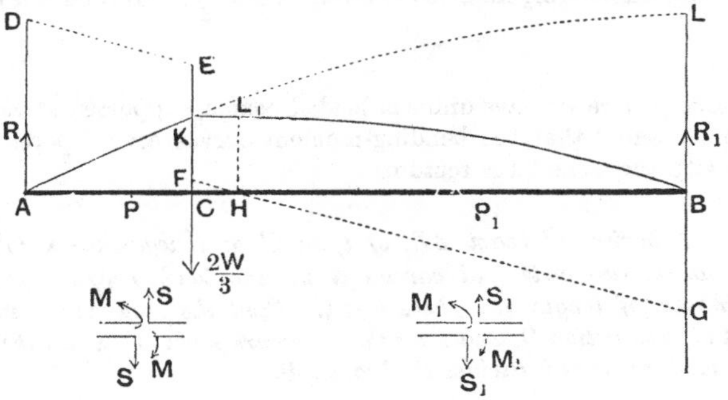

By taking moments about A and B, we easily have $R=W$ and $R_1=\frac{2W}{3}$.

For a point P between A and C, where $AP=x$, we have the bending moment M

$$=\text{moment} \curvearrowright \text{ of the forces acting on } AP \text{ about } P$$

$$=Rx-wx\,.\,\tfrac{1}{2}x=-\frac{W}{4a}[x^2-4ax] \qquad\qquad (1)$$

If S be the shearing stress there, then

$$R-S=wx,$$

so that

$$S=-\frac{W}{2a}(x-2a) \qquad\qquad (2).$$

For a point P_1 between C and B, where $AP_1 = x$, we have similarly

$$M_1 = R.x - wx.\tfrac{1}{2}x - \frac{2W}{3}\left(x - \frac{a}{2}\right) = -\frac{W}{4a}\left(x^2 - \frac{4ax}{3} - \frac{4a^2}{3}\right) \ldots\ldots(3),$$

and
$$R - S_1 = wx + \frac{2W}{3},$$

so that
$$S_1 = -\frac{W}{2a}\left(x - \frac{2a}{3}\right) \ldots\ldots\ldots\ldots(4).$$

Taking any suitable vertical distance to represent the unit of bending moment, the curve (1) represents the arc AK of a parabola whose vertex is at a point L vertically over B; whilst the curve (3) is an arc KB of an equal parabola whose vertex L_1 is on the ordinate

$$x = \frac{2a}{3}, \textit{ viz. } HL_1.$$

Again taking any suitable vertical distance to represent unit shear, then (2) is the straight line DE, and (4) is the straight line FHG.

As in the last article, the ordinate to the shear line measures the slope at the corresponding point of the bending-moment curve; each has discontinuity at the ordinate through C.

The maximum bending moment is given by $x = \frac{2a}{3}$, and then it is equal to $W.\frac{4a}{9}$.

In all cases where we have uniform loaded beams, supported at various points, we shall find that the bending-moment curves are all portions of parabolas with the same latus rectum.

Ex. 3. *A horizontal beam AB, of span $2l$ and negligible weight, is supported at its two ends and carries a moving load PQ, of uniform intensity w and of length $2a$, where $a < l$. Find the maximum bending moment at a cross section O, and shew that it occurs when O divides the load PQ in the same ratio as it divides the beam AB.*

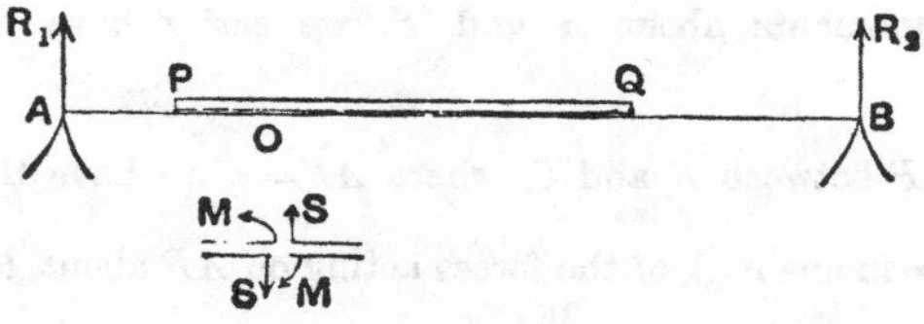

Let R_1 and R_2 be the reactions of the supports and let $AP = \xi$.

By taking moments about A and B, we have

$$R_1 = \frac{wa}{l}(2l - a - \xi) \text{ and } R_2 = \frac{wa}{l}(a + \xi).$$

Let $AO=x$ and let M be the bending moment at O. Taking moments for the part AO, we have

$$M=R_1 \cdot x-\frac{w}{2}(x-\xi)^2$$

$$=\frac{wa}{l}(2l-a-\xi)x-\frac{w}{2}(x-\xi)^2 \quad \text{......................(1).}$$

For a given position of the section O, M is a maximum when $\frac{dM}{d\xi}=0$,

i.e. when $$-\frac{ax}{l}+(x-\xi)=0,$$

i.e. when $$\xi=x\left(1-\frac{a}{l}\right) \quad \text{..............................(2),}$$

and then $$\frac{PO}{OQ}=\frac{x-\xi}{2a+\xi-x}=\frac{x}{2l-x}=\frac{AO}{OB}.$$

Substituting in (1) the value of ξ given by (2), we easily have

$$M_{\text{max.}}=\frac{wa}{l}\left(1-\frac{a}{2l}\right)(2l-x)x,$$

so that the curve of maximum bending moment is a parabola whose vertex is vertically above the middle point of AB.

Let S be the shearing stress at O. Resolving vertically for the part AO, we have

$$S=R_1-w(x-\xi)=\frac{w}{l}[2al-a^2-lx+(l-a)\xi]=\frac{dM}{dx}.$$

For a given position of the section O this clearly increases with ξ and it is thus a maximum when P is at O, and then $S_{\text{max.}}=\frac{w}{l}[2al-a^2-ax]$, and the curve of maximum shearing stress is thus a straight line.

Ex. 4. *A stiff horizontal beam AB is supported at one end A and at some other point C. If the greatest possible uniformly distributed weight is to be placed upon it without breaking it, shew that C must divide the beam in the ratio* $\sqrt{2}+1:1$.

Let $AB=2l$, $AC=y$, and let R_1 and R_2 be the reactions at A and C, so that, if w be the load per unit of length, then $R_1=2wl\frac{y-l}{y}$ and $R_2=\frac{2wl^2}{y}$.

If x be $<AC$, the bending moment for any section distant x from A

$$=R_1x-\frac{w}{2}x^2,$$

and is thus a maximum when $R_1-wx=0$.

Hence the maximum bending moment for the part AC

$$=\frac{1}{2}\frac{R_1^2}{w}=2wl^2\left(\frac{y-l}{y}\right)^2 \quad \text{..........................(1).}$$

Also the maximum bending moment for the part CB is clearly at C, and

$$=\frac{1}{2}w \cdot CB^2=\frac{w}{2}(2l-y)^2 \quad \text{.........................(2).}$$

If (1) and (2) are not equal we can lessen the greater, and hence lessen the maximum tendency to break, by altering y.

They are equal, and then the maximum bending moment is made as small as is possible, when

$$2l^2\left(\frac{y-l}{y}\right)^2=\tfrac{1}{2}(2l-y)^2 \qquad \text{.........................(3)},$$

and then $\qquad 2l-y=2l\frac{y-l}{y}, \quad i.e.\ y=l\sqrt{2}.$

In this case $\qquad \frac{AC}{CB}=\frac{y}{2l-y}=\frac{\sqrt{2}}{2-\sqrt{2}}=\sqrt{2}+1.$

The other solutions of (3) give impossible results since clearly y must be positive and greater than l and less than $2l$.

EXAMPLES

1. A beam AB is supported at its ends so as to be horizontal. Draw the curves of shearing stress and bending moment

(1) when the beam is uniformly loaded, shewing that the bending moment at P varies as $AP.PB$;

(2) when its weight is neglected but it supports a weight W at its middle point.

2. A beam AB is fixed at A so as to be horizontal there; draw the curves of shearing stress and bending moment

(1) when it is uniformly loaded;

(2) when its weight is neglected but it supports a weight W at its middle point.

3. Draw the shear and bending-moment diagrams in the case of a beam, 80 feet long, supported at the middle point, anchored at one end and loaded with 50 tons at the other end.

4. A beam, 25 feet long, is supported at one end and at a point 5 feet from the other end. It carries a distributed load of 500 pounds per foot run and a load of 10000 pounds at the overhanging end. Find the pressures on the supports, and the maximum bending moment and the section where this occurs. Draw also the curves of bending moment and of shearing stress.

[The pressures are 2187·5 and 20312·5 lbs. wt.; the greatest bending moment is at the second point of support.]

5. A beam AB, 10 feet long, is supported at two points 2 and 7 feet from A. Weights of 1 and 2 tons are placed at A and B, and in addition there is a uniformly distributed load of 2 tons per foot run between the supports. Sketch the diagrams of bending moments and shearing force and find, graphically or otherwise, where the bending moment is zero.

6. A beam AB, used as a cantilever, is anchored at A and supported at its middle point C which is at the same level as A. Draw the curves of bending moment and shearing stress

(1) when a weight of 10 tons is attached at B and a weight of 5 tons at the middle point of AC;

(2) when 10 tons is attached at B and there is a distributed weight of 5 tons on AC;

the weight of the beam being neglected in each case.

7. AB is a horizontal beam, 18 feet long, supported at A and B; C and D are points on it such that $AC = 6$ feet and $AD = 10$ feet. At C and D are placed loads of 4 and 5 tons, and there is a distributed load of 1 ton per foot from A to C; ·5 ton per foot from C to D and 2 tons per foot from D to B. Draw the curves of shearing stress and bending moment for the different portions of the beam.

8. An electric trolley-post is fixed vertically and has an overhanging arm which supports the wires at a distance of 10 feet out from the centre line of the post. Each post supports 200 lbs. of wire and the overhanging arm weighs 200 lbs. Assuming that the weight of the arm is evenly distributed along its length, sketch the curves shewing the shear and the bending moment along the length of the arm, and calculate the bending moment at the bottom of the post.

9. A horizontal beam, 25 feet long, is supported at one end A and at a point C 5 feet from the end B. The intensity of load gradually increases from $\frac{1}{2}$ ton per foot run at B to $1\frac{1}{2}$ tons per foot run at A. Find the maximum bending moment and shear force in the beam. Sketch diagrams of bending moment and shearing force.

10. AB is a stiff uniform beam, of weight W and length $2l$, and is supported at its ends so as to be horizontal; a man of weight W' stands on it at P where $AP = \xi$ $(< l)$. Shew that the curve of bending moment consists of two arcs of parabolas of equal latera recta, and that the bending moment is greatest at a point distant $l - \frac{W'}{W}\xi$ from A.

11. A beam, of 80 feet span and weighing one ton per foot run, carries a rolling load of two tons per foot run, and the rolling load covers a distance of 10 feet. Draw, roughly to scale, the curves of maximum positive and negative shearing force as the load crosses over.

12. A train equivalent to a rolling load of $1\frac{1}{4}$ tons per foot run traverses a girder of 120 foot span. Draw diagrams of maximum possible bending moment and maximum possible shear stress at every point if the rolling load is (1) greater than 120 feet in length, (2) only 60 feet in length.

13. Two rolling loads of 10 tons and 15 tons respectively, at an interval of $7\frac{1}{2}$ feet, cross a girder of 75 foot span, the larger load leading. Draw diagrams of maximum possible bending moment and shearing force for the whole girder.

14. A continuous load of w tons per foot is drawn slowly over a level bridge consisting of a single rigid girder a feet in length. The load is devoid of rigidity and is longer than the bridge, and the weight of the bridge itself is neglected. P is a point on the girder distant k feet from the nearer end. Shew that the maximum shearing stress at P is $w\frac{(a-k)^2}{2a}$ tons; and that if from the end O of the moving load in any position a vertical OS is drawn proportional to the shearing stress at P, the curve traced out by S consists of four parabolic arcs and a straight line.

15. A uniform girder is supported at its ends and the load is concentrated at its middle. Shew that the maximum bending moment is twice as great as it would be if the load were uniformly distributed.

16. The lower end A of a thin uniform rod is attached to a smooth hinge, its upper end B resting against a smooth vertical plane; shew that the tendency to break at any point P varies as the product of the distances of P from A and B.

17. A man, of weight W, can just walk across a certain horizontal plank, of weight nW and length l, without breaking it if the plank is supported at its two ends. Shew that if the plank, still horizontal, is clamped at one end and free at the other, the man can only venture along it for a distance

$$\tfrac{1}{4}l(1-\tfrac{3}{2}n).$$

18. A symmetrical arch, of span a and height h, is to be constructed of straight massless jointed rods to carry seven equal weights w at horizontal distances $\frac{1}{8}a$ apart, in such a way that there shall be no bending moment at any point of the rods. Shew how the form of the arch may be determined by graphical construction, and prove that the horizontal forces necessary to keep the ends in position are $\frac{wa}{h}$.

19. A uniform rod, of length a and weight w, has strings each of length b attached to its ends and to two fixed points, one of which is at a height a vertically above the other, so as to form a parallelogram. If the rod is made to rotate with uniform angular velocity ω about the vertical line, shew that the bending moment at a distance x from either end is $\frac{wx(a-x)\tan\alpha}{2a}$, where α is the inclination of each string to the vertical. Shew also that the tensions of the strings are equal and that

$$b\omega^2\cos\alpha = g.$$

20. A light horizontal rod, of length a, whose ends are supported, is loaded so that the tendency to break at any point is proportional to the load per unit length at the same point; shew that the load at any point varies as $\sin\frac{\pi x}{a}$, where x is the distance of the point from an end of the rod.

[If R be the reaction at an end, w_ξ the weight per unit of length at a point distant ξ from this end, and λw_x the bending moment at a distance x from the same end, the condition of the question gives

$$\lambda . w_x = R . x - \int_0^x (x-\xi) w_\xi d\xi.$$

Hence, on differentiating twice with respect to x, we have

$$\lambda \frac{d^2 w_x}{dx^2} = -w_x.$$

$$\therefore \quad w_x = A \sin\left[\frac{x}{\sqrt{\lambda}} + D\right], \text{ where } A \text{ and } D \text{ are constants.}$$

Also w_x must be zero when $x=0$ or a, since the bending moment must be zero at either end.

Hence $\qquad D=0, \text{ and } \dfrac{a}{\sqrt{\lambda}} = \pi.$]

21. A semi-circular wire, of weight W and radius a, hangs in a vertical plane from one end A about which it can turn freely. Find the bending moment at any point.

[In Art. 146, Ex. 1, it will be shewn that the centre of gravity G of the wire is at a distance $\frac{2a}{\pi}$ from the centre; also for equilibrium G must be vertically below A. Hence the diameter through A must be inclined at an angle α to the vertical, where $\tan\alpha = \frac{2}{\pi}$. If P be a point of the wire, such that AP is inclined at θ to the vertical, the bending moment at P = the sum of the moments of all the external forces on one side of P

$$= \frac{W}{\pi a}\int_\theta^\alpha 2a\, d\phi\, [2a \sin\phi \cos(\alpha-\phi) - 2a\sin\theta\cos(\alpha-\theta)]$$

$$= \frac{2aW}{\pi}[(\alpha-\theta)\sin(\alpha-2\theta) + \sin\theta\sin(\alpha-\theta)].]$$

22. A gipsy's tripod consists of three equal stiff uniform sticks freely hinged at one common end from which hangs the kettle. The other ends of the sticks rest on the ground and are prevented from slipping by a smooth circular hoop which encloses them and is fixed to the plane. Shew that the bending moment of each stick will be greatest at its middle point and that it will be independent of the length of the sticks and of the weight of the kettle.

23. A uniform rod, of length $2l$, rests symmetrically on two props in the same horizontal line at a distance $2c$ apart. If $l<2c$, shew that the tendency to break will be greatest at a prop or in the middle according as $2l \gtrless (2+\sqrt{2})c$; but that if $l>2c$, the tendency to break will be greatest at a prop. What happens if $2l=(2+\sqrt{2})c$?

[If $l<2c$, the bending moments at a prop and at the centre are of opposite signs, and the absolute magnitudes only must be compared.]

24. A beam, of length $2l$, is loaded uniformly, and the intensity of the load over the central half of the beam is double that over the two end quarters. It is to be supported by two props equidistant from the centre, and so situated that the bending moment on the beam is the least possible. Shew that the props should each be at a distance $\frac{l}{2}$ from the centre.

131. ***A beam, supported at the ends, is loaded with a number of concentrated loads; to shew that the funicular polygon for the loads is a diagram of bending moments for the beam.***

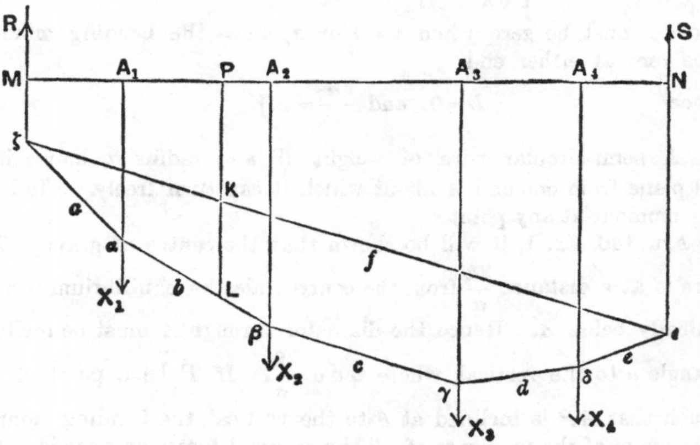

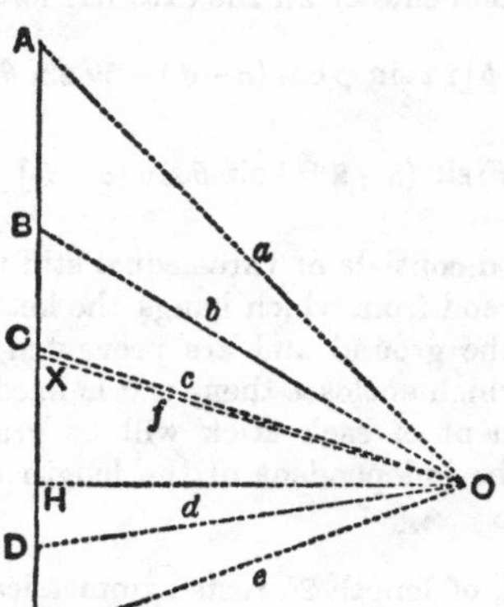

Let the beam MN be loaded at points A_1, A_2, A_3, A_4 with weights X_1, X_2, X_3, X_4 as in the figure. Let R and S be the reactions at M and N.

Draw a vertical line AE such that AB, BC, CD, DE represent X_1, X_2, X_3, X_4 respectively. Take any pole O, and draw OA, OB, OC, OD, OE.

Starting with any point α on X_1, draw lines $\zeta\alpha$, $\alpha\beta$ parallel to OA, OB and then $\beta\gamma$ parallel to OC, and so on as in Art. 113; we finally obtain the line $\epsilon\zeta$. Draw OX parallel to it, and then XA, EX represent R and S.

Also draw OH perpendicular to AE, so that $OH\,(=h)$ represents the constant horizontal component of $a, b, c, \ldots$.

Let P be any point on MN between A_1 and A_2, and draw PKL vertically to meet the funicular polygon in K and L. We shall shew that the bending moment at P is represented by KL.

From the force diagram it is clear that

$$X_2 \equiv b \text{ along } \alpha\beta \text{ and } c \text{ along } \gamma\beta,$$

$$X_3 \equiv c \text{ along } \beta\gamma \text{ and } d \text{ along } \delta\gamma,$$

$$X_4 \equiv d \text{ along } \gamma\delta \text{ and } e \text{ along } \epsilon\delta,$$

and

$$S \equiv e \text{ along } \delta\epsilon \text{ and } f \text{ along } \epsilon\zeta.$$

Hence

$$X_2, X_3, X_4, S \equiv b \text{ along } \alpha\beta \text{ and } f \text{ along } \epsilon\zeta.$$

Hence the sum of their moments about $P =$ moment of b along $\alpha\beta$ about P together with the moment of f along $\epsilon\zeta$ about P.

Now b along $L\beta$ is equal to a vertical component, which has no moment about P, and a horizontal component h whose moment about P is $h\,.\,PL$ ↻.

So the moment of f along $\epsilon\zeta$ about P is $h\,.\,PK$ ↺.

Hence the total moment about P of the forces X_2, X_3, X_4, S

$$= h\,.\,PL - h\,.\,PK = h\ KL \circlearrowright,$$

i.e. the bending moment about P is represented by KL, and is equal to this intercept multiplied by the force which is represented by the distance of the pole O from the line of loads.

Similarly any other case may be considered.

It will be noted that this proposition is really the same as that of Art. 117 for a system of forces which are all parallel.

Ex. Draw out the case of a beam, 60 feet long, supported at the ends and loaded at points distant 12, 28 and 48 feet from one end with weights 5, 6 and 4 tons respectively. Shew clearly the scale upon which the diagram must be read.

132 Another graphical construction for the bending moment may be given as follows.

Let $MN=l$, $MA_1=a_1$, $MA_2=a_2$, $MA_3=a_3$, and $MA_4=a_4$.

Draw MB_4 perpendicular to MN to represent $S.l$, and on it take B_3, B_2, B_1 so that B_4B_3, B_3B_2, B_2B_1, B_1M represent $X_4.a_4$, $X_3.a_3$, $X_2.a_2$, and $X_1.a_1$ respectively.

Join NB_4 meeting the vertical through A_4 in C_4; join C_4B_3, meeting the vertical through A_3 in C_3, and so on.

We thus obtain the polygon $MC_1C_2C_3C_4N$, the ordinate of which at any point P of the beam represents the bending moment at P.

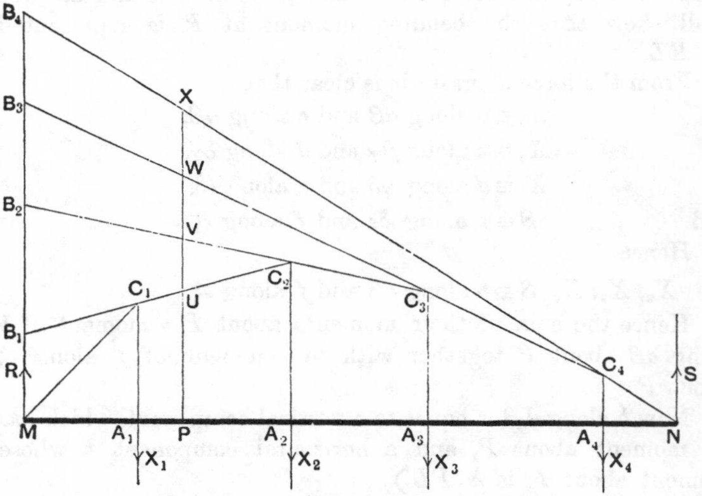

For if this ordinate PU meet C_2C_3, C_3C_4, C_4N in V, W, X, the bending moment at P

$$=S.PN-X_4.PA_4-X_3.PA_3-X_2.PA_2$$

$$=\frac{PN}{MN}.Sl-\frac{PA_4}{MA_4}.X_4a_4-\frac{PA_3}{MA_3}.X_3a_3-\frac{PA_2}{MA_2}.X_2a_2,$$

so that the bending moment at P is represented by

$$\frac{PN}{MN}.MB_4-\frac{PA_4}{MA_4}.B_4B_3-\frac{PA_3}{MA_3}.B_3B_2-\frac{PA_2}{MA_2}.B_2B_1,$$

i.e. by $\qquad PX-XW-WV-VU,$

i.e. by PU.

Similarly for any other point of the beam.

133. As in Art. 131, it is clear that the resultant of the loads X_2, X_3, X_4 is equivalent to b along $\alpha\beta$ and e along $\epsilon\delta$, *i.e.* the resultant of the loads X_2, X_3, X_4 passes through the intersection of the sides b and e of the funicular polygon. So for any other pair of sides.

Hence through the point of intersection of any two sides b and e of the funicular polygon passes the resultant of the loads that lie between b and e.

Now let the loading be continuous, so that the funicular polygon becomes a continuous curve and the sides b and e two tangents at points U and V of it.

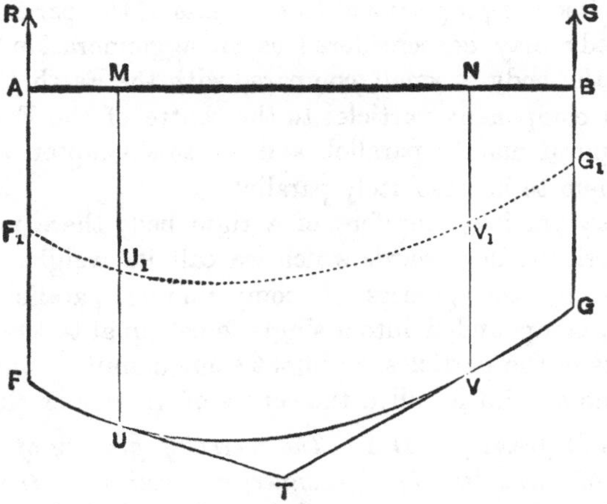

Then through T passes the resultant of all the loads on MN.

Draw the curve of loads $F_1U_1V_1G_1$.

The horizontal distance of the resultant load on MN from M

$$= \frac{w_1x_1 + w_2x_2 + \dots}{w_1 + w_2 + \dots} = \frac{\int y_1 dx_1 . x_1}{\int y_1 dx_1}$$

= distance from M of the centre of gravity of the load curve, as will be seen in Art. 145.

Hence the vertical through T passes through the centre of gravity of the load curve, *i.e.* any two tangents to the bending-moment curve intersect in a point which is vertically below the centre of gravity of the load curve.

CHAPTER VIII

CENTRE OF GRAVITY

134. Every particle of matter is attracted to the centre of the Earth, and the force with which the Earth attracts any particle to itself is proportional to the mass of the particle.

Any body may be considered as an agglomeration of particles. If the body be small, compared with the Earth, the lines joining its component particles to the centre of the Earth will be very approximately parallel, and in this chapter we shall consider them to be absolutely parallel.

On every particle, therefore, of a rigid body there is acting a force vertically downwards which we call its weight. These forces may by the process of compounding parallel forces, Art. 33, be compounded into a single force, equal to the sum of the weights of the particles, acting at some definite point of the body. Such a point is called the centre of gravity of the body.

Centre of gravity. Def. *The centre of gravity of a body, or system of particles rigidly connected together, is that point through which the line of action of the weight of the body always passes.*

135. Since the construction for the position of the resultant of parallel forces depends *only* on the point of application and magnitude, and *not* on the direction of the forces, the point we finally arrive at is the same if the body be turned through any angle; for the weights of the portions of the body are still parallel, although they have not the same direction, relative to the body, in the two positions.

We can hence shew that a body can only have one centre of gravity. For, if possible, let it have two centres of gravity G and G_1. Let the body be turned, if necessary, until GG_1 be horizontal. We shall then have the resultant of a system of vertical forces acting both through G and through G_1. But the resultant force, being itself necessarily vertical, cannot act in the horizontal line GG_1.

Hence there can be only one centre of gravity.

136. **If the body be not so small that the weights of its component parts may all be considered to be very approximately parallel, it has not necessarily a centre of gravity.**

In any case, the point of the body at which we arrive by the construction of Art. 33, has, however, very important properties and is called its Centre of Mass, or Centre of Inertia. If the body be of uniform density, its centre of mass coincides with its Centroid, or Mean Centre.

137. Thin uniform rod AB. The centre of gravity is clearly its middle point G; since, for every particle between G and A, there is an equal particle at an equal distance from G between G and B.

A uniform parallelogram ABCD. By dividing the parallelogram into a very large number of very thin strips, by lines parallel to AD, it is clear that all their centres of gravity, and hence that of the whole figure, lie on the line joining the middle points of AD and BC. Similarly, it lies on the line joining the middle points of AB and CD. Hence it is at the intersection of the diagonals.

Uniform triangular lamina ABC. Let D, E be the middle points of BC and CA. By dividing the triangle into a very large number of very thin strips by lines parallel to BC it is clear that all their centres of gravity, and therefore that of the whole figure, lie on AD. Similarly, it lies on BE. The centre of gravity G is therefore at the meet of AD and BE. By similar triangles GAB, GDE we easily have $GD = \frac{1}{2} GA$, and hence $GD = \frac{1}{3} DA$, giving G.

It is clear that G is also the centre of gravity of equal weights at A, B, C. For weights w at B and C are equivalent to $2w$ at D and this with w at A clearly gives $3w$ at G (Art. 31). Hence any uniform triangle may be replaced as far as its weight is concerned by particles each one-third of its weight placed at its angular points.

Uniform tetrahedron ABCD. Let G_4, G_3 be the centres of gravity of the faces ABC, DAB. By dividing the tetrahedron into a very large number of very thin slices by means of planes parallel to the face ABC, it can be shewn that the centre of gravity of each slice, and thus that of the whole tetrahedron, lies on DG_4. Similarly it lies on CG_3. Hence the required centre of gravity G is at their intersections. By similar triangles we then have, if E be the middle point of AB,

$$\frac{GG_4}{GD} = \frac{G_3G_4}{DC} = \frac{EG_4}{EC} = \frac{1}{3},$$

so that $G_4G = \frac{1}{4}G_4D$, giving the position of G.

The tetrahedron may be replaced by particles equal to one-quarter of its weight at each vertex. For particles w at A, B, C are, as in the case of the triangle, equivalent to $3w$ at G_4, and $3w$ at G_4 and w at D are equivalent to $4w$ at G (Art. 31).

Pyramid on any base. Solid Cone. If the base of the pyramid in the previous case be any plane figure $ABCLMN$... whose centre of gravity is G_1, it may be shewn, by a similar method of proof, that the centre of gravity must lie on the line joining D to G_1.

Also by drawing the planes DAG_1, DBG_1, ... the whole pyramid may be split into a number of pyramids on triangular bases, the centres of gravity of which all lie on a plane parallel to $ABCL$... and at a distance from D of three-quarters that of the latter plane. Hence the centre of gravity of the whole lies on the line G_1D, and divides it in the ratio 1 : 3.

Let now the sides of the plane base form a regular polygon, and let their number be indefinitely increased. Ultimately the plane base becomes a circle, and the pyramid becomes a solid cone having D as its vertex; also the point G_1 is now the centre of the circular base.

Hence the centre of gravity of a solid right circular cone is on the line joining the centre of the base to the vertex at a distance from the base equal to one-quarter of the distance of the vertex.

Surface of a right circular cone. Since the surface of a cone can be divided into an infinite number of triangular laminas, by joining the vertex of the cone to points on the circular base indefinitely close to one another, and since their centres of gravity all lie in a plane parallel to the base of the cone at a distance from the vertex equal to two-thirds of that of the base, the centre of gravity of the whole cone must lie in that plane.

But, by symmetry, the centre of gravity must lie on the axis of the cone.

Hence the required point is the point in which the above plane meets the axis, and therefore is on the axis at a point distant from the base one-third the height of the cone.

138. General formulae for the determination of the centre of gravity. If a system of particles whose weights are $w_1, w_2, w_3, \dots w_n$ be at points whose coordinates referred to fixed axes Ox, Oy, Oz in space are $(x_1, y_1, z_1), (x_2, y_2, z_2), \dots (x_n, y_n, z_n)$, then the coordinates $(\bar{x}, \bar{y}, \bar{z})$ of the centre of gravity G are given by

$$\bar{x} = \frac{w_1x_1 + w_2x_2 + \dots + w_nx_n}{w_1 + w_2 + \dots + w_n} = \frac{\Sigma(w_1x_1)}{\Sigma(w_1)},$$

$$\bar{y} = \frac{\Sigma(w_1y_1)}{\Sigma(w_1)} \text{ and } \bar{z} = \frac{\Sigma(w_1z_1)}{\Sigma(w_1)}.$$

These formulae are proved in Art. 34 since weights of particles are merely a particular case of a system of parallel forces.

If all the particles lie on a straight line, the first of these formulae give G; if they lie in a plane, only the first two are wanted.

139. *Given the centre of gravity of the two portions of a body, to find the centre of gravity of the whole body.*

Let the given centres of gravity be G_1 and G_2, and let the weights of the two portions be W_1 and W_2; the required point G, by Art. 31, divides G_1G_2 so that

$$G_1G : GG_2 :: W_2 : W_1.$$

The point G may also be obtained by the use of Art. 138.

Ex. *On the same base AB, and on opposite sides of it, isosceles triangles CAB and DAB are described whose altitudes are 12 inches and 6 inches respectively. Find the distance from AB of the centre of gravity of the quadrilateral CADB.*

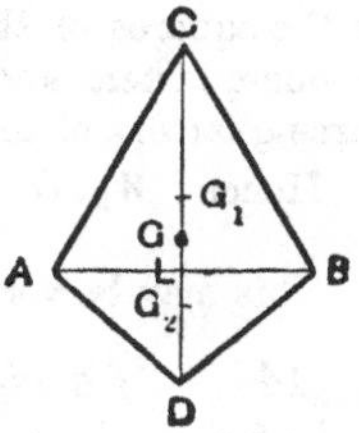

Let CLD be the perpendicular to AB, meeting it in L, and let G_1 and G_2 be the centres of gravity of the two triangles CAB and DAB respectively, so that

$$CG_1 = \tfrac{2}{3} . CL = 8,$$

and $$CG_2 = CL + LG_2 = 12 + 2 = 14.$$

The weights of the triangles are proportional to their areas, *i.e.* to $\frac{1}{2}AB . 12$ and $\frac{1}{2}AB . 6$.

If G be the centre of gravity of the whole figure, we have

$$CG = \frac{\Delta CAB \times CG_1 + \Delta DAB \times CG_2}{\Delta CAB + \Delta DAB} = \frac{12 \times 8 + 6 \times 14}{12 + 6} = 10.$$

Hence $$LG = CL - CG = 2 \text{ inches}.$$

This result may be verified experimentally by cutting the figure out of thin cardboard.

140. *Given the centre of gravity of the whole of a body and of a portion of the body, to find the centre of gravity of the remainder.*

Let G be the centre of gravity of a body $ABCD$, and G_1 that of the portion ADC. Let W be the weight of the whole body and W_1 that of the portion ACD, so that $W_2\ (= W - W_1)$ is the weight of the portion ABC.

Let G_2 be the centre of gravity of the portion ABC. Since W_1 at G_1 and W_2 at G_2 have their centre of gravity at G, therefore G must lie on G_1G_2 and be such that

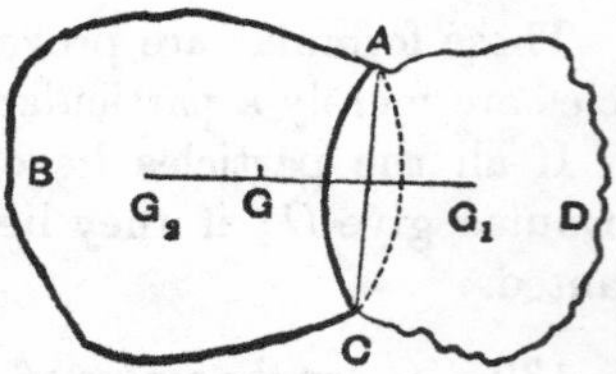

$$W_1 \,.\, GG_1 = W_2 \,.\, GG_2.$$

Hence, given G and G_1, we obtain G_2 by producing G_1G to G_2, so that

$$GG_2 = \frac{W_1}{W_2} \,.\, GG_1 = \frac{W_1}{W - W_1} \,.\, GG_1.$$

The required point may be also obtained by means of Art. 138.

Ex. ***From a circular disc, of radius r, is cut out a circle, whose diameter is a radius of the disc; find the centre of gravity of the remainder.***

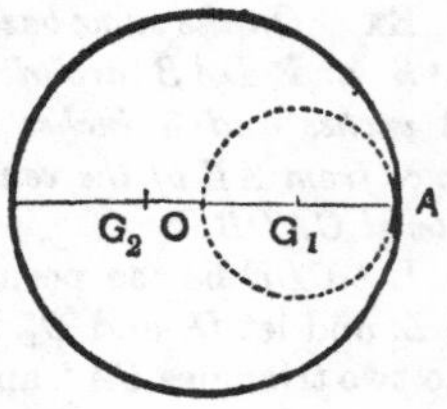

Since the areas of circles are to one another as the squares of their radii, the portion cut off is one-quarter, and the portion remaining is three-quarters, of the whole, so that $W_1 = \frac{1}{3} W_2$.

Hence $W_2 \,.\, OG_2 = W_1 \,.\, OG_1 = \frac{1}{3} W_2 \times \frac{1}{2} r$.

$\therefore \quad OG_2 = \frac{1}{6} r.$

This may be verified experimentally.

141. *If a rigid body be in equilibrium, one point only of the body being fixed, the centre of gravity of the body will be in the vertical line passing through the fixed point of the body.*

Let O be the fixed point of the body, and G its centre of gravity.

The forces acting on the body are the reaction at the fixed point of support of the body and the weights of the component parts of the body, which are equivalent to a single vertical force through the centre of gravity of the body.

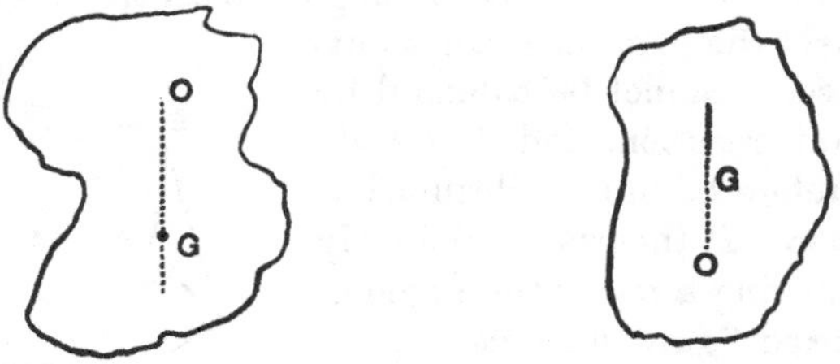

Also, when two forces keep a body in equilibrium, they must be equal and opposite and have the same line of action, so that the vertical line through G must pass through the point O.

Two cases arise; the first, in which the centre of gravity G is below the point of suspension O, and the second, in which G is above O. In the first case, the body, if slightly displaced from its position of equilibrium, will tend to return to this position; in the second case, the body will not tend to return to its position of equilibrium.

142. *If a body be placed with its base in contact with a horizontal plane, it will stand, or fall, according as the vertical line drawn through the centre of gravity of the body meets the plane within, or without, the base.*

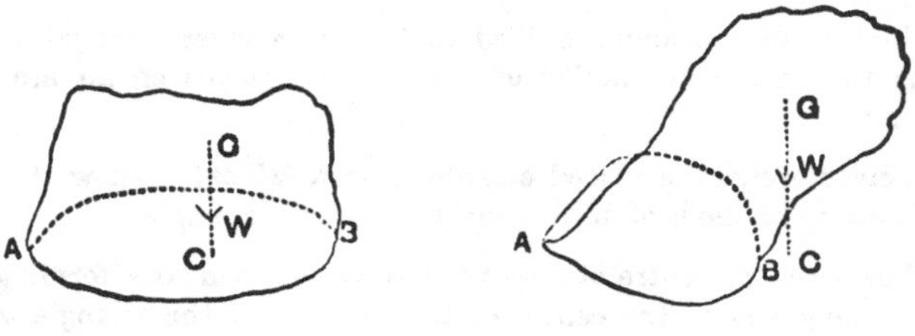

The forces acting on the body are its weight, which acts at its centre of gravity G, and the reactions of the plane, acting at different points of the base of the body. These reactions are all vertical, and hence they may be compounded into a single vertical force acting at some point of the base.

Since the resultant of two like parallel forces acts always at a point between the forces, it follows that the resultant of all the reactions on the base of the body cannot act through a point outside the base.

Hence, if the vertical line through the centre of gravity of the body meet the plane at a point outside the base, it cannot be balanced by the resultant reaction, and the body cannot therefore be in equilibrium, but must fall over. If the base of the body be a figure having a re-entrant angle as in the annexed figure, we must extend the meaning of the word "base" in the enunciation to mean the area included in the figure obtained by drawing a piece of thread tightly round the geometrical base. In the above figure the "base" therefore means the area *ABDEFA*.

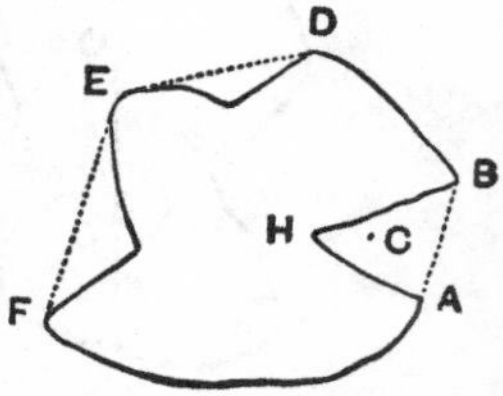

For example, the point *C*, at which the resultant reaction acts, may lie within the area *AHB*, but it cannot lie without the dotted line *AB*. If the point *C* were on the line *AB*, between *A* and *B*, the body would be on the point of falling over.

EXAMPLES

1. The base of a triangle is fixed, and its vertex moves on a given straight line; shew that the centre of gravity also moves on a straight line.

2. The base of a triangle is fixed, and it has a given vertical angle; shew that the centre of gravity of the triangle moves on an arc of a certain circle.

3. A given weight is placed anywhere on a triangle; shew that the centre of gravity of the system lies within a certain triangle.

4. Shew that the centre of gravity of three uniform rods forming the sides of a triangle is at the centre of the in-circle of the triangle whose angular points are the middle points of the three rods.

5. If three forces act on a point P which are represented by $\mu . PA$, $\mu . PB$, and $\mu . PC$ respectively, shew that their resultant is $3\mu . PG$, where G is the centre of gravity of the triangle ABC.

6. A particle P is acted upon by forces towards the points A, B, C, ... which are represented by $\lambda . PA$, $\mu . PB$, $\nu . PC$, ...; shew that their resultant is represented by $(\lambda+\mu+\nu+...) PG$, where G is the centre of gravity of weights placed at A, B, C, ... proportional to λ, μ, ν, ... respectively.

[This is the generalised form of Art. 25, and may be proved by successive applications of that article.]

7. *To find the centre of gravity of a quadrilateral lamina having two parallel sides.*

Let $ABCD$ be the quadrilateral, having the sides AB and CD parallel and equal to $2a$ and $2b$ respectively.

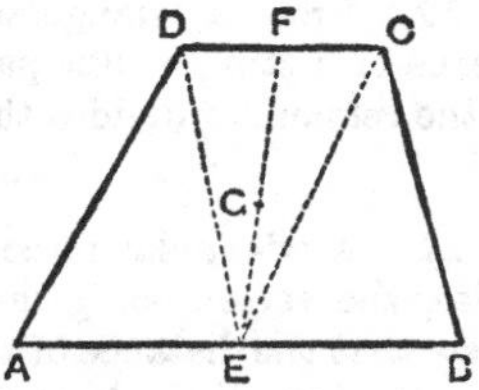

Let E and F be the middle points of AB and CD respectively. Join DE and EC; the areas of the triangles ADE, DEC, and BEC are proportional to their bases AE, DC, and EB, *i.e.* are proportional to a, $2b$, and a.

Replace them by particles equal to one-third of their weight placed at their angular points.

We thus have weights proportional to $\frac{a+2b}{3}$ at each of C and D, $\frac{a}{3}$ at each of A and B, and $\frac{2a+2b}{3}$ at E.

Again, replace the equal weights at C and D by a weight proportional to $\frac{2a}{3}+\frac{4b}{3}$ at the middle point F of CD, and the equal weights at A and B by a weight proportional to $\frac{2a}{3}$ at E. We thus have weights

$$\frac{2a}{3}+\frac{4b}{3} \text{ at } F, \text{ and } \frac{4a}{3}+\frac{2b}{3} \text{ at } E.$$

Hence the required centre of gravity G is on the straight line EF, and is such that

$$\frac{EG}{GF}=\frac{\text{weight at } F}{\text{weight at } E}=\frac{a+2b}{2a+b}.$$

8. *The distances of the angular points and intersection of the diagonals of a plane quadrilateral lamina from any line OX in its plane are a, b, c, d, and e; shew that the distance of the centre of inertia from the same line is $\frac{1}{3}(a+b+c+d-e)$.*

Let A, B, C, D be the angular points, and E the intersection of the diagonals. Then

$$\frac{\Delta ACD}{\Delta ACB}=\frac{\text{perpendicular from } D \text{ on } AC}{\text{perpendicular from } B \text{ on } AC}=\frac{DE}{EB}=\frac{d-e}{e-b}.$$

By Arts. 137 and 138 the distance of the centre of gravity of the ΔACD from OX is $\frac{a+c+d}{3}$ and that of the ΔACB is $\frac{a+c+b}{3}$.

Hence distance of required C.G. from OX

$$=\frac{\Delta ACD\times\frac{1}{3}(a+c+d)+\Delta ACB\times\frac{1}{3}(a+b+c)}{\Delta ACD+\Delta ACB}$$

$$=\frac{1}{3}\frac{(d-e)(a+c+d)+(e-b)(a+b+c)}{(d-e)+(e-b)}=\frac{1}{3}(a+b+c+d-e).$$

9. Prove the following construction for the centre of gravity G of a plane quadrilateral area $ABCD$; let L, M be the centres of gravity of the triangles ABC, ADC and let LM meet AC in N; then G lies on LM and is such that $MG=LN$.

10. From a triangular area ABC is cut off the nth of its area by means of a straight line parallel to BC; shew that the centre of gravity of the remainder divides the median through A in the ratio

$$n+\sqrt{n}-2:2(n+\sqrt{n}+1).$$

11. A triangular piece of paper is folded across the line bisecting two sides, the vertex being thus brought to lie on the base of the triangle. Shew that the distance of the centre of inertia of the paper in this position from the base of the triangle is three-quarters that of the centre of inertia of the unfolded paper from the same line.

12. A uniform rod is hung up by two strings attached to its ends, the other ends of the strings being attached to a fixed point; shew that the tensions of the strings are proportional to their lengths.

Prove that the same relation holds for a uniform triangular lamina hung up by three strings attached to its angular points.

13. The mass of the moon is ·013 times that of the earth. Taking the earth's radius as 4000 miles and the distance of the moon's centre from the earth's centre as 60 times the earth's radius, find the distance of the C.G. of the earth and moon from the centre of the earth.

[3080 miles nearly.]

14. Find the vertical angle of a cone in order that the centre of gravity of its whole surface, including its plane base, may coincide with the centre of gravity of its volume. [$2\sin^{-1}\frac{1}{3}$.]

15. A solid right circular cone has its base scooped out, so that the hollow is a right cone on the same base; how much must be removed so that the centre of gravity of the remainder may coincide with the vertex of the hollow? [Height of inner cone $=\frac{1}{3}$ height of outer cone.]

16. Shew how to cut out of a uniform cylinder a cone, whose base coincides with that of the cylinder, so that the centre of gravity of the remaining solid may coincide with the vertex of the cone.

[Height of cone $=(2-\sqrt{2})$ times the height of cylinder.]

17. A square hole is punched out of a circular lamina, the diagonal of the square being a radius of the circle. Shew that the centre of gravity of the remainder is at a distance $\frac{a}{8\pi-4}$ from the centre of the circle, where a is the diameter of the circle.

18. From a uniform triangular board a portion consisting of the area of the inscribed circle is removed; shew that the distance of the centre of gravity of the remainder from any side, a, is

$$\frac{S}{3as}\frac{2s^3-3\pi aS}{s^2-\pi S},$$

where S is the area and s the semi-perimeter of the board.

19. A circular hole of a given size is punched out of a uniform circular plate; shew that the centre of gravity lies within a certain circle.

20. A triangular lamina ABC, obtuse-angled at C, stands with the side AC in contact with a table. Shew that the least weight, which suspended from B will overturn the triangle, is

$$\tfrac{1}{3}W\frac{a^2+3b^2-c^2}{c^2-a^2-b^2}, \text{ where } W \text{ is the weight of the triangle.}$$

Interpret the above if $c^2>a^2+3b^2$.

21. A cone, whose height is equal to four times the radius of its base, is hung from a point in the circumference of its base; shew that it will rest with its base and axis equally inclined to the vertical.

22. If A and B be the positions of two masses, m and n, and if G be their centre of gravity, shew that, if P be any point, then

$$m\,.\,AP^2+n\,.\,BP^2=m\,.\,AG^2+n\,.\,BG^2+(m+n)\,PG^2.$$

Similarly, if there be any number of masses, $m, n, p, \ldots$ at points $A, B, C, \ldots$, and G be their centre of gravity, shew that

$$m\,.\,AP^2+n\,.\,BP^2+p\,.\,CP^2+\ldots$$
$$=m\,.\,AG^2+n\,.\,BG^2+p\,.\,CG^2+\ldots+(m+n+p+\ldots)\,PG^2.$$

23. A frustum of a solid right cone is placed with its base on a rough inclined plane whose inclination is gradually increased; if R and r be the radii of the larger and smaller sections, and h be the height of the frustum, shew that the frustum will ultimately either tumble, or slide, according as the coefficient of friction

$$\gtrless \frac{4R}{h}\cdot\frac{R^2+Rr+r^2}{R^2+2Rr+3r^2}.$$

24. The top of a right cone, of vertical angle 2α, is cut off by a plane making an angle β with the axis, and is placed on a perfectly rough inclined plane with the major axis of the base along a line of greatest slope; in this position it is on the point of tumbling over; shew that the tangent of the inclination of the inclined plane to the horizon has one of the values

$$\frac{4\sin 2\alpha\pm\sin 2\beta}{\cos 2\alpha-\cos 2\beta}.$$

25. Into a thin cylindrical vase, of weight W and cross section s, whose centre of gravity is distant b from its base, is poured liquid of density ρ. When the height of the centre of gravity of the whole is a minimum, shew that the weight of the liquid is

$$\sqrt{W(W+2sb\rho)}-W.$$

143. The formulae of Art. 138 may be used to give the centre of gravity of any arc, area, or solid of known shape.

Centre of gravity of an arc. If P be any point (x, y) of a plane arc, PQ an elementary arc δs, whose density at the point P is ρ and whose weight is therefore proportional to ρds, these formulae give

$$\bar{x} = \frac{\Sigma(\rho\,\delta s\,.\,x)}{\Sigma(\rho\,.\,\delta s)} = \frac{\int\rho x\,ds}{\int\rho\,ds},$$

the limits of the integrals extending from one end to the other of the arc considered.

Similarly, $$\bar{y} = \frac{\int\rho y\,ds}{\int\rho\,ds}.$$

Also $\dfrac{ds}{dx} = \sqrt{1 + \left(\dfrac{dy}{dx}\right)^2}$, and $\dfrac{dy}{dx}$ is known from the equation to the curve.

If, as is usually the case, the arc is of uniform density, ρ is constant and divides out from both numerator and denominator; if it be of variable density, then ρ must be given as a function of x or y.

Similar formulae hold for a curve in three dimensions; but now

$$\frac{ds}{dx} = \sqrt{1 + \left(\frac{dy}{dx}\right)^2 + \left(\frac{dz}{dx}\right)^2}.$$

If the arc have its equation given in polar coordinates, say $r = f(\theta)$, then

$$\delta s = \sqrt{\delta r^2 + r^2\delta\theta^2},$$

and the formulae give

$$\bar{x} = \frac{\int\rho r\cos\theta\,ds}{\int\rho\,ds}, \text{ and } \bar{y} = \frac{\int\rho r\sin\theta\,ds}{\int\rho\,ds}.$$

Similar formulae hold for three dimensions.

144. Ex. 1. *Find the centre of gravity of the arc of the parabola $y^2 = 4ax$ included between the vertex and an ordinate at a distance at^2 from the vertex.*

Here $y\dfrac{dy}{dx} = 2a$. $\therefore$ $\dfrac{ds}{dx} = \sqrt{1 + \left(\dfrac{dy}{dx}\right)^2} = \sqrt{1 + \dfrac{4a^2}{y^2}} = \sqrt{\dfrac{x+a}{x}}$.

$$\therefore\ \bar{x} = \frac{\int_0^{at^2} ds\,.\,x}{\int_0^{at^2} ds} = \frac{\int_0^{at^2}\sqrt{x(x+a)}\,dx}{\int_0^{at^2}\sqrt{\dfrac{x+a}{x}}\,dx},$$

and
$$\bar{y}=\frac{\int_0^{at^2} ds\,.\,y}{\int_0^{at^2} ds}=2\sqrt{a}\,.\,\frac{\int_0^{at^2}\sqrt{x+a}\,dx}{\int_0^{at^2}\sqrt{\frac{x+a}{x}}\,dx}.$$

Now
$$\int_0^{at^2}\sqrt{x(x+a)}\,dx=\int_0^{at^2}\sqrt{\left(x+\frac{a}{2}\right)^2-\frac{a^2}{4}}\,dx$$
$$=\left[\frac{1}{2}\left(x+\frac{a}{2}\right)\sqrt{x^2+ax}-\frac{a^2}{8}\log\left(x+\frac{a}{2}+\sqrt{x^2+ax}\right)\right]_0^{at^2}$$
$$=\frac{a^2}{4}\left[t\,(2t^2+1)\sqrt{1+t^2}-\log\,(t+\sqrt{1+t^2})\right].$$

Also $\int_0^{at^2}\sqrt{\frac{x+a}{x}}\,dx=\int_0^t 2a\sqrt{1+v^2}\,dv$, on putting $x=av^2$,
$$=a\left[t\sqrt{1+t^2}+\log\,(t+\sqrt{1+t^2})\right],$$

and
$$\int_0^{at^2}\sqrt{x+a}\,dx=\left[\frac{2}{3}(x+a)^{\frac{3}{2}}\right]_0^{at^2}=\frac{2}{3}a^{\frac{3}{2}}\left[(1+t^2)^{\frac{3}{2}}-1\right].$$

On making these substitutions, we have
$$\bar{x}=\frac{a}{4}\cdot\frac{t\,(1+2t^2)\sqrt{1+t^2}-\log\left[t+\sqrt{1+t^2}\right]}{t\sqrt{1+t^2}+\log\left[t+\sqrt{1+t^2}\right]},$$

and
$$\bar{y}=\frac{4a}{3}\cdot\frac{(1+t^2)^{\frac{3}{2}}-1}{t\sqrt{1+t^2}+\log\left[t+\sqrt{1+t^2}\right]}.$$

Ex. 2. *Find the centroid of the arc of the catenary*
$$y=\frac{c}{2}\left(e^{\frac{x}{c}}+e^{-\frac{x}{c}}\right)$$
which is included between the origin and any point (x, y).

Here
$$\left(\frac{ds}{dx}\right)^2=1+\left(\frac{dy}{dx}\right)^2=1+\frac{1}{4}\left(e^{\frac{x}{c}}-e^{-\frac{x}{c}}\right)^2=\frac{1}{4}\left(e^{\frac{x}{c}}+e^{-\frac{x}{c}}\right)^2$$

so that
$$\frac{ds}{dx}=\frac{1}{2}\left(e^{\frac{x}{c}}+e^{-\frac{x}{c}}\right)=\cosh\frac{x}{c},\ \text{and } s=c\sinh\frac{x}{c}.$$

$$\therefore\ \bar{x}=\frac{\int_0^x x\,ds}{\int_0^x ds}=\frac{\int_0^x x\cosh\frac{x}{c}\,dx}{\int_0^x\cosh\frac{x}{c}\,dx}=\frac{cx\sinh\frac{x}{c}-c^2\cosh\frac{x}{c}+c^2}{c\sinh\frac{x}{c}}$$
$$=x-\frac{c\,(y-c)}{s}.$$

Also
$$\bar{y}=\frac{\int y\,ds}{\int ds}=\frac{\int_0^x c\cosh^2\frac{x}{c}\,dx}{\int_0^x \cosh\frac{x}{c}\,dx}=\frac{c}{2}\frac{\int_0^x\left(1+\cosh\frac{2x}{c}\right)dx}{\int_0^x \cosh\frac{x}{c}\,dx}$$
$$=\frac{x+\frac{c}{2}\sinh\frac{2x}{c}}{2\sinh\frac{x}{c}}=\frac{x+c\sinh\frac{x}{c}\cosh\frac{x}{c}}{2\sinh\frac{x}{c}}=\frac{y}{2}+\frac{cx}{2s}.$$

EXAMPLES

Find the positions of the centroids of the arcs of the following curves:

1. Cycloid $x=a(\theta+\sin\theta)$, $y=a(1-\cos\theta)$ which is in the positive quadrant. $\left[\bar{x}=\left(\pi-\frac{4}{3}\right)a;\ \bar{y}=\frac{2a}{3}.\right]$

2. $x^{\frac{2}{3}}+y^{\frac{2}{3}}=a^{\frac{2}{3}}$ between two successive cusps. $\left[\bar{x}=\bar{y}=\frac{2a}{5}.\right]$

3. Helix $x=a\cos\theta$, $y=a\sin\theta$, $z=b\theta$ included between the point $\theta=0$ and the point $\theta=a$. $\left[\bar{x}=\frac{a\sin a}{a};\ \bar{y}=\frac{a(1-\cos a)}{a};\ \bar{z}=\frac{1}{2}ba.\right]$

4. If the density of a complete circular arc varies as the square of the distance from a point O on the arc, shew that its centroid divides the diameter through O in the ratio 3 : 1.

5. Find the centre of gravity of the wire of a cork-screw, of given length, radius, and pitch, supposing the thickness of the wire at any point to be $b+nz$, where z is the distance of a point, measured parallel to the axis, from one end.

145. Centre of gravity of any plane area. In Cartesian coordinates the element of area is $\delta x\,\delta y$, and if ρ be its density the weight w is proportional to $\rho\,\delta x\,\delta y$; the fundamental formulae thus become

$$\bar{x}=\frac{\Sigma\rho\,\delta x\,\delta y\,.\,x}{\Sigma\rho\,\delta x\,\delta y}=\frac{\iint\rho x\,dx\,dy}{\iint\rho\,dx\,dy},\text{ and }\bar{y}=\frac{\iint\rho y\,dx\,dy}{\iint\rho\,dx\,dy},$$

the limits being chosen so as to include all the area considered.

If the area be of uniform density, and the ordinate at distance x cut the curve in points whose ordinates are y_1 and y_2, we may take as the element of area $(y_1-y_2)\,\delta x$, whose centre of gravity has as coordinates x and $\frac{y_1+y_2}{2}$, in the limit when δx is very small.

The fundamental formulae then give

$$\bar{x} = \frac{\Sigma (y_1 - y_2)\,\delta x \,.\, x}{\Sigma (y_1 - y_2)\,\delta x} = \frac{\int x (y_1 - y_2)\,dx}{\int (y_1 - y_2)\,dx},$$

and
$$\bar{y} = \frac{\Sigma (y_1 - y_2)\,\delta x \,.\, \dfrac{y_1 + y_2}{2}}{\Sigma (y_1 - y_2)\,\delta x} = \frac{1}{2}\frac{\int (y_1^2 - y_2^2)\,dx}{\int (y_1 - y_2)\,dx}.$$

The values of y_1 and y_2 are known from the equation to the curve, and the limits for x are such as to include all the area considered.

If the curve be given in polar coordinates, $r = f(\theta)$, referred to O as pole, and if P and Q be points whose vectorial angles are θ and $\theta + \delta\theta$, the polar element of area is $\frac{1}{2} r^2 . \delta\theta$ and its centre of gravity is, when $\delta\theta$ is very small, at the point whose polar coordinates are $\frac{2}{3} r$ and θ, and whose Cartesian coordinates are $\frac{2}{3} r \cos\theta$ and $\frac{2}{3} r \sin\theta$.

Hence
$$\bar{x} = \frac{\Sigma \frac{1}{2} r^2 \delta\theta \,.\, \frac{2}{3} r \cos\theta}{\Sigma \frac{1}{2} r^2 . \delta\theta} = \frac{2}{3}\frac{\int r^3 \cos\theta\, d\theta}{\int r^2\, d\theta},$$

where $r = f(\theta)$, and a similar equation for $\bar{y}$, giving the centre of gravity of a sectorial area AOB, the limits for θ being the vectorial angles of A and B.

If this sectorial area be of variable density, or thickness, we must take the element of area as $r\,\delta\theta\;\delta r$ of density ρ and then, for the sectorial area AOB,

$$\bar{x} = \frac{\Sigma r\,\delta\theta\,\delta r \rho \,.\, r\cos\theta}{\Sigma r \delta\theta\,\delta r \rho \,.\, \theta} = \frac{\iint \rho r^2 \cos\theta\, d\theta\, dr}{\iint \rho r\, d\theta\, dr},$$

and similarly for $\bar{y}$. The value of ρ is given as a function of r and θ; the limits of integration for r are 0 to $f(\theta)$, and the limits for θ the vectorial angles of A and B.

146. Ex. 1. *To find the centre of gravity of the arc, sector and segment of a circle.*

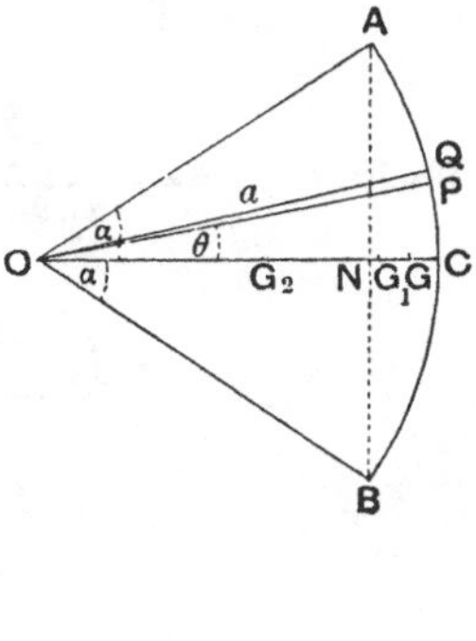

Let the arc ACB subtend an angle 2α at the centre O of the circle, and let C be the middle point of the arc.

Taking OC as the axis of x, then for the arc we have

$$\bar{x} = \frac{\int ds \,.\, x}{\int ds} = \frac{\int_{-\alpha}^{+\alpha} a\,d\theta \,.\, a\cos\theta}{\int_{-\alpha}^{+\alpha} a\,d\theta} = a \,.\, \frac{\sin\alpha}{\alpha}$$

For the sector AOB,

$$\bar{x} = \frac{\int \frac{1}{2} a^2 d\theta \,.\, \frac{2}{3} a \cos\theta}{\int \frac{1}{2} a^2 d\theta} = \frac{2}{3} a \; \frac{\sin\alpha}{\alpha} = OG_1.$$

Let G_2 be the centre of gravity of the triangle AOB, so that

$$OG_2 = \frac{2}{3} ON = \frac{2}{3} a \cos\alpha.$$

Then the weight of the segment $ANBC$ acting at its centre of gravity G together with that of the triangle AOB balance about G_1.

$$\therefore \; \frac{2}{3} \frac{a \sin\alpha}{\alpha} = OG_1 = \frac{\Delta\, AOB \times OG_2 + [\text{sector } AOB - \Delta\, AOB] \,.\, OG}{\text{sector } AOB}$$

$$= \frac{\frac{1}{2} a^2 \sin 2\alpha \,.\, \frac{2}{3} a \cos\alpha + [\frac{1}{2} a^2 \,.\, 2\alpha - \frac{1}{2} a^2 \sin 2\alpha] \,.\, OG}{\frac{1}{2} a^2 \,.\, 2\alpha}.$$

This gives $$OG = \frac{2a}{3} \frac{\sin^3\alpha}{\alpha - \sin\alpha\cos\alpha}.$$

Cor. By putting $\alpha = \frac{\pi}{2}$, we see that the centre of gravity of the arc of a semicircle is distant $\frac{2a}{\pi}$ from the centre, and that the distance of the centre of gravity of a semi-circular area is distant $\frac{4a}{3\pi}$ from the centre

Ex. 2. *Find the centre of gravity of the area bounded by the axis of y, the cycloid, $x = a(\theta + \sin\theta)$, $y = a(1 - \cos\theta)$, and its base.*

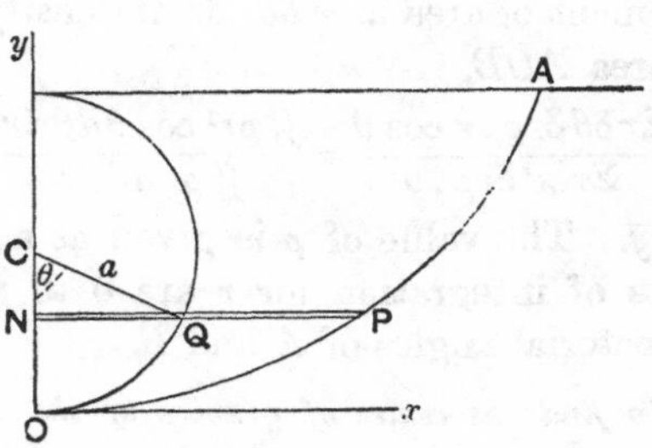

We have $$\bar{x} = \frac{\int xdy \,.\, \frac{x}{2}}{\int xdy} = \frac{a}{2} \frac{\int_0^\pi (\theta + \sin\theta)^2 \,.\, \sin\theta \,.\, d\theta}{\int_0^\pi (\theta + \sin\theta) \,.\, \sin\theta \,.\, d\theta}$$

$$= \frac{a}{2} \frac{\int_0^\pi \{\theta^2 \sin\theta + \theta(1 - \cos 2\theta) + \frac{1}{4}(3\sin\theta - \sin 3\theta)\}\, d\theta}{\int_0^\pi (\theta\sin\theta + \frac{1}{2} - \frac{1}{2}\cos 2\theta)\, d\theta}.$$

Now, by the extended rule for integration by parts,

$$\int \theta^2 \sin\theta\, d\theta = -\theta^2 \cos\theta + 2\theta \sin\theta + 2\cos\theta,$$

$$\int \theta \cos 2\theta\, d\theta = \tfrac{1}{2}\theta \sin 2\theta + \tfrac{1}{4}\cos 2\theta,$$

$$\int \theta \sin\theta\, d\theta = -\theta\cos\theta + \sin\theta,$$

and
$$\int \theta \sin 2\theta\, d\theta = -\tfrac{1}{2}\theta\cos 2\theta + \tfrac{1}{4}\sin 2\theta.$$

$$\therefore\ \bar{x} = \frac{a}{2}\cdot\frac{\left[\theta^2(\tfrac{1}{2} - \cos\theta) + \theta(2\sin\theta - \tfrac{1}{2}\sin 2\theta) + \tfrac{5}{4}\cos\theta - \tfrac{1}{4}\cos 2\theta + \tfrac{1}{12}\cos 3\theta\right]_0^{\pi}}{\left[\theta(\tfrac{1}{2} - \cos\theta) + \sin\theta - \tfrac{1}{4}\sin 2\theta\right]_0^{\pi}}$$

$$= \frac{a}{2}\cdot\frac{\pi^2 \cdot \frac{3}{2} - \frac{8}{3}}{\pi \cdot \frac{3}{2}} = \frac{9\pi^2 - 16}{18\pi}\cdot a.$$

Also
$$\bar{y} = \frac{\int x\,dy \cdot y}{\int x\,dy} = a\,\frac{\int_0^{\pi} (\theta + \sin\theta)\sin\theta\,(1 - \cos\theta)\,d\theta}{\int_0^{\pi} (\theta + \sin\theta)\sin\theta\,d\theta}$$

$$= a\,\frac{\int_0^{\pi} [\theta(\sin\theta - \tfrac{1}{2}\sin 2\theta) + \tfrac{1}{2} - \tfrac{1}{2}\cos 2\theta - \sin^2\theta\cos\theta]\,d\theta}{\int_0^{\pi} [\theta\sin\theta + \tfrac{1}{2} - \tfrac{1}{2}\cos 2\theta]\,d\theta}$$

$$= a\,\frac{\left[\theta(\tfrac{1}{4}\cos 2\theta + \tfrac{1}{2} - \cos\theta) + \sin\theta - \tfrac{3}{8}\sin 2\theta - \frac{\sin^3\theta}{3}\right]_0^{\pi}}{\left[\theta(\tfrac{1}{2} - \cos\theta) + \sin\theta - \tfrac{1}{4}\sin 2\theta\right]_0^{\pi}} = \frac{7a}{6}.$$

Ex. 3. *Find the centre of gravity of the loop of the curve $r = a\cos 3\theta$ containing the initial line.*

The values of θ giving the loop are from $-\frac{\pi}{6}$ to $\frac{\pi}{6}$.

Hence
$$\bar{x} = \frac{\int \frac{1}{2}r^2 d\theta \cdot \frac{2}{3} r\cos\theta}{\int \frac{1}{2} r^2 d\theta},\ \text{between limits } -\frac{\pi}{6} \text{ and } \frac{\pi}{6},$$

$$= \frac{2a}{3}\,\frac{\int_0^{\frac{\pi}{6}} \cos^3 3\theta \cdot \cos\theta \cdot d\theta}{\int_0^{\frac{\pi}{6}} \cos^2 3\theta \cdot d\theta} = \frac{a}{3}\,\frac{\int_0^{\frac{\pi}{6}} (3\cos 3\theta + \cos 9\theta)\cos\theta\,d\theta}{\int_0^{\frac{\pi}{6}} (1 + \cos 6\theta)\,d\theta}$$

$$= \frac{a}{6}\,\frac{\int_0^{\frac{\pi}{6}} [3\cos 2\theta + 3\cos 4\theta + \cos 8\theta + \cos 10\theta]}{\int_0^{\frac{\pi}{6}} (1 + \cos 6\theta)\,d\theta} = \frac{81\sqrt{3}}{80\pi}\cdot a.$$

Ex. 4. ***In a semi-circular disc, bounded by a diameter OA, the density at any point varies as the distance from O; find the position of its centre of gravity.***

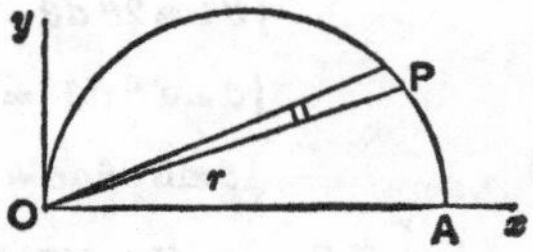

Here the density $\rho=\lambda r$.

Hence

$$\bar{x}=\frac{\iint r\,d\theta\,dr\,.\,\lambda r\,.\,r\cos\theta}{\iint r\,d\theta\,dr\,.\,\lambda r}$$

$$=\frac{\iint r^3\cos\theta\,dr\,d\theta}{\iint r^2\,dr\,d\theta};\ \text{and}\ \bar{y}=\frac{\iint r^3\sin\theta\,dr\,d\theta}{\iint r^2\,dr\,d\theta}.$$

The limits of the integrations are from $r=0$ to $r=2a\cos\theta$, where a is the radius, and from $\theta=0$ to $\theta=\frac{\pi}{2}$.

$$\therefore\ \bar{x}=\frac{\int_0^{\frac{\pi}{2}}\cos\theta\left[\frac{r^4}{4}\right]_0^{2a\cos\theta}d\theta}{\int_0^{\frac{\pi}{2}}\left[\frac{r^3}{3}\right]_0^{2a\cos\theta}d\theta}=\frac{3a}{2}\frac{\int_0^{\frac{\pi}{2}}\cos^5\theta\,d\theta}{\int_0^{\frac{\pi}{2}}\cos^3\theta\,d\theta}=\frac{3a}{2}\cdot\frac{\frac{4\,.\,2}{5\,.\,3}\,.\,1}{\frac{2}{3}\,.\,1}=\frac{6a}{5}.$$

Also
$$\bar{y}=\frac{\int_0^{\frac{\pi}{2}}\sin\theta\left[\frac{r^4}{4}\right]_0^{2a\cos\theta}d\theta}{\int_0^{\frac{\pi}{2}}\left[\frac{r^3}{3}\right]_0^{2a\cos\theta}d\theta}=\frac{3a}{2}\frac{\int_0^{\frac{\pi}{2}}\cos^4\theta\sin\theta\,d\theta}{\int_0^{\frac{\pi}{2}}\cos^3\theta\,d\theta}=\frac{9a}{20}.$$

EXAMPLES

Find the positions of the centroids of the areas of the following curves:

1. The parabola $y^2=4ax$ between the axis of x and the ordinate $x=h$.
$$\left[\bar{x}=\frac{3h}{5};\ \bar{y}=\frac{3}{4}\sqrt{ah}.\right]$$

2. The part of the ellipse $\frac{x^2}{a^2}+\frac{y^2}{b^2}=1$ which lies in the positive quadrant.
$$\left[\frac{\bar{x}}{a}=\frac{\bar{y}}{b}=\frac{4}{3\pi}.\right]$$

3. The parabola $\left(\frac{x}{a}\right)^{\frac{1}{2}}+\left(\frac{y}{b}\right)^{\frac{1}{2}}=1$ between the curve and the axes.
$$\left[\frac{\bar{x}}{a}=\frac{\bar{y}}{b}=\frac{1}{5}.\right]$$

4. $\left(\frac{x}{a}\right)^{\frac{2}{3}}+\left(\frac{y}{b}\right)^{\frac{2}{3}}=1$, lying in the positive quadrant. $\left[\frac{\bar{x}}{a}=\frac{\bar{y}}{b}=\frac{256}{315\pi}.\right]$

5. $y=\sin x$ between $x=0$ and $x=\pi$. $\left[\bar{x}=\frac{\pi}{2};\ \bar{y}=\frac{\pi}{8}.\right]$

6. $ay^2=x^3$ between the origin and $x=b$. $\left[\bar{x}=\frac{5}{7}b.\right]$

7. $y^2(2a-x)=x^3$ and its asymptote. $\left[\bar{x}=\frac{5a}{3}\right]$

8. A loop of the curve $y^2(a+x)=x^2(a-x)$. $\left[\bar{x}=\frac{a}{3}\cdot\frac{3\pi-8}{4-\pi}.\right]$

9. The cardioid $r=a(1+\cos\theta)$. $\left[\bar{x}=\frac{5a}{6}.\right]$

10. One loop of $r=a\cos 2\theta$. $\left[\bar{x}=\frac{128\sqrt{2}}{105}\cdot\frac{a}{\pi}.\right]$

11. One loop of the lemniscate of Bernoulli $r^2=a^2\cos 2\theta$. $\left[\bar{x}=\frac{\pi a\sqrt{2}}{8}.\right]$

12. One loop of $r=a\cos n\theta$. $\left[\bar{x}=\frac{16an^4\cos\frac{\pi}{2n}}{(n^2-1)(9n^2-1)\pi}.\right]$

13. $\left(\frac{x}{a}\right)^n+\left(\frac{y}{b}\right)^n=1$, lying in the positive quadrant.

$$\left[\frac{\bar{x}}{a}=\frac{\bar{y}}{b}=\frac{2}{3}\frac{\left\{\Gamma\left(\frac{2}{n}\right)\right\}^2}{\Gamma\left(\frac{1}{n}\right).\Gamma\left(\frac{3}{n}\right)}.\right]$$

Find the positions of the centroids of the areas enclosed by the following curves:

14. $y^2=ax$ and $x^2=by$. $\left[\bar{x}.a^{\frac{1}{3}}=\bar{y}.b^{\frac{1}{3}}=\frac{9}{20}a^{\frac{2}{3}}b^{\frac{2}{3}}.\right]$

15. $y^2=ax$ and $y^2=2ax-x^2$ on the positive side of the axis of x.

$$\left[\frac{5\bar{x}}{15\pi-44}=\bar{y}=\frac{a}{3\pi-8}.\right]$$

16. $x^2+y^2-2ax=0$ and $x^2+y^2-2bx=0$ on the positive side of the axis of x. $\left[\bar{x}=\frac{a^2+ab+b^2}{a+b};\ \bar{y}=\frac{4}{3}\frac{a^2+ab+b^2}{\pi(a+b)}\right]$

17. $y^2=4ax$ and $y=mx$. $\left[\bar{x}=\frac{8a}{5m^2};\ \bar{y}=\frac{2a}{m}.\right]$

18. $y^2=4ax$, $y^2=4bx$, $x^2=4cy$ and $x^2=4dy$. $\left[\bar{x}=\frac{9}{5}\frac{(b^{\frac{4}{3}}-a^{\frac{4}{3}})(d^{\frac{5}{3}}-c^{\frac{5}{3}})}{(b-a)(d-c)}.\right]$

19. The density at any point of a circular lamina varies as the nth power of the distance from a point O on the circumference; shew that the centre of gravity of the lamina divides the diameter through O in the ratio $n+2:2$.

20. A circular disc, of radius a, whose density is proportional to the distance from the centre, has a hole cut in it bounded by a circle, of diameter b, which passes through the centre. Shew that the distance from the centre of the disc of the centre of gravity of the remaining portion is $\dfrac{6b^4}{15\pi a^3 - 10b^3}$.

21. The centre of inertia of a circular disc, the density of which varies inversely as the fourth power of the distance from an external point O in the plane, is the inverse point of O with respect to the boundary of the disc.

22. The distance from the cusp of the centroid of the area of the cardioid $r = a(1+\cos\theta)$, when the density at any point varies as the nth power of the distance from the cusp, is $\dfrac{(n+2)(2n+5)}{(n+3)(n+4)}a$.

23. Find the centre of gravity of a plate in the form of a quadrant AOB of an ellipse, the thickness at any point of the plate varying as the product of the distances of the point from OA and OB. $\left[\dfrac{\bar{x}}{a} = \dfrac{\bar{y}}{b} = \dfrac{8}{15}.\right]$

24. Find the coordinates of the centre of gravity of a lamina in the shape of a quadrant of the curve $\left(\dfrac{x}{a}\right)^{\frac{2}{3}} + \left(\dfrac{y}{b}\right)^{\frac{2}{3}} = 1$, the density being given by $\sigma = kxy$. $\left[\dfrac{\bar{x}}{a} = \dfrac{\bar{y}}{b} = \dfrac{128}{429}.\right]$

25. A chord of an ellipse cuts off a segment of constant area; shew that the locus of its centre of gravity is a similar, similarly situated, and concentric ellipse.

26. The locus of the centre of gravity of all equal segments cut off from a parabola is an equal parabola.

27. If G is the centre of gravity of any arc PQ of the lemniscate $r^2 = a^2\cos 2\theta$, shew that OG bisects the angle POQ.

28. *A curve is such that the centroid of the arc of it intercepted between two radii vectores drawn from a fixed point always lies on the straight line bisecting the angle between these radii; shew that the curve is either a lemniscate of Bernouilli or a circle.*

[For all values of β, we are given that $\tan\dfrac{\beta}{2} = \dfrac{\int_0^\beta ds \, . \, r\sin\theta}{\int_0^\beta ds \, . \, r\cos\theta}$.

Hence, if $\dfrac{r\,ds}{d\theta} = F(\theta)$, we have $\int_0^\beta F(\theta)\sin\theta\,d\theta = \tan\dfrac{\beta}{2}\int_0^\beta F(\theta)\cos\theta\,d\theta$.

Differentiating with respect to β,

$$F(\beta)\sin\beta = \tan\frac{\beta}{2}\,.\,F(\beta)\cos\beta + \frac{1}{2}\sec^2\frac{\beta}{2}\int_0^\beta F(\theta)\cos\theta\,d\theta.$$

$$\therefore\quad \sin\beta\,.\,F(\beta) = \int_0^\beta F(\theta)\cos\theta\,d\theta.$$

Differentiating again, we have

$$\cos\beta\,.\,F(\beta) + \sin\beta\,.\,F'(\beta) = F(\beta)\cos\beta.$$

Hence $F'(\beta) = 0$ and hence $F(\beta) = \text{constant}$.

$$\therefore\quad r\sqrt{r^2 + \left(\frac{dr}{d\theta}\right)^2} = \text{constant} = a^2,$$

from which we easily have that either r is constant or $r^2 = a^2\cos(2\theta + \gamma)$.]

29. Shew that the circle is the only curve in which the centroid of the area included between the curve and two radii drawn from a fixed point always lies on the straight line bisecting the angle between the radii.

147. *Centre of gravity of a solid and surface of revolution.*

Let the curve AB revolve round the axis of x. The volume generated by the element of area $PMNQ$ between two ordinates at distances x and $x + \delta x$ from Oy is $\pi y^2\,.\,\delta x$, and its centre of gravity is at a distance x from O, when δx is indefinitely small.

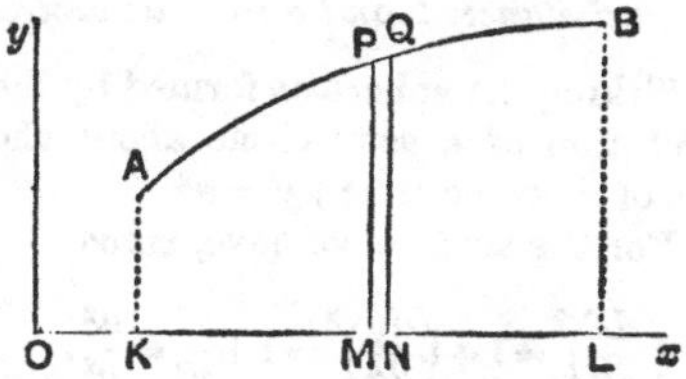

Hence, if the solid be of uniform density,

$$\bar{x} = \frac{\Sigma\,\pi y^2\delta x\,.\,x}{\Sigma\,\pi y^2\delta x} = \frac{\int y^2 x\,dx}{\int y^2\,dx},$$

where y is known in terms of x from the equation to the curve and the limits of x are OK and OL.

The surface generated by the revolution of the arc PQ $(= \delta s)$ about Ox is $2\pi y\,.\,\delta s$. Hence, for the surface,

$$\bar{x} = \frac{\Sigma\,2\pi y\,\delta s\,.\,x}{\Sigma\,2\pi y\,\delta s} = \frac{\int yx\,ds}{\int y\,ds}.$$

Now $\frac{ds}{dx}$, $\left[= \sqrt{1 + \left(\frac{dy}{dx}\right)^2}\right]$, and y are known from the equation to the curve and hence the integrations can be performed.

If the generating curve be given by the equation $r=f(\theta)$ in polar coordinates, the element $r\delta\theta\,.\,\delta r$ describes a circle of radius $r\sin\theta$, and $\bar{x}$ for the volume

$$=\frac{\Sigma r\delta\theta\,.\,\delta r\,.\,2\pi r\sin\theta\,.\,r\cos\theta}{\Sigma r\delta\theta\,\delta r\,.\,2\pi r\sin\theta}=\frac{\iint r^3\sin\theta\cos\theta\,dr\,d\theta}{\iint r^2\sin\theta\,dr\,d\theta},$$

the limits for r being 0 to $f(\theta)$, and those of θ depending on the part of the curve considered.

So, for the surface,

$$\bar{x}=\frac{\Sigma\,\delta s\,.\,2\pi r\sin\theta\,.\,r\cos\theta}{\Sigma\,\delta s\,.\,2\pi r\sin\theta}=\frac{\int r^2\sin\theta\cos\theta\,.\,ds}{\int r\sin\theta\,.\,ds},$$

where $r=f(\theta)$, and $\left(\frac{ds}{d\theta}\right)^2=r^2+\left(\frac{dr}{d\theta}\right)^2$.

148. Ex. 1. *Find the centre of gravity of the surface and volume of the part of a sphere, of radius a, included between parallel planes which are at distances b and c from its centre.*

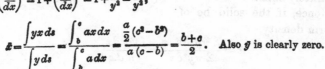

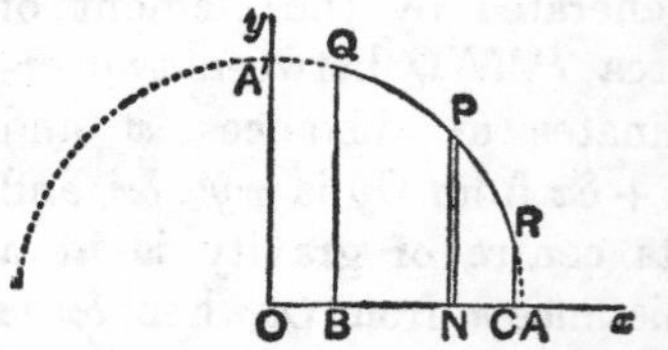

Taking the sphere as formed by the revolution of a semi-circle about the axis of x, we have $x^2+y^2=a^2$.

For the surface we have, since

$$\left(\frac{ds}{dx}\right)^2=1+\left(\frac{dy}{dx}\right)^2=1+\frac{x^2}{y^2}=\frac{a^2}{y^2},$$

$$\bar{x}=\frac{\int yx\,ds}{\int y\,ds}=\frac{\int_b^c ax\,dx}{\int_b^c a\,dx}=\frac{\frac{a}{2}(c^2-b^2)}{a(c-b)}=\frac{b+c}{2}.$$ Also $\bar{y}$ is clearly zero.

Hence the centre of gravity of a zone of a sphere is half-way between its plane ends.

For the volume,

$$\bar{x}=\frac{\int \pi y^2\,dx\,.\,x}{\int \pi y^2\,dx}=\frac{\int_b^c x(a^2-x^2)\,dx}{\int_b^c (a^2-x^2)\,dx}=\tfrac{3}{4}(b+c)\frac{2a^2-b^2-c^2}{3a^2-b^2-bc-c^2}.$$

Cor. If we put $c=a$ and $b=0$, we have the case of a hemisphere. If it be hollow, then $\bar{x}=\frac{a}{2}$; if it be solid, $\bar{x}=\frac{3a}{8}$.

Ex. 2. ***Find the centre of gravity of the volume formed by revolving the area bounded by the parabolas $y^2=4ax$ and $x^2=4by$ about the axis of x.***

We easily see that the point of intersection P of the two curves is given by

$$x'=4a^{\frac{1}{3}}b^{\frac{2}{3}},\ y'=4a^{\frac{2}{3}}b^{\frac{1}{3}}.$$

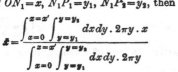

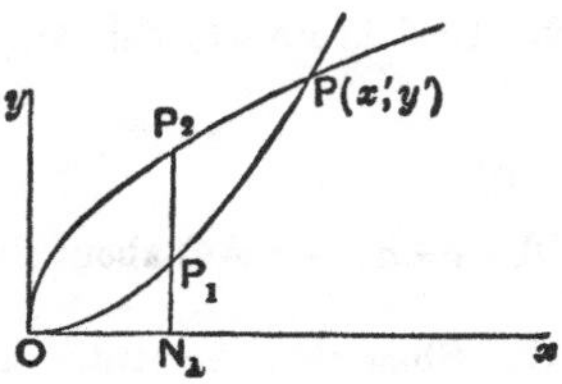

If $ON_1=x$, $N_1P_1=y_1$, $N_1P_2=y_2$, then

$$\bar{x}=\frac{\displaystyle\int_{x=0}^{x=x'}\int_{y=y_1}^{y=y_2} dx\,dy\,.\,2\pi y\,.\,x}{\displaystyle\int_{x=0}^{x=x'}\int_{y=y_1}^{y=y_2} dx\,dy\,.\,2\pi y}$$

$$=\frac{\displaystyle\int_0^{x'}(y_2^2-y_1^2)\,.\,x\,dx}{\displaystyle\int_0^{x'}(y_2^2-y_1^2)\,dx}=\frac{\displaystyle\int_0^{x'}\left[4ax-\frac{x^4}{16b^2}\right]x\,dx}{\displaystyle\int_0^{x'}\left[4ax-\frac{x^4}{16b^2}\right]dx}=\frac{20}{9}a^{\frac{1}{3}}b^{\frac{2}{3}}.$$

EXAMPLES

Find the centroids of the surfaces formed by the revolution of the following curves:

1. Parabola $y^2=2ax$ cut off by $x=c$ about its axis.

$$\left[\bar{x}=\frac{(3c-a)(a+2c)^{\frac{3}{2}}+a^{\frac{5}{2}}}{5\{(a+2c)^{\frac{3}{2}}-a^{\frac{3}{2}}\}}.\right]$$

2. Cycloid $x=a(\theta+\sin\theta)$, $y=a(1-\cos\theta)$ about the axis of y.

$$\left[\bar{y}=\frac{2a}{15}\cdot\frac{15\pi-8}{3\pi-4}.\right]$$

3. Cardioid $r=a(1+\cos\theta)$ about its axis. $\left[\bar{x}=\frac{50}{63}a.\right]$

4. One loop of $r^2=a^2\cos 2\theta$ about the initial line. $\left[\bar{x}=\frac{a}{6}(2+\sqrt{2}).\right]$

Find the centroids of the volumes formed by the revolution of the following curves:

5. The portion of the parabola $y^2=4ax$, cut off by the ordinate $x=h$, about the axis of x. $\left[\bar{x}=\frac{2h}{3}.\right]$

6. $y^{m+n}=a^mx^n$ about the axis of x. $\left[\bar{x}=\frac{m+3n}{m+2n}\cdot\frac{x}{2}.\right]$

7. $y^4-axy^2+x^4=0$ about the axis of x. $\left[\bar{x}=\frac{3a\pi}{32}.\right]$

8. A circle of radius a through two right angles about a tangent line.

$$\left[\bar{x}=\frac{5a}{2\pi}\right]$$

9. Cycloid $x=a(\theta+\sin\theta)$, $y=a(1-\cos\theta)$ about the axis of y.

$$\left[\bar{y}=\frac{63\pi^2-64}{9\pi^2-16}\cdot\frac{a}{6}.\right]$$

10. $r=a(1+\cos\theta)$ about its axis. $\left[\bar{x}=\frac{4a}{5}.\right]$

11. Shew that the centre of gravity of a lune of a sphere, of angle 2α, is at a distance $\frac{\pi}{4}\cdot\frac{a\sin\alpha}{\alpha}$ from its axis.

12. Shew that the distance from the centre of the centre of gravity of a sector of a sphere, of radius a, is $\frac{3}{8}a(1+\cos\alpha)$, where α is the angle which the radius to any point of the spherical base of the sector makes with the axis of the sector.

13. If the density at any point of a sphere of radius a varies directly as the distance from the centre, and a sphere described on the radius as diameter be cut out, shew that the centre of gravity of the remainder is at a distance $\frac{4}{63}a$ from the centre.

14. Shew that the centre of gravity of a sphere, the density at any point of which varies inversely as the square of the distance from a fixed point on the surface of the sphere, bisects the radius through the fixed point.

15. The density of a hemisphere varies as the nth power of the distance from the centre; shew that the centre of gravity divides the radius perpendicular to its plane surface in the ratio $n+3:n+5$.

16. Taking Laplace's law for the density of the earth, assumed to be a sphere of radius a, *viz.* $\rho=\rho_0\frac{\sin\theta}{\theta}$, where $\theta=\frac{\mu x}{a}$, for the density ρ at a distance x from the centre, shew that the distance from the centre of the centre of gravity of half the earth, bounded by a plane through its centre, is

$$a\frac{(2-\mu^2)\cos\mu+2\mu\sin\mu-2}{2\mu(\sin\mu-\mu\cos\mu)}.$$

17. A portion of an anchor ring is formed by the revolution of a circular area of radius a about a line in its plane at a distance c from its centre, where $c>a$. If 2α be the angle through which it revolves, prove that the centre of gravity of the solid is at a distance from the line equal to

$$\frac{4c^2+a^2}{4c}\cdot\frac{\sin\alpha}{\alpha}.$$

18. A uniform solid is bounded by the surface formed by the revolution of a cycloid about its base and is then cut in halves by a plane through the axis of revolution. Shew that the centre of gravity of each half is at a distance $\frac{7a}{3\pi}$ from the plane face, where a is the radius of the generating circle of the cycloid.

19. A solid is bounded by half the surface formed by the revolution of the cardioid $r=a(1+\cos\theta)$ about its axis, and by a plane base through its axis; shew that the distance of its centre of gravity from the axis is $\frac{63}{128}a$ and its distance from the pole, measured parallel to the axis, is $\frac{4a}{5}$.

20. A quadrant of a circle, of radius a, makes a complete revolution about a straight line which is parallel to one of the bounding radii of the quadrant at a distance b from it and which does not cut it. Shew that the distances of the centres of gravity of the curved surface and volume so generated from its plane surface are $\frac{a(2b\pm a)}{\pi b\pm 2a}$ and $\frac{a}{2}\frac{8b\pm 3a}{3\pi b\pm 4a}$.

149. General formulae for the centre of gravity of any volume. If P be any point (x, y, z) of the volume and $(x+\delta x,\ y+\delta y,\ z+\delta z)$ a close point Q, the volume of the elementary parallelopiped bounded by planes through P and Q parallel to the planes yOz, zOx, xOy is $\delta x\,.\,\delta y\,.\,\delta z$. Its density being ρ, the fundamental formulae become

$$\bar{x}=\frac{\Sigma\,\delta x\,.\,\delta y\,.\,\delta z\,.\,\rho\,.\,x}{\Sigma\,\delta x\,\delta y\,\delta z\,\rho}=\frac{\iiint \rho x\,dx\,dy\,dz}{\iiint \rho\,dx\,dy\,dz}.$$

So $$\bar{y}=\frac{\iiint \rho y\,dx\,dy\,dz}{\iiint \rho\,dx\,dy\,dz}\quad\text{and}\quad \bar{z}=\frac{\iiint \rho z\,dx\,dy\,dz}{\iiint \rho\,dx\,dy\,dz}.$$

The limits are such as to include all the volume considered.

Ex. 1. *Find the centre of gravity of the positive octant of the ellipsoid*

$$\frac{x^2}{a^2}+\frac{y^2}{b^2}+\frac{z^2}{c^2}=1,$$

which is of constant density.

Here $\bar{x}=\frac{\iiint x\,dx\,dy\,dz}{\iiint dx\,dy\,dz}$...(1).

The limits for z are from 0 to MR, i.e. $c\sqrt{1-\frac{x^2}{a^2}-\frac{y^2}{b^2}}$, for x from 0 to NS, i.e. $a\sqrt{1-\frac{y^2}{b^2}}$, and for y from 0 to b.

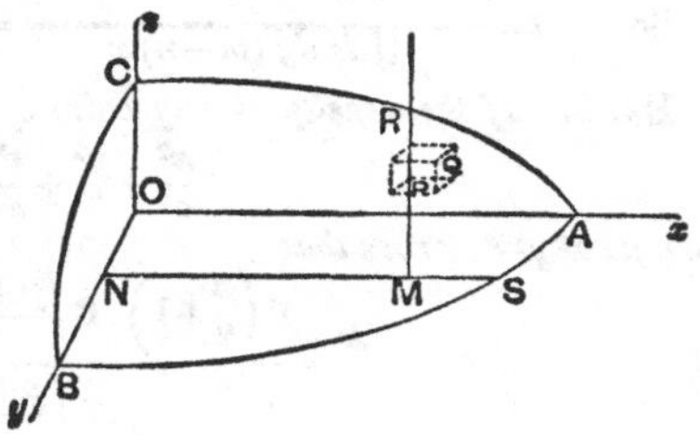

Hence the numerator of (1)

$$=\iint cx\sqrt{1-\frac{x^2}{a^2}-\frac{y^2}{b^2}}\,dx\,dy=\int\left[-\frac{a^2c}{3}\left(1-\frac{x^2}{a^2}-\frac{y^2}{b^2}\right)^{\frac{3}{2}}\right]_0^{NS}dy$$

$$=\int_0^b\frac{a^2c}{3}\left(1-\frac{y^2}{b^2}\right)^{\frac{3}{2}}dy=\int_0^{\frac{\pi}{2}}\frac{a^2c}{3}\cos^3\theta\,.\,b\cos\theta\,d\theta,\qquad(\text{if } y=b\sin\theta),$$

$$=\frac{a^2bc}{3}\cdot\frac{3\,.\,1}{4\,.\,2}\,\frac{\pi}{2}=\frac{\pi}{16}a^2bc.$$

Also the denominator of (1) = volume of the octant $=\frac{1}{8}\cdot\frac{4}{3}\pi abc.$

Hence $\bar{x}=\frac{3a}{8}$; similarly, $\bar{y}=\frac{3b}{8}$ and $\bar{z}=\frac{3c}{8}$.

Ex. 2. *Find the centre of gravity of the volume cut off from the cylinder* $2x^2+y^2=2ax$ *by the planes* $z=mx$, $z=nx$.

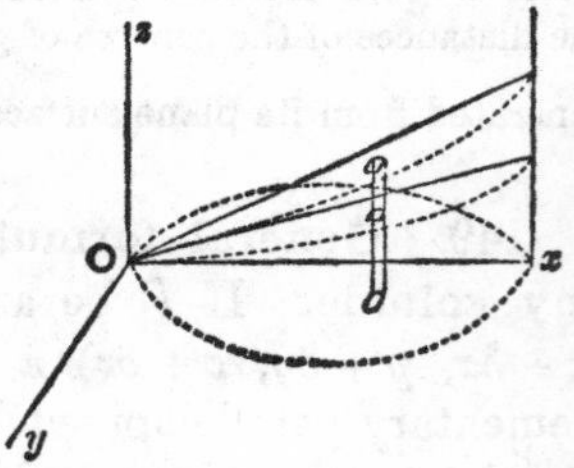

If we take any element $\delta x\,\delta y$ of the section by the plane xy, the volume above it is clearly $\delta x\,.\,\delta y\times(m-n)\,x$, and the height of its centre of gravity above the plane of xy is $\frac{m+n}{2}x$.

Hence $\bar{x}=\frac{\iint dx\,dy\,(m-n)\,x\,.\,x}{\iint dx\,dy\,(m-n)\,x}$,

the limits for y being $-\sqrt{2ax-2x^2}$ to $\sqrt{2ax-2x^2}$, and for x from 0 to a.

$$\text{Hence}\quad \bar{x}=\frac{\int_0^a x^{\frac{5}{2}}\sqrt{a-x}\,dx}{\int_0^a x^{\frac{3}{2}}\sqrt{a-x}\,dx}=a\frac{\int_0^{\frac{\pi}{2}}\sin^6\phi\cos^2\phi\,d\phi}{\int_0^{\frac{\pi}{2}}\sin^4\phi\cos^2\phi\,d\phi},\ \text{if } x=a\sin^2\phi,$$

$$=a\frac{\int_0^{\frac{\pi}{2}}(\sin^6\phi-\sin^8\phi)\,d\phi}{\int_0^{\frac{\pi}{2}}(\sin^4\phi-\sin^6\phi)\,d\phi}=\frac{5a}{8}.$$

$$\text{So}\qquad \bar{z}=\frac{\iint dx\,dy\,(m-n)\,x\times\frac{m+n}{2}x}{\iint dx\,dy\,(m-n)\,x}=\frac{m+n}{2}\,.\,\bar{x}=\frac{5a}{16}(m+n).$$

Ex. 3. *If the density at any point of an octant of the ellipsoid*

$$\frac{x^2}{a^2}+\frac{y^2}{b^2}+\frac{z^2}{c^2}=1$$

vary as $x^py^qz^r$, *prove that*

$$\frac{\bar{x}}{a}=\frac{\Gamma\left(\frac{p}{2}+1\right).\,\Gamma\left(\frac{p+q+r}{2}+\frac{5}{2}\right)}{\Gamma\left(\frac{p}{2}+\frac{1}{2}\right).\,\Gamma\left(\frac{p+q+r}{2}+3\right)}.$$

Consider the cases $p=q=r=0$, $p=q=r=1$ *and* $p=q=r=2$.

The density being $\lambda x^p y^q z^r$, we have $\bar{x}=\dfrac{\iiint dx\,dy\,dz\,.\,\lambda x^p y^q z^r\,.\,x}{\iiint dx\,dy\,dz\,.\,\lambda x^p y^q z^r}$,

where x, y, z have any positive values subject to the condition

$$\frac{x^2}{a^2}+\frac{y^2}{b^2}+\frac{z^2}{c^2}\leqq 1.$$

Put $x=a\xi^{\frac{1}{2}}$, $y=b\eta^{\frac{1}{2}}$, $z=c\zeta^{\frac{1}{2}}$. $\therefore\ \bar{x}=a\,.\,\dfrac{\iiint \xi^{\frac{p}{2}}\eta^{\frac{q}{2}-\frac{1}{2}}\zeta^{\frac{r}{2}-\frac{1}{2}}\,.\,d\xi d\eta d\zeta}{\iiint \xi^{\frac{p}{2}-\frac{1}{2}}\eta^{\frac{q}{2}-\frac{1}{2}}\zeta^{\frac{r}{2}-\frac{1}{2}}d\xi d\eta d\zeta}$,

where ξ, η, ζ have any positive values subject to the condition $\xi+\eta+\zeta\leqq 1$.

Hence, by Dirichlet's integrals,

$$\bar{x}=a\,.\,\frac{\Gamma\left(\frac{p}{2}+1\right)\Gamma\left(\frac{q}{2}+\frac{1}{2}\right)\Gamma\left(\frac{r}{2}+\frac{1}{2}\right)}{\Gamma\left(\frac{p+q+r}{2}+3\right)}\div\frac{\Gamma\left(\frac{p}{2}+\frac{1}{2}\right)\Gamma\left(\frac{q}{2}+\frac{1}{2}\right)\Gamma\left(\frac{r}{2}+\frac{1}{2}\right)}{\Gamma\left(\frac{p+q+r}{2}+\frac{5}{2}\right)}$$

$$=a\,.\,\frac{\Gamma\left(\frac{p}{2}+1\right)\Gamma\left(\frac{p+q+r}{2}+\frac{5}{2}\right)}{\Gamma\left(\frac{p}{2}+\frac{1}{2}\right)\Gamma\left(\frac{p+q+r}{2}+3\right)}.$$

If $p=q=r=0$, then $\dfrac{\bar{x}}{a}=\dfrac{\Gamma(1)\,\Gamma(\frac{5}{2})}{\Gamma(\frac{1}{2})\,\Gamma(3)}=\dfrac{\Gamma(1).\frac{3}{2}.\frac{1}{2}.\Gamma(\frac{1}{2})}{\Gamma(\frac{1}{2}).2.1.\Gamma(1)}=\dfrac{3}{8}$, as in Ex. 1.

If $p=q=r=1$, then $\dfrac{\bar{x}}{a}=\dfrac{\Gamma(\frac{3}{2})\,\Gamma(4)}{\Gamma(1)\,\Gamma(\frac{9}{2})}=\dfrac{\Gamma(\frac{3}{2}).3.2.1.\Gamma(1)}{\Gamma(1).\frac{7}{2}.\frac{5}{2}.\frac{3}{2}.\Gamma(\frac{3}{2})}=\dfrac{16}{35}$.

If $p=q=r=2$, then $\dfrac{\bar{x}}{a}=\dfrac{\Gamma(2)\,\Gamma(\frac{11}{2})}{\Gamma(\frac{3}{2})\,\Gamma(6)}=\dfrac{\Gamma(2).\frac{9}{2}.\frac{7}{2}.\frac{5}{2}.\frac{3}{2}.\Gamma(\frac{3}{2})}{\Gamma(\frac{3}{2}).5.4.3.2.\Gamma(2)}=\dfrac{63}{128}$.

150. If the solid have its equation given in polar coordinates so that the coordinates of a point P are (r, θ, ϕ), where $OP=r$, $zOP=\theta$, and ϕ is the angle the plane zOP makes with zOx, then the element of volume is $\delta r\,.\,r\delta\theta\,.\,r\sin\theta\,\delta\phi$, *i.e.* $r^2\sin\theta\,\delta r\,.\,\delta\theta\,.\,\delta\phi$, and we have

$$\bar{x}=\frac{\Sigma\, r^2\sin\theta\,\delta r\,.\,\delta\theta\,.\,\delta\phi\,.\,r\cos\phi\sin\theta}{\Sigma\, r^2\sin\theta\,\delta r\,.\,\delta\theta\,.\,\delta\phi}=\frac{\iiint r^3\sin^2\theta\cos\phi\,dr\,d\theta\,d\phi}{\iiint r^2\sin\theta\,dr\,d\theta\,d\phi},$$

$$\bar{y}=\frac{\Sigma\, r^2\sin\theta\,\delta r\,.\,\delta\theta\,.\,\delta\phi\,.\,r\sin\phi\sin\theta}{\Sigma\, r^2\sin\theta\,\delta r\,.\,\delta\theta\,.\,\delta\phi}=\frac{\iiint r^3\sin^2\theta\sin\phi\,dr\,d\theta\,d\phi}{\iiint r^2\sin\theta\,dr\,d\theta\,d\phi},$$

and $$\bar{z}=\frac{\Sigma\, r^2\sin\theta\,\delta r\,.\,\delta\theta\,.\,\delta\phi\,.\,r\cos\theta}{\Sigma\, r^2\sin\theta\,\delta r\,.\,\delta\theta\,.\,\delta\phi}=\frac{\iiint r^3\sin\theta\cos\theta\,dr\,d\theta\,d\phi}{\iiint r^2\sin\theta\,dr\,d\theta\,d\phi}$$

the limits being such as to include all the solid.

Ex. *Find the centre of gravity of a hemisphere whose density varies as the distance from a point on its plane edge.*

The equation to the hemisphere is

$$(x-a)^2+y^2+z^2=a^2,$$

or, in polar coordinates, $r=2a\cos\phi\sin\theta$.

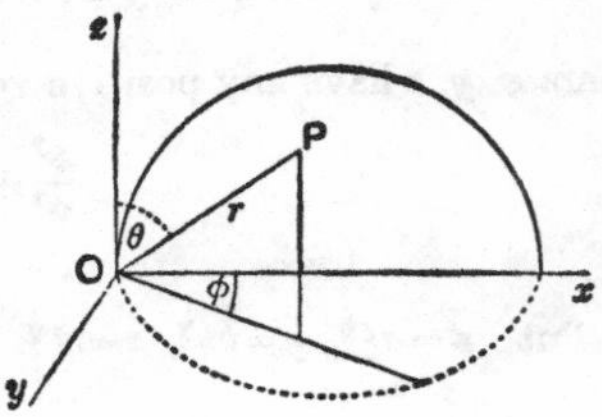

The limits for r are thus 0 to

$$2a\cos\phi\sin\theta.$$

Those for ϕ are $-\frac{\pi}{2}$ to $\frac{\pi}{2}$, and those for θ are zero to $\frac{\pi}{2}$.

If the density at any point is λr, the element of mass is

$$\delta r\,.\,r\,\delta\theta\,.\,r\sin\theta\,\delta\phi\,.\,\lambda r.$$

Hence

$$\bar{x}=\frac{\iiint\lambda r^3\sin\theta\,dr\,d\theta\,d\phi\,.\,r\cos\phi\sin\theta}{\iiint\lambda r^3\sin\theta\,.\,dr\,d\theta\,d\phi}$$

$$=\frac{\frac{1}{5}\iint(2a\cos\phi\sin\theta)^5\sin^2\theta\cos\phi\,d\theta\,d\phi}{\frac{1}{4}\iint(2a\cos\phi\sin\theta)^4\sin\theta\,d\theta\,d\phi}=\frac{8a}{5}\,\frac{\iint\cos^6\phi\,.\,\sin^7\theta\,d\theta\,d\phi}{\iint\cos^4\phi\,.\,\sin^5\theta\,d\theta\,d\phi}$$

$$=\frac{8a}{5}\cdot\frac{\frac{5.3.1}{6.4.2}\,\frac{\pi}{2}\times\frac{6.4.2}{7.5.3}}{\frac{3.1}{4.2}\,\frac{\pi}{2}\times\frac{4.2}{5.3}}=\frac{8a}{7}.$$

Clearly $\bar{y}=0$, by symmetry.

Also

$$\bar{z}=\frac{\iiint\lambda r^3\sin\theta\,dr\,d\theta\,d\phi\,.\,r\cos\theta}{\iiint\lambda r^3\sin\theta\,dr\,d\theta\,d\phi}$$

$$=\frac{\frac{1}{5}\iint(2a\cos\phi\sin\theta)^5\sin\theta\cos\theta\,d\theta\,d\phi}{\frac{1}{4}\iint(2a\cos\phi\sin\theta)^4\sin\theta\,d\theta\,d\phi}=\frac{8a}{5}\,\frac{\iint\cos^5\phi\,.\,\sin^6\theta\cos\theta\,d\theta\,d\phi}{\iint\cos^4\phi\,.\,\sin^5\theta\,d\theta\,d\phi}$$

$$=\frac{8a}{5}\cdot\frac{\frac{4.2}{5.3.1}\cdot\frac{1}{7}}{\frac{3.1}{4.2}\,\frac{\pi}{2}\times\frac{4.2}{5.3.1}}=\frac{128}{105}\,\frac{a}{\pi}.$$

EXAMPLES

Find the centroids of the volumes included between the following surfaces:

1. $x^2+y^2=2ax$, $z=mx$ and $z=nx$. $\left[\frac{5a}{4},\ 0,\ \frac{5a\,(m+n)}{8}\right]$

2. $x^2+y^2=a^2$, $z=0$ and $z=x\tan\alpha+h$. $\left[\frac{a^2}{4h}\tan\alpha,\ 0,\ \frac{a^2}{8h}\tan^2\alpha+\frac{h}{2}.\right]$

3. $\frac{x^2}{a^2}+\frac{y^2}{b^2}=1$, $z=0$ and $lx+my+nz=1$.

$$\left[-\frac{a^2l}{4}, -\frac{b^2m}{4}, \frac{l^2a^2+m^2b^2+4}{8n}.\right]$$

4. $\frac{x^2}{a^2}+\frac{y^2}{b^2}-\frac{z^2}{c^2}=1$, $x=0$ and $z=\pm c$ $\left[\frac{a}{8\pi}\{7\sqrt{2}+3\log(1+\sqrt{2})\}.\right]$

5. $\frac{y^2}{b^2}+\frac{z^2}{c^2}-\frac{2x}{a}=0$, $x=2a$, $y=0$ and $z=0$. $\left[\frac{4a}{3}, \frac{32b}{15\pi}, \frac{32c}{15\pi}.\right]$

6. $\left(\frac{x}{a}\right)^{\frac{2}{3}}+\left(\frac{y}{b}\right)^{\frac{2}{3}}+\left(\frac{z}{c}\right)^{\frac{2}{3}}=1$ contained in the positive octant.

$$\left[\frac{21a}{128}, \frac{21b}{128}, \frac{21c}{128}.\right]$$

7. A body is formed of the portion of a uniform solid sphere, of radius a, which is cut off by a circular cylinder, of diameter a, passing through the centre of the sphere. Shew that the centre of gravity of the portion that lies within the cylinder is at a distance $\frac{12a}{15\pi-20}$ from the centre of the sphere.

151. *Centre of gravity of any spherical triangle.*

Let ABC be the triangle, and O the centre of the sphere. Let OC be the axis of z and Ox, Oy two perpendicular axes, Ox being in the plane COA.

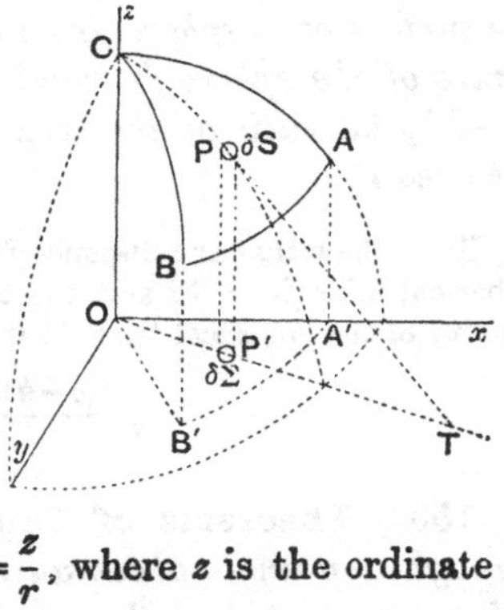

Let P be any point on the triangle; let the tangent at P to the circle CP meet the plane xOy in T, and let PP' be the ordinate to this plane. Take a small element δS of the triangle at P, and let its projection on the plane xOy be $\delta\Sigma$.

Then $\frac{\delta\Sigma}{\delta S}=\cos P'TP=\sin POT=\frac{z}{r}$, where z is the ordinate of P and r is the radius of the sphere.

$$\therefore\ z\,.\,\delta S=r\,.\,\delta\Sigma.$$

Hence, if $\bar{z}$ be the ordinate of the required centre of gravity,

$$\bar{z}=\frac{\Sigma z\,.\,\delta S}{\Sigma\delta S}=\frac{\int r\,.\,d\Sigma}{\int dS}=r\frac{\Sigma}{S}\quad\text{..............(1)},$$

where S is the area of the triangle and Σ is the area of its projection on xOy.

Now

Σ = the projection of the area ACB on xOy

= the projection of the area AOB on the same plane

$= \frac{1}{2}r^2 . \angle AOB \times$ cosine of the angle between AOB and xOy

$= \frac{1}{2}r^2 . c \times$ sine of the inclination of OC to AOB

$= \frac{1}{2}r^2 . c \sin p_3$, where p_3 is the arc drawn from C perpendicular to AB,

$= \frac{1}{2}r^2 . c . \sin b . \sin A$.

Also $S = r^2 . E$, where E is the spherical excess;

$$\therefore \bar{z} = \frac{1}{2}\frac{rc \sin b \sin A}{E}.$$

This gives the distance from O along OC of the projection of the centre of gravity upon OC; similar formulae give the projections on OA and OB. Hence its position is known.

152. The relation (1) of the previous article is clearly true for any area on the sphere, whether a triangle or not.

Hence *the distance of the centre of gravity of any area S on the surface of a sphere from any plane xOy passing through the centre of the sphere is equal to the radius of the sphere multiplied by the ratio of the area of the projection of S upon xOy to the area S.*

Ex. Shew that the distance from the plane, through the side AB of a spherical triangle ABC and the centre O of the sphere, of the centre of gravity of the spherical triangle is

$$\frac{1}{2}\frac{r(c - b \cos A - a \cos B)}{E}.$$

153. Theorems of Pappus. *If any plane area revolve through any angle about an axis in its own plane, then* (1) *the volume generated by the area is equal to the product of the area and the length of the path described by the centroid of the area, and* (2) *the surface generated by the area is equal to the product of the perimeter of the area and the length of the path described by the centroid of the perimeter.*

Let A be the area and S the perimeter of the curve; $\bar{y}$ the distance of the centroid of the area and $\bar{y}'$ that of the perimeter of the curve from the axis of rotation which is taken to be the axis of x.

(1) Let P be any point of the area of the curve whose ordinate is y; then, if the rotation be through an angle θ, the length of the arc described by $P = y.\theta$.

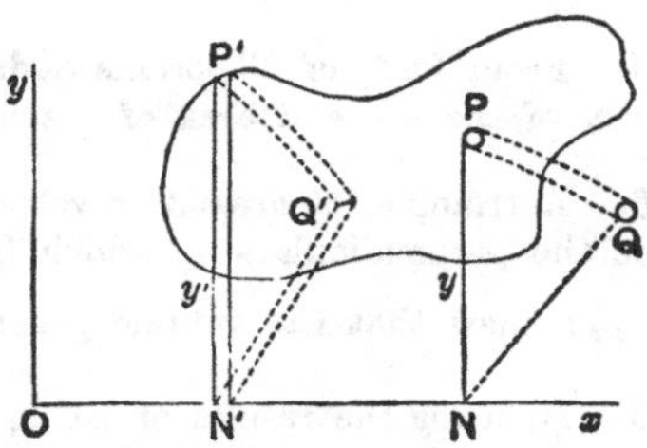

Hence the volume described by an element dA of area at P $= y\theta.dA$.

The whole volume described by the area

$= \Sigma y\theta.dA = \theta.\Sigma y.dA = \theta.\bar{y}A$ (by Art. 145) $= A.\bar{y}\theta$

= the area of the curve multiplied by the length of the arc described by the centroid of the area.

(2) Let P' be a point on the perimeter of the curve whose ordinate is y'; during the rotation the length of the curve described by $P' = y'.\theta$.

Hence the surface described by an element δs of the perimeter at $P' = y'\theta.\delta s$.

Hence the whole surface traced out by the perimeter

$= \Sigma y'\theta.\delta s = \theta.\Sigma y'.\delta s = \theta.\bar{y}'S$ (by Art. 143) $= S.\bar{y}'\theta$

= the perimeter of the curve multiplied by the length of the arc described by the centroid of the perimeter.

EXAMPLES

1. *Find the volume and surface of an anchor-ring or tore.*

An anchor-ring is the surface generated by the revolution of a circle about an axis in its own plane. If a is the radius of the circle, and b the distance of its centre from the axis of rotation, then in a complete revolution the distance described by the centre $= 2\pi b$.

Hence the volume of the anchor-ring $= \pi a^2 \times 2\pi b = 2\pi^2 a^2 b$, and its surface $= 2\pi a \times 2\pi b = 4\pi^2 ab$.

2. A solid sector of a sphere of radius a stands on a base whose rim is circular, and the diameter of this rim subtends an angle 2α at the centre of the sphere; shew that the volume of the sector is $\frac{4\pi}{3}a^3 \sin^2\frac{\alpha}{2}$ and that its curved surface is $4\pi a^2 \sin^2\frac{\alpha}{2}$.

[Revolve a sector of a circle about one of its bounding radii.]

3. Apply Pappus' Theorems to find the surface and volume of a frustum of a right cone in terms of its height and the radii of its plane ends.

4. From Pappus' Theorems deduce the position of the centres of gravity of the arc and area of a semi-circle.

5. A triangle, of area Δ, revolves about a straight line in its own plane the perpendiculars on which from its angular points are p_1, p_2, and p_3; shew that the volume generated is $\frac{2\pi}{3}\Delta(p_1+p_2+p_3)$.

6. By using the results of Ex. 1, page 146, and Ex. 2, page 148, find the volume and surface of the solid formed by a complete revolution of a cycloid about its base.

CHAPTER IX

STABLE AND UNSTABLE EQUILIBRIUM

154. WE have pointed out in Art. 141 that the body in the first figure of that article would, if slightly displaced, tend to return to its position of equilibrium, and that the body in the second figure would not tend to return to its original position of equilibrium, but would recede still further from that position.

These two bodies are said to be in stable and unstable equilibrium respectively.

Consider, again, the case of a heavy sphere, resting on a horizontal plane, whose centre of gravity is not at its centre.

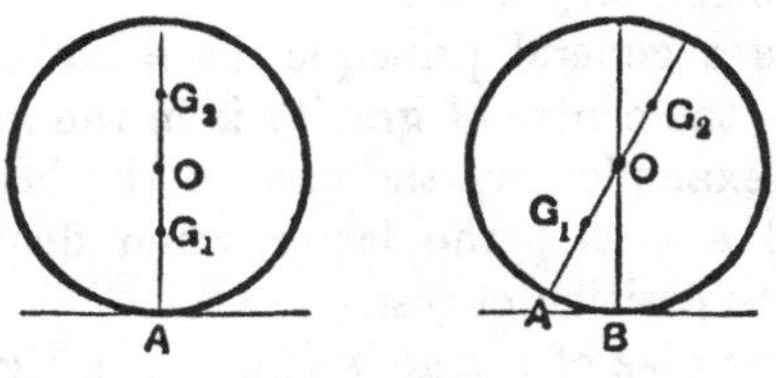

Let the first figure represent the position of equilibrium, the centre of gravity being either below the centre O, as G_1, or above, as G_2. Let the second figure represent the sphere turned through a small angle, so that B is now the point of contact with the plane. The reaction of the plane still acts through the centre of the sphere.

If the weight of the body act through G_1, it is clear that the body will return towards its original position of equilibrium, and therefore the body was originally in stable equilibrium.

If the weight act through G_2, the body will move still

further from its original position of equilibrium, and therefore it was originally in unstable equilibrium.

If however the centre of gravity of the body had been at O, then, in the case of the second figure, the weight would still be balanced by the reaction of the plane; the body would thus remain in the new position, and the equilibrium would be called neutral.

155. Def. A body is said to be in **stable** equilibrium when, if it be slightly displaced from its position of equilibrium, the forces acting on the body tend to make it return towards its position of equilibrium; it is in **unstable** equilibrium when, if it be slightly displaced, the forces tend to move it still further from its position of equilibrium; it is in **neutral** equilibrium, if the forces acting on it in its displaced position are in equilibrium.

In general bodies which are "top-heavy," or which have small bases, are unstable.

Thus in theory a pin might be placed upright with its point on a horizontal table so as to be in equilibrium; in practice the "base" would be so small that the slightest displacement would bring the vertical through its centre of gravity outside its base and it would fall. So with a billiard cue placed vertically with its end on the table.

A body is, as a general principle, in a stable position of equilibrium when the centre of gravity is in the lowest position it can take up; examples are the case of the last article, and the pendulum of a clock; the latter when displaced always returns towards its position of rest.

Again, take the case of a man walking on a tight rope. He generally carries a pole heavily weighted at one end, so that the centre of gravity of himself and the pole is always below his feet. When he feels himself falling in one direction, he shifts his pole so that this centre of gravity shall be on the other side of his feet, and then the resultant weight pulls him back again towards the upright position.

If a body has more than one theoretical position of equilibrium, the one in which its centre of gravity is lowest will in general be the stable position, and that in which the centre of gravity is highest will be the unstable one.

156. ***A body rests in equilibrium upon another fixed body, the portions of the two bodies in contact being spheres of radii r and R respectively, and the straight line joining the centres of the spheres being vertical; if the first body be slightly displaced, to find whether the equilibrium is stable or unstable, the bodies being rough enough to prevent sliding.***

Let O be the centre of the spherical surface of the lower body, and O_1 that of the upper body. Let A_1G_1 be h.

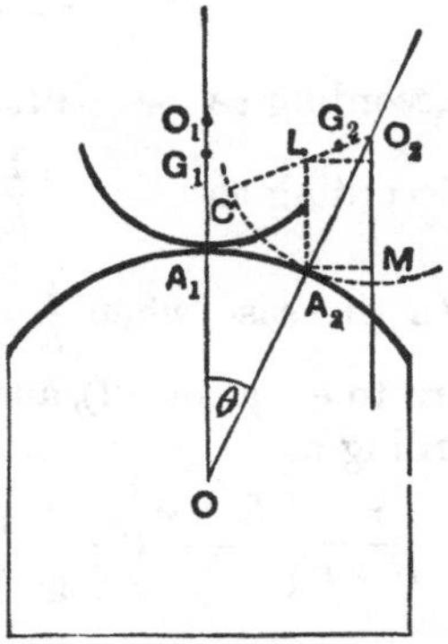

Let the upper body be slightly displaced, by rolling, so that the new position of the centre of the upper body is O_2, the new point of contact is A_2, the new position of the centre of gravity is G_2, and the new position of the point A_1 is C. Hence CG_2 is h.

Draw A_2L vertically to meet O_2C in L, and O_2M vertically to meet a horizontal line through A_2 in M.

Let $\angle A_2OA_1 = \theta$, and $\angle A_2O_2C = \phi$, so that $\angle CO_2M = (\theta + \phi)$.

Since the upper body has rolled into its new position, the arc A_1A_2 = the arc CA_2, so that

$$R \cdot \theta = r \cdot \phi \quad \text{.......................(1)},$$

where R and r are respectively the radii of the lower and upper surfaces.

The equilibrium is stable, or unstable, according as G_2 lies to the left, or right, of the line A_2L,

i.e. according as the distance of G_2 from O_2M is > or < A_2M,

i.e. according as $(r-h)\sin(\theta+\phi)$ is > or < $r\sin\theta$,

i.e. according as $(r-h)\sin\left[\dfrac{R+r}{r}\theta\right]$ > or < $r\sin\theta$,

i.e. according as $(r-h)\left[\dfrac{R+r}{r}\theta - \dfrac{1}{\underline{|3}}\left(\dfrac{R+r}{r}\right)^3\theta^3 + \dots\right]$

$$> \text{ or } < r\left[\theta - \frac{\theta^3}{\underline{|3}} + \dots\right] \dots(2),$$

by substituting the expansion for the sine of an angle in terms of the angle,

i.e. according as

$$(r-h)\left[\frac{R+r}{r}-\frac{1}{\lfloor 3}\left(\frac{R+r}{r}\right)^3\theta^2+\ldots\right]>\text{ or }<r\left[1-\frac{\theta^2}{\lfloor 3}+\ldots\right],$$

i.e., when θ is made indefinitely small, according as

$$(r-h)\frac{R+r}{r}>\text{ or }<r,$$

i.e. according as $\qquad rR>\text{ or }<h(R+r)$,

i.e. according as $\qquad \frac{1}{h}>\text{ or }<\frac{1}{R}+\frac{1}{r}$ ....(3).

In the case when $\frac{1}{h}=\frac{1}{R}+\frac{1}{r}$, *i.e.* when $h=\frac{Rr}{R+r}$, we must return to equation (2), and the equilibrium is stable or unstable according as

$$\frac{r^2}{R+r}\left[\frac{R+r}{r}\theta-\frac{1}{\lfloor 3}\left(\frac{R+r}{r}\right)^3\theta^3+\ldots\right]$$
$$>\text{ or }<r\left(\theta-\frac{\theta^3}{\lfloor 3}+\ldots\right),$$

i.e. according as

$$-\tfrac{1}{6}(R+r)^2\theta^3+\text{higher powers of }\theta>\text{ or }<-\tfrac{1}{6}r^2\theta^3+\ldots,$$

i.e. according as

$$(R+r)^2-\text{sqs. etc. of }\theta<\text{ or }>r^2-\text{sqs. etc. of }\theta,$$

i.e. according as $(R+r)^2<$ or $>r^2$, when θ is made indefinitely small.

Hence the equilibrium is unstable in this case. The equilibrium is thus only stable when $\frac{1}{h}>\frac{1}{r}+\frac{1}{R}$. In all other cases it is unstable.

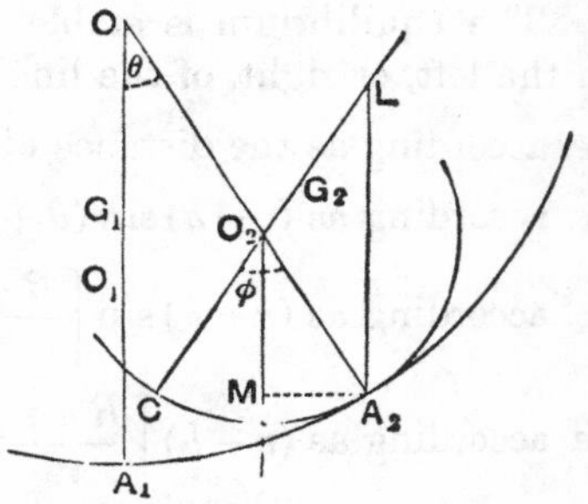

If the curvature of the lower surface be in the other direction as in the second figure, then in this case the angle

$$CO_2M=\phi-\theta=\frac{R-r}{r}\theta,$$

and the equilibrium is stable or unstable according as G_2 is to the left or right of A_2L,

i.e. according as $\quad O_2G_2 \sin(\phi - \theta) \lesseqgtr MA_2$,

i.e. according as $\quad (h - r) \sin \dfrac{R-r}{r}\theta \lesseqgtr r \sin \theta$,

i.e. according as

$$(h-r)\left[\frac{R-r}{r}\theta - \frac{1}{\underline{|3}}\left(\frac{R-r}{r}\theta\right)^3 + \dots\right] \lesseqgtr r\left(\theta - \frac{\theta^3}{\underline{|3}} + \dots\right) \dots (4),$$

i.e. according as

$$(h-r)\left[\frac{R-r}{r} - \frac{1}{6}\left(\frac{R-r}{r}\right)^3\theta^2 + \dots\right] \lesseqgtr r\left(1 - \frac{\theta^2}{6} + \dots\right),$$

i.e. according as $(h-r)\dfrac{R-r}{r} \lesseqgtr r$, if θ be indefinitely small,

i.e. according as $h \lesseqgtr \dfrac{Rr}{R-r}$, *i.e.* according as $\dfrac{1}{h} \gtreqless \dfrac{1}{r} - \dfrac{1}{R}$.

In the critical case when $h = \dfrac{Rr}{R-r}$ we must return to (4) and proceed to higher powers.

(4) then becomes

$$\frac{r^2}{R-r}\left[\frac{R-r}{r}\theta - \frac{1}{\underline{|3}}\left(\frac{R-r}{r}\theta\right)^3 + \dots\right] \lesseqgtr r\left(\theta - \frac{\theta^3}{\underline{|3}} + \dots\right),$$

i.e.

$$-\frac{1}{6}\frac{(R-r)^2}{r^2}\theta^3 + \dots \lesseqgtr -\frac{\theta^3}{6} + \dots,$$

i.e. $\dfrac{(R-r)^2}{r^2}$ − powers of $\theta \gtreqless 1$ − powers of θ.

Hence, when θ is indefinitely small, we see that the equilibrium is stable or unstable according as $(R-r)^2 \gtreqless r^2$,

i.e. according as $R \gtreqless 2r$.

In the case when $R = 2r$ and hence $h = \dfrac{Rr}{R-r} = 2r$, then

$$\phi = \frac{R}{r}\theta = 2\theta,$$

and $O_2G_2 \sin(\phi - \theta) = (h-r)\sin(\phi - \theta) = r \sin \theta = MA_2$ always.

Hence, in this particular case, G_2 always coincides with L; and the upper body will always rest through whatever angle it be rolled since its centre of gravity is now always vertically above the point of contact.

Cor. 1. If the upper body have a plane face in contact with the lower body, as in the following figure, r is infinite in value. Hence the equilibrium is stable if

$$\frac{1}{h} \text{ be} > \frac{1}{R}; \textit{ i.e.}, \text{ if } h \text{ be} < R.$$

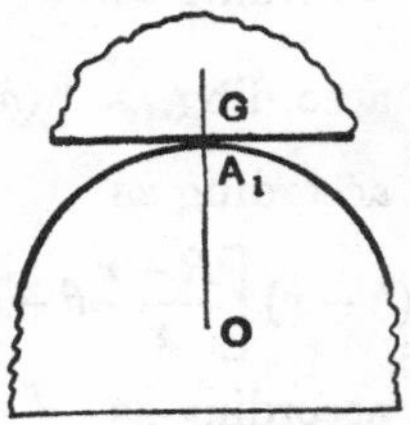

Hence the equilibrium is stable, if the distance of the centre of gravity of the upper body from its plane face be less than the radius of the lower body; otherwise the equilibrium is unstable.

Cor. 2. If the lower body be a plane, so that R is infinite, the equilibrium is stable if

$$\frac{1}{h} \text{ be} > \frac{1}{r}, \textit{ i.e.}, \text{ if } h \text{ be} < r.$$

Hence, if a body of spherical base be placed on a horizontal table, it is in stable equilibrium, if the distance of its centre of gravity from the point of contact be less than the radius of the spherical surface.

157. If the portions of the surfaces in contact are not spheres, but surfaces whose radii of curvature are R and r, it is similarly found that the equilibrium is stable or unstable according as $\frac{1}{h} \gtrless \frac{1}{R} + \frac{1}{r}$.

In the neutral or critical case, when $\frac{1}{h} = \frac{1}{R} + \frac{1}{r}$, the determination of the stability is a question of some difficulty. For its consideration the student may refer to Routh's *Analytical Statics*, or Minchin's *Statics*.

It is there shewn that the equilibrium is stable or unstable according as

$$\frac{d}{ds}\left(\frac{1}{R}\right) + \frac{d}{ds}\left(\frac{1}{r}\right)$$

is negative or positive.

If this condition fails, as it does when the points of contact are points of maximum or minimum curvature, the equilibrium is found to be stable or unstable according as

$$\frac{d^2}{ds^2}\left(\frac{1}{R}\right) + \frac{d^2}{ds^2}\left(\frac{1}{r}\right) + \frac{(R+r)(R+2r)}{R^3r^3}$$

is negative or positive.

EXAMPLES

1. A body, consisting of a cone and a hemisphere on the same base, rests on a rough horizontal table, the hemisphere being in contact with the table; shew that the greatest height of the cone, so that the equilibrium may be stable, is $\sqrt{3}$ times the radius of the hemisphere.

2. A hemisphere rests in equilibrium on a sphere of equal radius; shew that the equilibrium is unstable when the curved, and stable when the flat, surface of the hemisphere rests on the sphere.

3. A uniform beam, of thickness $2b$, rests symmetrically on a perfectly rough horizontal cylinder of radius a; shew that the equilibrium of the beam will be stable or unstable according as b is less or greater than a.

4. A heavy uniform cube balances on the highest point of a sphere, whose radius is r. If the sphere be rough enough to prevent sliding, and if the side of the cube be $\frac{\pi r}{2}$, shew that the cube can rock through a right angle without falling.

5. A lamina in the form of an isosceles triangle, whose vertical angle is α, is placed on a sphere, of radius r, so that its plane is vertical and one of its equal sides is in contact with the sphere; shew that, if the triangle be slightly displaced in its own plane, the equilibrium is stable if $\sin\alpha$ be less than $\frac{3r}{a}$, where a is one of the equal sides of the triangle.

6. A solid homogeneous hemisphere of radius r has a solid right cone of the same substance constructed on its base; the hemisphere rests on the convex side of a fixed sphere of radius R, the axis of the cone being vertical. Shew that the greatest height of the cone consistent with stability for a small rolling displacement is

$$\frac{r}{R+r}[\sqrt{(3R+r)(R-r)}-2r].$$

7. A weight W is supported on a smooth inclined plane by a given weight P, connected with W by means of a string passing round a fixed pulley whose position is given. Find the position of equilibrium of W on the plane, and shew that it is stable.

8. A rough uniform circular disc, of radius r and weight p, is movable about a point distant c from its centre. A string, rough enough to prevent any slipping, hangs over the circumference and carries unequal weights W and w at its ends. Find the positions of equilibrium, and determine whether they are stable or unstable.

9. A solid sphere rests inside a fixed rough hemispherical bowl of twice its radius. Shew that, however large a weight is attached to the highest point of the sphere, the equilibrium is stable.

10. A thin hemispherical bowl, of radius b and weight W, rests in equilibrium on the highest point of a fixed sphere, of radius a, which is rough enough to prevent any sliding. Inside the bowl is placed a small smooth sphere of weight w. Shew that the equilibrium is not stable unless

$$w < W . \frac{a-b}{2b}$$

11. Shew that a sphere partially immersed in a basin of water cannot rest in stable equilibrium on the summit of any convex part of the base.

12. Three equal particles repelling each other with forces proportional to the nth power of the distance are connected together by three equal elastic strings. Find the position of equilibrium and shew that it is stable if $n < \frac{p}{p-a}$, where a is the unstretched and p the stretched length of any string.

13. A solid ellipsoid, whose axes are of lengths $2a$, $2b$, $2c$, rests with the "c-axis" vertical on a rough horizontal plane. The centre of gravity is on the vertical axis at a distance h from the bottom vertex. Shew that the equilibrium is stable if h is less than both $\frac{a^2}{c}$ and $\frac{b^2}{c}$.

14. A heavy cone rests with the centre of its base on the vertex of a fixed paraboloid of revolution, and the height of the cone is equal to twice the latus rectum of the generating parabola. Prove that the equilibrium is neutral to a first approximation, but that it is really stable.

15. A heavy body, the section of which is a cycloid, rests on a rough horizontal plane and has its centre of gravity at the centre of curvature of the curve at the point of contact; shew that the equilibrium is unstable.

16. A solid frustum of a paraboloid of revolution, of height h and latus rectum $4a$, rests with its vertex on the vertex of a paraboloid of revolution, whose latus rectum is $4b$; shew that the equilibrium is stable if

$$h < \frac{3ab}{a+b}.$$

17. A lamina in the form of a cycloid, whose generating circle is of radius a, rests on the top of another cycloid whose generating circle is of radius b, their vertices being in contact and their axes vertical. If h be the height of the centre of gravity of the upper cycloid above its vertex, shew that the equilibrium is stable only if $h < \frac{4ab}{a+b}$ and is unstable if $h \geqq \frac{4ab}{a+b}$.

18. A parabolical cup, whose weight is W, stands on a horizontal table and contains a quantity of water, of weight nW; if h be the height of the centre of gravity of the cup and the contained water, shew that the equilibrium is stable provided that the latus rectum of the parabola is $> 2(n+1)h$.

158. Suppose that we have a body, or system of bodies, under the influence of no forces except their weights, and supported by reactions with smooth fixed surfaces or by other forces which do not appear in the equation of virtual work. Then, if $w_1, w_2, \ldots$ be the weights of the different bodies and $z_1, z_2, \ldots$ the heights of their centres of gravity above some fixed plane, the equation of virtual work becomes

$$-w_1 . \delta z_1 - w_2 . \delta z_2 - w_3 . \delta z_3 + \ldots = 0.$$

If W be the total weight of the system and $\bar{z}$ the height of its centre of gravity, this becomes

$$-W . \delta\bar{z} = 0.$$

But $\delta\bar{z} = 0$ is the first condition that the height of the centre of gravity may be a maximum or minimum.

If the height of the centre of gravity is a true maximum, then, for any small displacement of the system, the centre of gravity is lowered; if, after any such displacement, the system be momentarily held at rest and then let go, it is clear that it would not go back to its position of equilibrium, for that would be contrary to the dynamical principle that the kinetic energy of any such system must be equal to the work done. The system would therefore not go back to its position of equilibrium but would depart still more from it. The equilibrium in this case is said to be *unstable*.

If the height of the centre of gravity is a true minimum, then, for any small displacement, the height of the centre of gravity is increased; in this case, if the system be momentarily held at rest after any such small displacement and then be let go, it would return to its position of equilibrium; in this case the equilibrium is said to be *stable*.

Hence the equilibrium of a body or system of bodies can often be found as follows: let the height $\bar{z}$ of its centre of gravity above a fixed plane be expressed as a function of some one independent variable θ. Solve the equation $\frac{d\bar{z}}{d\theta} = 0$ in the form $\theta = \alpha, \beta, \gamma, \ldots$. If the value $\theta = \alpha$, when substituted in the value of $\frac{d^2\bar{z}}{d\theta^2}$, makes it positive, so that $\bar{z}$ is a true minimum, then the value $\theta = \alpha$ gives a position of *stable* equilibrium.

If the value $\theta=\alpha$, when substituted in the value of $\frac{d^2\bar{z}}{d\theta^2}$, makes it negative, so that $\bar{z}$ is a true maximum, then the corresponding position is one of *unstable* equilibrium.

As the system moves into its different positions the centre of gravity will describe some curve, and we know that in such a curve the maximum and minimum ordinates occur alternately. It follows that the positions of unstable and stable equilibrium occur alternately.

159. **Ex. 1.** *A square lamina rests in a vertical plane on two smooth pegs which are in the same horizontal line. Shew that there is only one position of equilibrium unless the distance between the pegs is greater than one-quarter of the diagonal of the square, but that, if this condition is satisfied, there may be three positions of equilibrium and the symmetrical position will be stable, but the other two positions of equilibrium will be unstable.*

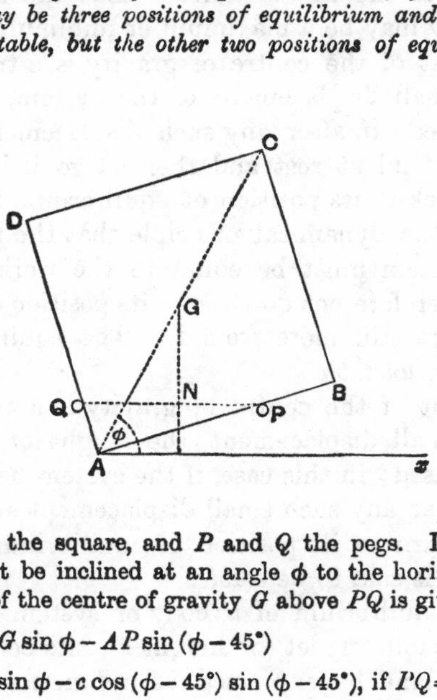

Let $ABCD$ be the square, and P and Q the pegs. Let the diagonal $AC=2d$, and let it be inclined at an angle ϕ to the horizontal Ax. The height $GN\ (=\bar{z})$ of the centre of gravity G above PQ is given by

$$\bar{z}=AG\sin\phi-AP\sin(\phi-45^\circ)$$

$$=d\sin\phi-c\cos(\phi-45^\circ)\sin(\phi-45^\circ),\text{ if } PQ=c,$$

i.e. $$\bar{z}=d\sin\phi+\frac{c}{2}\cos 2\phi \quad\text{.......................................(1)};$$

$$\therefore\ \frac{d\bar{z}}{d\phi}=d\cos\phi-c\sin 2\phi \quad\text{.......................................(2)},$$

and $$\frac{d^2\bar{z}}{d\phi^2}=-d\sin\phi-2c\cos 2\phi \quad\text{.......................................(3)}.$$

Now, since the pegs are smooth, the equation of virtual work reduces to $W . \delta\bar{z}=0$. Hence, by (2), the positions of equilibrium are given by

$$\cos\phi\,(d-2c\sin\phi)=0 \quad \text{.......................(4).}$$

The solutions of this equation are $\phi=90°$ and $\sin\phi=\dfrac{d}{2c}$.

This latter equation has real roots only when $2c>d$, *i.e.* when $PQ>\frac{1}{4}AC$.

Take the case when $2c>d$.

There are then three positions; the first when AC is vertical and the other two when AC is inclined at either side of the vertical at an angle $\sin^{-1}\dfrac{d}{2c}$ to the horizontal.

When $\phi=90°$, then, by (3), $\dfrac{d^2\bar{z}}{d\phi^2}=-d+2c=\text{positive}$.

Therefore $\bar{z}$ is a minimum and the equilibrium is stable.

When

$$\sin\phi=\frac{d}{2c},\ \text{then}\ \frac{d^2\bar{z}}{d\phi^2}=-d\sin\phi-2c+4c\sin^2\phi=\frac{d^2-4c^2}{2c}=\text{negative}.$$

In this case $\bar{z}$ is a maximum and the equilibrium is unstable.

Next take the case when $2c<d$.

In this case there is only one position of equilibrium given by $\phi=90°$, and then $\dfrac{d^2\bar{z}}{d\phi^2}=-d+2c=\text{negative}$.

$\bar{z}$ is now a maximum and the equilibrium is unstable.

Ex. 2. *A rod SH, of length* $2c$ *and whose centre of gravity G is at a distance d from its centre, has a string, of length* $2c\sec\alpha$, *tied to its two ends and the string is then slung over a small smooth peg P; find the position of equilibrium and shew that the position which is not vertical is unstable.*

Since $SP+PH=2c\sec\alpha$, the peg P must be somewhere on an ellipse of foci S and H and semi-major axis $c\sec\alpha$.

Also its semi-minor axis

$$=\sqrt{c^2\sec^2\alpha-CH^2}=c\tan\alpha.$$

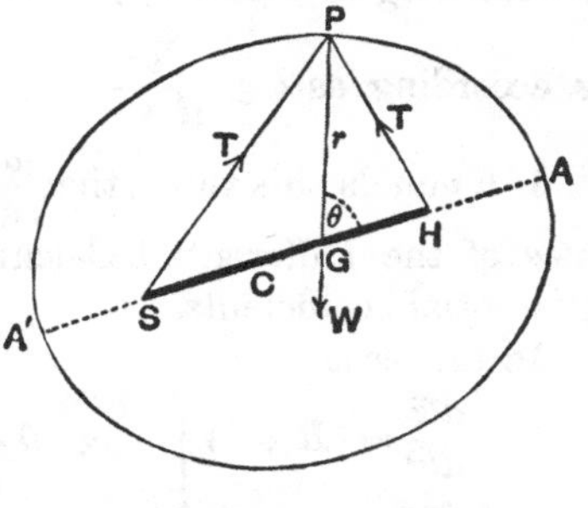

Hence the equation to the ellipse is $x^2\sin^2\alpha+y^2=c^2\tan^2\alpha$, or, referred to polar coordinates through G,

$$\sin^2\alpha\,(r\cos\theta+d)^2+r^2\sin^2\theta=c^2\tan^2\alpha \quad \text{...(1).}$$

If we find the value of θ for which r is a maximum or minimum, and take the corresponding point P of the ellipse for the position of the peg, and make PG vertical, we shall have the slant position of equilibrium.

(1) gives

$$\cos^2\theta \,.\, r^2\cos^2 a - 2\cos\theta \,.\, dr\sin^2 a = r^2 - c^2\tan^2 a + d^2\sin^2 a.$$

$$\therefore\ \cos\theta = \frac{d\sin a\tan a + \sqrt{r^2 - (c^2 - d^2)\tan^2 a}}{r\cos a}.$$

The least value of r is clearly $\sqrt{c^2 - d^2}\tan a$, and then $\cos\theta = \frac{d\tan a}{\sqrt{c^2 - d^2}}$.

Since in this case r is a minimum the centre of gravity is at its minimum depth below the peg, and therefore at the maximum height above the horizontal, and the equilibrium is unstable. The other two positions of equilibrium are when P is at A or A', and the rod is then clearly vertical.

If GP is a minimum it is clear that GP must be a normal at P; so that P may also be found from the fact that its normal passes through a known point G on the major axis.

160. The stability of the question of Art. 156 may also be easily considered by this method. For if $\bar{z}$ be the height, in the first case, of G_2 above O, we have

$$\bar{z} = (R + r)\cos\theta - (r - h)\cos\frac{R + r}{r}\theta \text{..............(1)};$$

$$\therefore\ \frac{d\bar{z}}{d\theta} = -(R + r)\sin\theta + (r - h)\frac{R + r}{r}\sin\frac{R + r}{r}\theta \text{ ...(2)},$$

and $$\frac{d^2\bar{z}}{d\theta^2} = -(R + r)\cos\theta + (r - h)\left(\frac{R + r}{r}\right)^2\cos\frac{R + r}{r}\theta \text{ ...(3)}.$$

A maximum or minimum value of $\bar{z}$ is clearly given by $\theta = 0$. The corresponding value of $\bar{z}$ is then a minimum or maximum, and the equilibrium stable or unstable, according as $\frac{d^2\bar{z}}{d\theta^2}$ is positive or negative,

i.e. according as $-(R + r) + (r - h)\left(\frac{R + r}{r}\right)^2$ is positive or negative,

i.e. according as $h \lessgtr \frac{Rr}{R + r}$.

If h equals this value, then $\frac{d^2\bar{z}}{d\theta^2}$ is zero when $\theta = 0$ and, by the rules of the Differential Calculus, we must consider the higher differential coefficients.

In this case

$$\frac{d^2\bar{z}}{d\theta^2} = (R + r)\left[-\cos\theta + \cos\left(\frac{R + r}{r}\theta\right)\right];$$

$$\therefore\ \frac{d^3\bar{z}}{d\theta^3} = (R + r)\left[\sin\theta - \left(\frac{R + r}{r}\right)\sin\left(\frac{R + r}{r}\theta\right)\right],$$

and $$\frac{d^4\bar{z}}{d\theta^4} = (R+r)\left[\cos\theta - \left(\frac{R+r}{r}\right)^2 \cos\left(\frac{R+r}{r}\theta\right)\right].$$

When $\theta = 0$, $\frac{d^3\bar{z}}{d\theta^3}$ is zero and $\frac{d^4\bar{z}}{d\theta^4}$ is negative.

Hence $\bar{z}$ is a maximum and the equilibrium unstable.

In the second case if $\bar{z}$ be the *depth* of G_2 *below* O we have

$$\bar{z} = (R-r)\cos\theta - (h-r)\cos\frac{R-r}{r}\theta.$$

The equilibrium is then stable or unstable according as $\bar{z}$ is a maximum or a minimum,

i.e. according as $\frac{d^2\bar{z}}{d\theta^2}$ is negative or positive when $\theta = 0$,

i.e., as before, according as $h \lessgtr \frac{rR}{R-r}$.

If h equals this value, then

$$\frac{d^2\bar{z}}{d\theta^2} = (R-r)\left[-\cos\theta + \cos\left(\frac{R-r}{r}\theta\right)\right].$$

Then $\frac{d^3\bar{z}}{d\theta^3}$ is zero when $\theta = 0$, and $\frac{d^4\bar{z}}{d\theta^4}$ is negative or positive according as $1 - \left(\frac{R-r}{r}\right)^2$ is negative or positive,

i.e. according as $R \gtrless 2r$.

Hence $\bar{z}$ is a maximum or minimum, and thus the equilibrium stable or unstable, according as $R \gtrless 2r$.

161. If in the question of Art. 156 the common normal in the position of equilibrium is not vertical, the problem may be treated as follows, in the case where the displacement is such that the centre of gravity G moves in the vertical plane through the common normal.

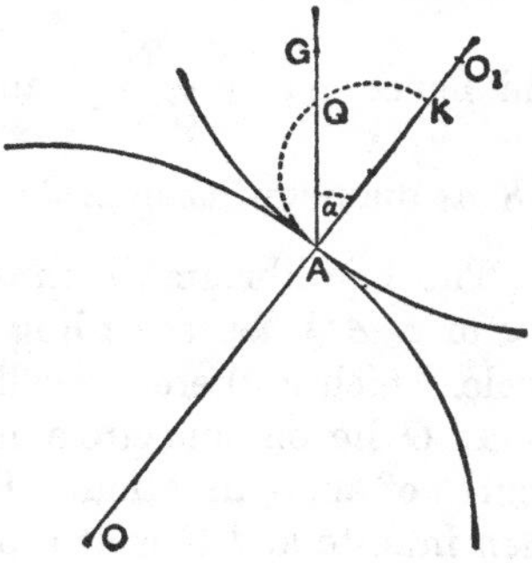

Let R and r be the radii of curvature of the lower and upper bodies at the point of contact A. Since there is equilibrium, G must be vertically over A; let $AG = h$, and $\angle GAO_1 = \alpha$.

The equilibrium will be stable or unstable according as G moves upwards or downwards when the upper body is slightly displaced, *i.e.* according as the concavity of the path of G is turned upwards or downwards, *i.e.* according as the centre of curvature of the path of G is above or below G.

Now, by the theory of the Curvature of Roulettes, the radius of curvature ρ of the path of G is given by

$$\frac{\cos\alpha}{\rho-h}+\frac{\cos\alpha}{h}=\frac{1}{R}+\frac{1}{r},$$

so that
$$\rho=\frac{h\left(\frac{1}{R}+\frac{1}{r}\right)}{\frac{1}{R}+\frac{1}{r}-\frac{\cos\alpha}{h}},$$

where ρ is measured positively from G towards A.

Hence ρ is positive or negative, *i.e.* the centre of curvature of the path of G is below or above G,

according as $\frac{1}{R}+\frac{1}{r}$ is $>$ or $<$ $\frac{\cos\alpha}{h}$,

i.e. according as $h >$ or $< \frac{Rr}{R+r}\cos\alpha$.

Hence the equilibrium is stable or unstable according as

$$h < \text{or} > \frac{Rr}{R+r}\cos\alpha.$$

If along AO_1 we measure off AK such that $\frac{1}{AK}=\frac{1}{R}+\frac{1}{r}$, and hence $AK=\frac{Rr}{R+r}$, and if AG meet in Q the circle on AK as diameter, then $AQ=AK\cos\alpha=\frac{Rr}{R+r}\cos\alpha$.

The equilibrium is thus stable or unstable according as $h <$ or $> AQ$, *i.e.* according as G lies within or without this circle, which is therefore called the *circle of stability*.

If G lie on this circle, its equilibrium is neutral to the first degree of approximation. The radius of curvature of its path is then infinite and G is at a point of inflexion of its path. This circle is therefore often known as the *circle of inflexions*.

EXAMPLES

1. A heavy uniform rod rests with one end against a smooth vertical wall and with a point in its length resting on a smooth peg. Find the position of equilibrium and shew that it is unstable.

2. Two equal uniform rods are firmly jointed at one end so that the angle between them is α, and they rest in a vertical plane on a smooth sphere of radius r. Shew that they are in stable or unstable equilibrium according as the length of either rod is $\gtrless 4r \operatorname{cosec} \alpha$.

3. A beam rests with its ends upon two smooth inclined planes, which are inclined at angles α and β to the horizon and which intersect in a horizontal line; find the position of equilibrium and shew that it is unstable.

4. A uniform heavy bar AB can move freely in a vertical plane about a hinge at A, and has a string attached to its end B which after passing over a small pulley at a point C vertically above A is attached to a weight. Shew that the position of equilibrium in which AB is inclined to the vertical is an unstable one.

5. A smooth beam AB, of weight W, rests with one end A on a smooth horizontal plane AC and the other end B against a smooth vertical wall BC. The end A is connected by a string which passes over a smooth pulley at C and is attached to a weight W'. A, B, C being in one vertical plane, find the position of equilibrium and shew that it is unstable.

6. Shew that the equilibrium of the rod in Ex. 2 of Art. 56 is stable.

7. Four uniform rods, each of length $2a$, are hinged at their ends so as to form a rhombus and the system is hung over two smooth pegs in the same horizontal line at a distance $a\sqrt{2}$, the pegs being in contact with different rods. Shew that the system is in equilibrium when the rhombus is a square, but that the equilibrium is not stable for all displacements.

8. A square lamina rests with its plane perpendicular to a smooth wall one corner being attached to a point in the wall by a fine string of length equal to the side of the square. Find the position of equilibrium and shew that it is stable.

9. A uniform isosceles triangular lamina ABC rests in equilibrium with its equal sides AB and AC in contact with two smooth pegs in the same horizontal line at a distance c apart. If the perpendicular AD upon BC is h, shew that there are three positions of equilibrium, of which the one with AD vertical is stable and the other two are unstable, if $h < 3c \operatorname{cosec} A$; whilst, if $h \geqq 3c \operatorname{cosec} A$, there is only one position of equilibrium, which is unstable.

10. A square board is hung flat against a wall, by means of a string fastened to the two extremities of the upper edge and hung round a perfectly smooth rail; when the length of the string is less than the diagonal of the board, shew that there are three positions of equilibrium.

Shew that the position of symmetry is unstable.

11. A rectangular picture hangs in a vertical position by means of a string, of length l, which after passing over a smooth nail has its ends attached to two points symmetrically situated in the upper edge of the picture at a distance c apart. If the height of the picture be a, shew that there is no position of equilibrium in which a side of the picture is inclined to the horizon if $la > c\sqrt{c^2+a^2}$, whilst if $la < c\sqrt{c^2+a^2}$ there are two such positions which are both stable.

Shew also that in the latter case the position in which the side is vertical is stable for some and unstable for other displacements.

12. A smooth ellipse is fixed with its axis vertical and in it is placed a beam with its ends resting on the arc of the ellipse; if the length of the beam be not less than the latus rectum of the ellipse, shew that when it is in stable equilibrium it will pass through the focus.

13. A uniform rod, of length $2l$, is attached by smooth rings at both ends to a parabolic wire, fixed with its axis vertical and vertex downwards, and of latus rectum $4a$. Shew that the angle θ which the rod makes with the horizontal in a slanting position of equilibrium is given by $\cos^2\theta = \frac{2a}{l}$; and that, if these positions exist, they are stable.

Shew also that the positions in which the rod is horizontal are stable or unstable according as the rod is below or above the focus.

14. A solid hemisphere rests on a plane inclined to the horizon at an angle α, $< \sin^{-1}\frac{3}{8}$, and the plane is rough enough to prevent any sliding. Find the position of equilibrium and shew that it is stable.

15. If a body rest with a plane face in contact with a perfectly rough sphere of radius a at a point at which the normal makes an angle θ with the vertical, and if the centre of gravity be at a distance h vertically above the point of contact, prove that the equilibrium will be stable if $h < a\cos\theta$, and unstable if $h \geqq a\cos\theta$.

16. Shew that the half of an ellipse cut off by any diameter will always have one position of stable equilibrium when resting with its curved surface in contact with a horizontal plane, if the eccentricity be less than $\frac{2}{\sqrt{3\pi}}$.

17. An elliptic cylinder is placed with its axis horizontal on a rough plane inclined to the horizon at an angle less than the angle of friction;

prove that the cylinder cannot rest if the inclination of the plane exceeds $\sin^{-1}\left(\frac{a^2-b^2}{a^2+b^2}\right)$; and if the inclination is equal to $\sin^{-1}\left(\frac{a^2-b^2}{a^2+b^2}\right)$ the equilibrium is neutral to a first approximation.

18. An elliptic disc, of semi-axes a and b, slides in a vertical plane so as always to be in contact with two smooth rods, OP and OQ, which are in the same vertical plane and at right angles. Shew that in the stable positions of equilibrium the major axis of the ellipse is parallel to one or other of the rods, whilst in the unstable position it is inclined at an angle θ to OP given by $\sin^2\theta = \frac{a^2\sin^2\alpha - b^2\cos^2\alpha}{a^2-b^2}$, where α is the inclination of OP to the vertical.

19. A smooth elliptic cylinder of semi-axes a and b slides between two planes each inclined at an angle α to the vertical. If $\tan\alpha$ lie between $\sqrt{\frac{b}{a}}$ and $\sqrt{\frac{a}{b}}$, shew that in the position of stable equilibrium the major axis of the cylinder is inclined to the vertical at an angle $\tan^{-1}\sqrt{\frac{a\sin^2\alpha - b\cos^2\alpha}{a\cos^2\alpha - b\sin^2\alpha}}$, and when $\tan\alpha$ does not lie between these limits find the positions of stable and unstable equilibrium.

20. Four equal masses, attracting according to the law of the inverse square of the distance, are placed at the corners of a rectangle; shew that a particle at the centre of the rectangle will be in unstable equilibrium for all displacements in the plane of the rectangle if the ratio of the length to the breadth of the rectangle is less than $\sqrt{2}$.

If this ratio is equal to $\sqrt{2}$, shew that there is stable equilibrium for a displacement parallel to the breadth of the rectangle.

CHAPTER X

FORCES IN THREE DIMENSIONS

162. ***To find the resultant of any given system of forces acting at given points of a rigid body.***

Take any convenient origin or base point, O, and axes of coordinates Ox, Oy, Oz.

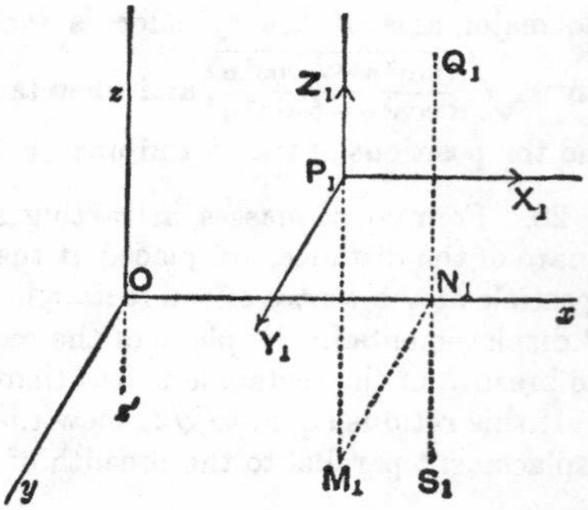

Let (x_1, y_1, z_1) be the coordinates of any point P_1 of the body at which acts one of the given forces R, whose components parallel to the axes are X_1, Y_1, Z_1.

Draw P_1M_1 perpendicular to the plane xOy, M_1N_1 perpendicular to Ox, and $Q_1N_1S_1$ parallel to Oz.

Along the lines Oz, Oz', N_1Q_1, N_1S_1 let forces, each equal to Z_1, be introduced. These, being in equilibrium among themselves, do not alter the effect of the given forces.

Now the forces Z_1 along P_1Z_1 and N_1S_1 form a couple of moment $Z_1 . M_1N_1$, *i.e.* $Z_1 . y_1$, in a plane perpendicular to Ox and in the positive direction about Ox; they are therefore equivalent to a couple whose axis is along Ox and is positive.

The forces Z_1 along N_1Q_1 and Oz' form a couple whose moment is $Z_1 . ON_1$, *i.e.* Z_1x_1, in a plane perpendicular to Oy and in the negative direction about Oy.

Hence the force Z_1 at P_1 is equivalent to

a force Z_1 at O along Oz,

a couple of moment $+ y_1Z_1$ about Ox,

and a couple of moment $- x_1Z_1$ about Oy.

Similarly the force X_1 at P_1 is equivalent to

a force X_1 at O along Ox,

a couple of moment $+z_1X_1$ about Oy,

and a couple of moment $-y_1X_1$ about Oz.

Also the force Y_1 at P_1 is equivalent to

a force Y_1 at O along Oy,

a couple of moment $+x_1Y_1$ about Oz,

and a couple of moment $-z_1Y_1$ about Ox.

Hence finally the three component forces X_1, Y_1, Z_1 acting at P_1 are equivalent to

forces X_1, Y_1, Z_1 along Ox, Oy, Oz respectively,

a couple $y_1Z_1 - z_1Y_1$ about Ox,

a couple $z_1X_1 - x_1Z_1$ about Oy,

and a couple $x_1Y_1 - y_1X_1$ about Oz.

In a similar manner we may replace the force, acting at another point (x_2, y_2, z_2) and whose components are X_2, Y_2, Z_2, by forces along Ox, Oy, Oz and couples about these lines as axes.

Hence finally the whole system of forces is equivalent to

a force along $Ox = X_1 + X_2 + \ldots = \Sigma(X_1) = X$,

a force along $Oy = Y_1 + Y_2 + \ldots = \Sigma(Y_1) = Y$,

a force along $Oz = Z_1 + Z_2 + \ldots = \Sigma(Z_1) = Z$,

a couple about $Ox = \Sigma(y_1Z_1 - z_1Y_1) = L$,

a couple about $Oy = \Sigma(z_1X_1 - x_1Z_1) = M$,

and a couple about $Oz = \Sigma(x_1Y_1 - y_1X_1) = N$

These three forces are equivalent to a single force R acting through O, such that $R^2 = X^2 + Y^2 + Z^2$, along a line whose direction cosines are $\frac{X}{R}, \frac{Y}{R}, \frac{Z}{R}$. [Art. 26.]

The three component couples are, by Art. 49, equivalent to a couple of moment G, such that $G^2 = L^2 + M^2 + N^2$, whose axis is along a line whose direction cosines are $\frac{L}{G}, \frac{M}{G}, \frac{N}{G}$.

Hence the system of forces has been reduced to a single force acting through an arbitrarily chosen point O, and a couple whose axis passes through O.

163. This combination of a force and a couple is often called a dyname, and the quantities X, Y, Z, L, M, N are its components.

The system of forces and couples along and about the axes of coordinates may, for brevity, be called the system $(X, Y, Z; L, M, N)$.

164. *General definition of the moment of a force about a line.*

The moment of a force P about a given line is obtained thus; resolve P into two components, Q parallel to the line and S perpendicular to it; the product of S and the shortest distance between the line of action of S and the given line is the required moment about the given line.

In the figure of Art. 162 the moment of the given force R about the axis of x is equal to the component $\sqrt{Y_1^2 + Z_1^2}$ multiplied by the shortest distance between its line of action and Ox, and is thus equal to the moment of the component $\sqrt{Y_1^2 + Z_1^2}$ about N_1, which again, by Art. 38, is equal to the sum of the moments of its two components Y_1 and Z_1 about N_1, and this sum finally is equal to $y_1 Z_1 - z_1 Y_1$.

165. *General conditions of equilibrium of a rigid body.*

A force R and a couple G together cannot produce equilibrium. For the couple G can be replaced by two equal and opposite forces one of which acts through the point O where R meets the plane of the couple. This force and R can be compounded into a single force which passes through O and does not meet the other force of the couple; and hence we cannot have equilibrium.

Hence there can be equilibrium only when the force R and the couple G separately vanish.

But, by Art. 162, $R^2 = X^2 + Y^2 + Z^2$ and $G^2 = L^2 + M^2 + N^2$.

Hence for equilibrium we must have

$$X = 0, \qquad Y = 0, \qquad Z = 0;$$
$$L = 0, \qquad M = 0, \text{ and } N = 0;$$

i.e. the sums of the resolved parts of the system of forces parallel to any three axes of coordinates must separately vanish, and also the sums of their moments about the three axes must separately vanish.

EXAMPLES

1. Two equal forces R act on a cube, whose centre is fixed and whose edge is $2a$, along diagonals of adjacent faces which do not meet; shew that the moment of the couple which will keep the cube at rest is either $Ra\sqrt{3}$ or Ra according to the directions of the forces.

2. Six forces, each equal to P, act along the edges of a cube, taken in order, which do not meet a given diagonal. Shew that their resultant is a couple of moment $2\sqrt{3}.Pa$, where a is the edge of the cube.

3. OA, OB, OC are edges of a cube of side a and OO', AA', BB', CC' are its diagonals; along OB', $O'A$, BC and $C'A'$ act forces equal to P, $2P$, $3P$ and $4P$; shew that they are equivalent to a force $\sqrt{35}P$ at O along a line whose direction-cosines are proportional to -3, -5, 6 together with a couple $\frac{Pa}{2}\sqrt{114}$ about a line whose direction-cosines are proportional to 7, -2, 2.

4. Forces act through the angular points of a tetrahedron perpendicular to the opposite faces and proportional to them. Shew that they are in equilibrium if they act either all inwards or all outwards.

5. In any rectilinear solid figure couples, whose axes are all drawn outwards, act one in each face proportional to the area of that face; shew that they are in equilibrium.

6. Four forces act along generators of the same system of a hyperboloid. Their magnitudes are such that if they were transferred parallel to themselves to act at one point they would be in equilibrium; shew that they are in equilibrium when acting along the generators.

[The equation to any generator of the hyperboloid

$$\frac{x^2}{a^2}+\frac{y^2}{b^2}-\frac{z^2}{c^2}=1 \text{ is } \frac{x-a\cos\theta}{a\sin\theta}=\frac{y-b\sin\theta}{-b\cos\theta}=\frac{z}{c}.]$$

166. Constrained bodies. A body is said to be constrained when one or more points of the body are fixed. For example, a rod attached to a wall by a ball-socket has one point fixed and is constrained.

If a rigid body have two points A and B fixed, all the points of the body in the line AB are fixed, and the only way in which the body can move is by turning round AB as an axis. For example, a door attached to the door-post by two hinges can only turn about the line joining the hinges.

If a rigid body have three points in it fixed, the three points not being in the same straight line, it is plainly immovable.

167. *Conditions of equilibrium of a rigid body with one point fixed.*

Take the fixed point as the origin O and any three perpendicular lines through it as the axes. Let the external forces acting on the body, apart from the force of constraint at A, reduce to component forces X, Y, Z parallel to the axes and component couples L, M, N about the axes, as in Art. 162.

Let the force of constraint at O, which gives equilibrium, have as components X', Y', Z' parallel to the axes.

Then for equilibrium we have, by Art. 165,

$$X + X' = 0, \quad Y + Y' = 0, \quad Z + Z' = 0 \quad \text{......(1)},$$
$$L = 0, \quad M = 0, \quad N = 0 \quad \text{......(2)}.$$

The equations (1) give only the component reactions at O in terms of the external forces.

The equations (2) give the conditions of equilibrium, *viz.* that *the sums of the moments of the external forces about any three perpendicular lines passing through the fixed point O must be separately zero.*

If the external forces all act in one plane passing through O, the preceding conditions reduce to the simpler condition that the sum of the moments about O must vanish.

168. *Conditions of equilibrium of a rigid body which has two points, A and B, fixed so that the body can turn about the fixed axis AB.*

Take the straight line AB as the axis of z and any point O on it as the origin.

Let $OA = z'$, $OB = z''$, and let the components at A and B of the forces of constraint be X', Y', Z' and X'', Y'', Z''.

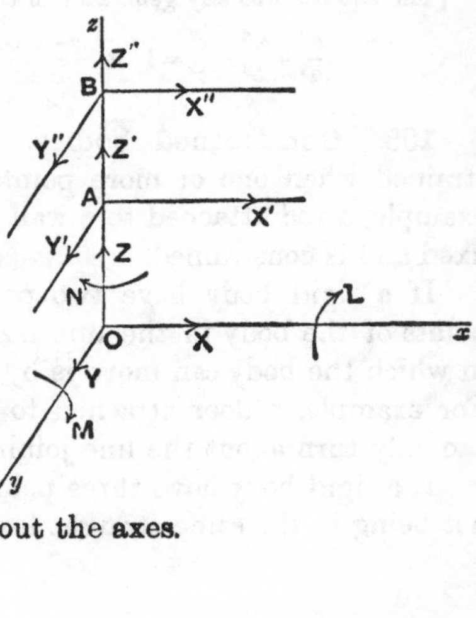

Let the external forces acting on the body apart from the forces of constraint reduce, as in Art. 162, to component forces X, Y, Z at O parallel to the axes and component couples L, M, N about the axes.

Then the component forces for the whole system of forces are $X + X' + X''$, $Y + Y' + Y''$, $Z + Z' + Z''$ and the component couples are

$$L - Y'z' - Y''z'', \quad M + X'z' + X''z'' \text{ and } N.$$

Hence the conditions of equilibrium are

$$X + X' + X'' = 0 \quad \text{..................(1)},$$
$$Y + Y' + Y'' = 0 \quad \text{..................(2)},$$
$$Z + Z' + Z'' = 0 \quad \text{..................(3)},$$
$$L - Y'z' - Y''z'' = 0 \quad \text{..................(4)},$$
$$M + X'z' + X''z'' = 0 \quad \text{..................(5)},$$

and

$$N = 0 \quad \text{..................(6)}.$$

(1) and (5) give X' and X''; (2) and (4) give Y' and Y'', the only relation between Z' and Z'' is equation (3), so that their values are indeterminate.

Finally the only relation between the external forces is equation (6), so that the condition of equilibrium is that *the sum of the moments of the external forces about the fixed axis AB must be zero.*

169. The reactions Z', Z'' may be expected to be indeterminate. For suppose the body to be a gate supported in the usual way by supports at A and B. If the staple at A be moved a very little higher than the proper position it will carry all the weight of the gate; if on the other hand it be placed a very little lower than the proper position, then all the weight will fall on to B. We should therefore expect the distribution of the weight to be an indeterminate one when the distance between the two staples of the post is exactly equal to the distance between the rings of the gate.

170. Ex. 1. *A circular uniform table, of weight 80 lbs., rests on four equal legs placed symmetrically round its edge; find the least weight which hung upon the edge of the table will just overturn it.*

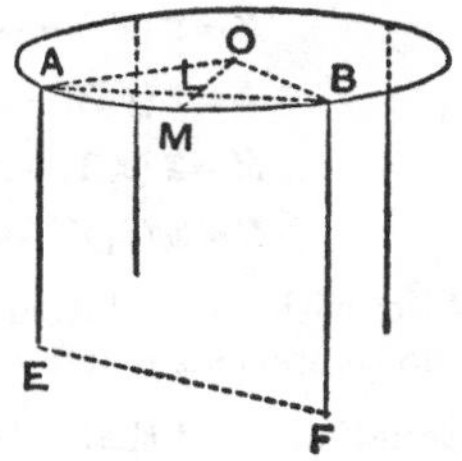

Let AE and BF be two of the legs of the table, whose centre is O.

If the weight be hung on the portion of the table between A and B the table will, if it turn at all, turn about the line joining the points E and F. Also it will be just on the point of turning when the weight and the weight of the table have equal moments about EF. Now the weight will clearly have the greatest effect when placed at M, the middle point of the arc AB.

Let OM meet AB in L, and let x be the required weight. Taking moments about EF, we have $x \,.\, LM = 80 \,.\, OL$.

$$\therefore\ x\left(1-\frac{1}{\sqrt{2}}\right)OA = 80\,.\,\frac{1}{\sqrt{2}}\,.\,OA, \text{ i.e. } x = 193{\cdot}1 \text{ lbs. wt.}$$

Ex. 2. *A heavy door is hung so that the line joining its hinges is of length $2h$ and is inclined at θ to the vertical; it is kept in a position inclined at an angle ϕ to the vertical plane through the line of hinges by a force P, perpendicular to the door, acting at a point whose distances from the line of hinges and from the lower edge of the door are b and c; find P and the actions at the hinges as far as they can be found, the weight of the door being W.*

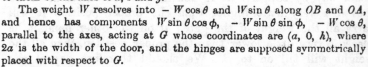

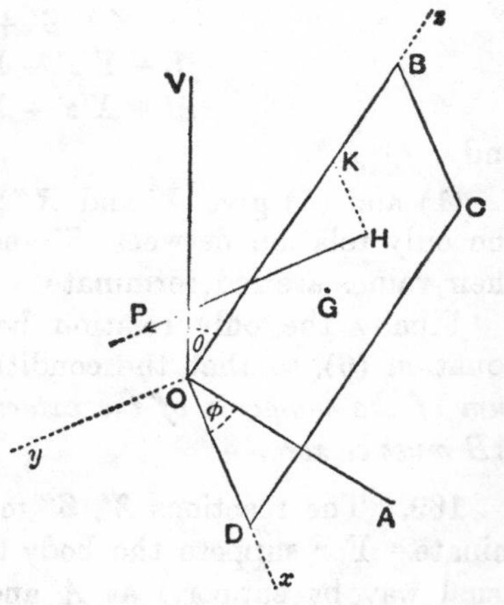

Let OA be perpendicular to the line joining the hinges, O and B, in a plane through OB and the vertical OV, and let $OBCD$ be the door so that $\angle AOD = \phi$.

Let OD, OB and a perpendicular to them be the axes of x, z and y.

The weight W resolves into $-W\cos\theta$ and $W\sin\theta$ along OB and OA, and hence has components $W\sin\theta\cos\phi$, $-W\sin\theta\sin\phi$, $-W\cos\theta$, parallel to the axes, acting at G whose coordinates are $(a, 0, h)$, where $2a$ is the width of the door, and the hinges are supposed symmetrically placed with respect to G.

The supporting force P at H has components 0, P, 0 parallel to the axes acting at a point whose coordinates are $(b, 0, c)$.

Hence, by Art. 162,

$$X = \Sigma(X_1) = W\sin\theta\cos\phi,$$
$$Y = \Sigma(Y_1) = -W\sin\theta\sin\phi + P,$$
$$Z = \Sigma(Z_1) = -W\cos\theta,$$
$$L = \Sigma(y_1 Z_1 - z_1 Y_1) = Wh\sin\theta\sin\phi - cP,$$
$$M = \Sigma(z_1 X_1 - x_1 Z_1) = Wh\sin\theta\cos\phi + aW\cos\theta,$$

and $$N = \Sigma(x_1 Y_1 - y_1 X_1) = -Wa\sin\theta\sin\phi + bP.$$

Also, with the notation of Art. 168, $z'=0$ and $z''=2h$, so that X', Y', Z' are the component reactions at O, and X'', Y'', Z'' those at the hinge B.

Equation (6) of that article gives $P = W\dfrac{a}{b}\sin\theta\sin\phi$.

Equations (1) and (5) give

$$X' = \frac{W}{2}\left[\frac{a}{h}\cos\theta - \sin\theta\cos\phi\right], \text{ and } X'' = -\frac{W}{2}\left[\frac{a}{h}\cos\theta + \sin\theta\cos\phi\right].$$

Equations (2) and (4) give

$$Y' = \frac{W}{2}\left[1 - \frac{2a}{b} + \frac{ca}{bh}\right]\sin\theta\sin\phi, \text{ and } Y'' = \frac{W}{2}\left[1 - \frac{ca}{bh}\right]\sin\theta\sin\phi.$$

Also, (3) gives $Z' + Z'' = W\cos\theta$.

EXAMPLES

1. A square table stands on four legs placed respectively at the middle points of its sides; find the greatest weight that can be put at one of the corners without upsetting the table.

2. A round table stands upon three equidistant weightless legs at its edge, and a man sits upon its edge opposite a leg. It just upsets and falls upon its edge and two legs. He then sits upon its highest point and just tips it up again. Shew that the radius of the table is $\sqrt{2}$ times the length of a leg.

3. A door, of weight W, is free to turn about an axis AB which is inclined at an angle α to the vertical; shew that the couple necessary to keep it in a position in which it is inclined at an angle β to the vertical plane through AB is $Wa \sin\alpha \sin\beta$, where a is the distance of its centre of gravity from AB.

4. A rectangular gate is hung in the ordinary way on two hinges so that the line joining the hinges makes an angle α with the vertical. Shew that the work which must be done to move it through an angle θ from its position of equilibrium is $Wa \sin\alpha (1 - \cos\theta)$, where W is the weight and $2a$ the breadth of the gate.

5. A rectangular table is supported in a horizontal position by four legs at its four angles, and a given weight is placed at a given point of it; shew that the thrust on each leg is indeterminate, and find the greatest and least value it can have for a given position of the weight.

6. A rigid rectangular table has equal legs at the four corners which are slightly compressible, and the compression in each leg is assumed to be proportional to the thrust on that leg. If the centre of gravity of the table lie within the parallelogram formed by joining the middle points of the sides, find the thrust in each leg. If the centre of gravity does not lie within this parallelogram, shew that the table rests on three legs only.

[Since the diagonals of the table remain straight, the mean of the compressions of each pair of opposite legs is equal to the depth through which the centre of the table moves. Hence the sum of the thrusts of each pair of opposite corners is the same.]

171. ***If three forces acting on a body keep it in equilibrium, they must lie in a plane.***

Let the three forces be P, Q, and R, and let P_1 and Q_1 be any two points on the lines of action of P and Q respectively.

Since the forces are in equilibrium, they can, taken together, have no effect to turn the body about the line P_1Q_1. But the forces P and Q meet this line, and therefore separately have no effect to turn the body about P_1Q_1. Hence the third force R can have no effect to turn the body about P_1Q_1. Therefore the line P_1Q_1 must meet R.

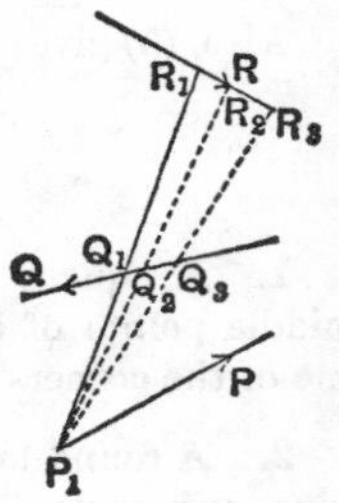

Similarly, if Q_2, Q_3, ... be other points on the line of action of Q, the lines P_1Q_2, P_1Q_3, ... must meet R.

Hence R must lie in the plane through P_1 and the line of action of Q, *i.e.* the lines of action of Q and R must be in a plane which passes through P_1.

But P_1 is any point on the line of action of P; and hence the above plane passes through any point on the line of action of P, *i.e.* it contains the line of action of P.

Cor. From Art. 54 it now follows that the three forces must also meet in a point or be parallel.

172. ***If four forces acting on a body are in equilibrium, shew that they are generators of the same hyperboloid.***

Let the lines of action of the four forces be P, Q, R, S. From any point on P draw a line L to meet both Q and R. Since the four forces are in equilibrium the sum of their moments about L must vanish; hence S must meet L. Starting with any two other points on P we obtain two other lines M and N which meet P, Q, R, S.

Now three non-intersecting lines L, M, N determine a hyperboloid of one sheet of which L, M, N are generators of the same system. The lines P, Q, R, S which meet L, M, N are then generators of this hyperboloid which belong to the other system.

It will be noted that it is assumed that no two of the four forces are either parallel or meet in a point; for, if so, they could be compounded into a single force and we should fall back on the case of the last article.

173. *If five forces acting on a body are in equilibrium, they can be intersected by two straight lines.*

Let the lines of action of the forces be P, Q, R, S and T. Through P, Q, R draw a hyperboloid of one sheet, and let S meet it in the points A and B.

Through A there passes one generator of the hyperboloid which is of the opposite system to that of P, Q, R and which therefore meets P, Q, R; since this generator meets P, Q, R, S the sum of their moments about it is zero.

But, by taking moments about it for the whole system, we see that the sum of the moments of P, Q, R, S and T about it is zero. Hence the moment of T about it must be zero, *i.e.* it meets T also.

Similarly, through B there passes a generator which meets all five forces.

It will be noted that these straight lines are real, coincident, or imaginary according as S meets the hyperboloid in real, coincident, or imaginary points.

174. Ex. 1. *A heavy rod OA can turn freely about a point O, whose distance from a rough wall is k, the height of the wall being h; the rod rests with a point of itself upon the top edge of the wall. Shew that the greatest angle which the rod can make with the perpendicular drawn from O to the top edge of the wall is*

$$\sin^{-1}\left(\frac{\mu k}{h}\right).$$

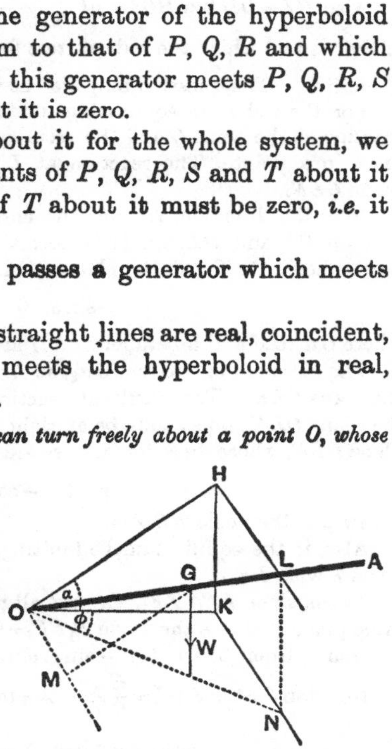

Let L be the point of the rod in contact with the wall, OK the perpendicular to the wall and OH the perpendicular to the top of the wall, so that $OK = k$, $KH = h$, and $\angle HOK = \alpha$.

Let OM be parallel to HL and GM be perpendicular to OM.

The direction of the normal reaction R at L is perpendicular to both OL and HL and is thus perpendicular to the plane OLH.

The friction μR is opposite to the direction in which L would move and is thus perpendicular to OL in the plane OLH.

Since GM is inclined at α to the horizon, W is equivalent to $W \sin \alpha$ along GM and $W \cos \alpha$ perpendicular to the plane OLH through the point G.

Taking moments about a perpendicular through O to the plane OHL, we thus have

$$\mu R \,.\, OL = W \sin \alpha \,.\, OG \sin HOL \;\ldots\ldots\ldots\ldots\ldots\ldots (1).$$

Also, taking moments about a line perpendicular to OL in the plane HOL, we have

$$R \, . \, OL = W \cos a \, . \, OG \quad \text{.........................(2).}$$

$$\therefore \quad \sin HOL = \mu \cot a = \frac{\mu k}{h}.$$

Also $HL = OH \tan HOL = \mu k \sqrt{\dfrac{h^2+k^2}{h^2-\mu^2 k^2}}$, and twice HL is the length of the wall on which the rod can rest for equilibrium to be possible.

Aliter. The question may also be solved by the use of Art. 171.

For the rod is in equilibrium under three forces, *viz.* its weight, the reaction at the point O and the resultant reaction at L, which must thus meet in a point. The reaction at L must therefore lie in the vertical plane OLN.

Now the direction, LF_1, of the normal reaction at L is perpendicular to both OL and HL, *i.e.* it is normal to the plane OHL, so that it is perpendicular to OH in the plane OHK and thus has as direction cosines

$$(-\sin a, \ 0, \ \cos a) \quad \text{.............................(3),}$$

where OK, OM and a parallel to KH are the axes of coordinates.

The direction of friction, LF_2, at L is along the perpendicular to OL in the plane OLH. The resultant reaction, which we have seen must lie in the plane OLN, must thus be at right angles to LF_3, the normal to the plane OLN, whose direction cosines are

$$(\sin \phi, \ -\cos \phi, \ 0) \quad \text{........................(4),}$$

where ϕ is the angle KON.

Also, if the equilibrium be limiting, this resultant reaction makes an angle λ with LF_1.

Again, since LF_1, LF_2, LF_3 are all perpendicular to OL, they lie in the same plane. Hence the angle $F_1LF_3 = 90° + \lambda$.

Hence, from (3) and (4), $-\sin \phi \sin a = \cos(90° + \lambda) = -\sin \lambda$.

Now $\tan \theta = \tan HOL = \dfrac{HL}{OH} = \cos a \tan \phi = \dfrac{\cos a \sin \lambda}{\sqrt{\sin^2 a - \sin^2 \lambda}}$,

$$\therefore \quad \sin \theta = \cot a \tan \lambda = \mu \cot a, \text{ as before.}$$

Ex. 2. *A heavy plug in the shape of a frustum of a cone exactly fits a conical hole of the same size, the common axis being vertical. The vertical angle of the cone is* $2a$, *and the radii of the circular bases of the frustum are* a *and* b. *The normal reaction per unit of area being supposed constant, shew that the moment of the least couple that will twist the plug is* $\dfrac{2}{3}\mu W \dfrac{a^2+ab+b^2}{a+b}$ *cosec* a, *where* W *is the weight of the plug and* μ *is the coefficient of friction.*

The area of the slant surface of the hole

$$= \pi(a+b) \times \text{slant side} = \pi \frac{a^2-b^2}{\sin a}.$$

Hence, if R be the constant normal reaction per unit of area, we have

$$W=\pi\frac{a^2-b^2}{\sin\alpha}\,.\,R\sin\alpha=\pi\,(a^2-b^2)\,R\;\ldots\ldots\ldots\ldots\ldots\ldots(1).$$

If x be the radius of any section of the plug, the area of the hole between it and the section of radius $x+\delta x=2\pi x\,.\,\frac{\delta x}{\sin\alpha}$, and hence the moment of the friction on it about the axis

$$=2\pi x\,.\,\frac{\delta x}{\sin\alpha}\,.\,\mu R\,.\,x=\frac{2\mu W}{(a^2-b^2)\sin\alpha}\,.\,x^2\delta x.$$

Hence the moment of the required couple

$$=\frac{2\mu W}{(a^2-b^2)\sin\alpha}\int_b^a x^2dx=\frac{2\mu W}{3\sin\alpha}\frac{a^3-b^3}{a^2-b^2}=\frac{2\mu W}{3}\frac{a^2+ab+b^2}{a+b}\operatorname{cosec}\alpha.$$

EXAMPLES

1. Two smooth planes, each inclined at an angle α to the vertical, intersect in a horizontal line. A uniform rod, of weight W and length $2a$, is placed between them in a horizontal position making an angle θ with their line of intersection. Shew that the horizontal couple required to maintain equilibrium is $Wa\cos\theta\cot\alpha$.

2. A uniform straight rod, of length $2c$, is placed in a horizontal position as high as possible within a hollow rough sphere, of radius a. Shew that the line joining the middle point of the rod to the centre of the sphere makes with the vertical an angle $\tan^{-1}\frac{\mu a}{\sqrt{a^2-c^2}}$.

3. A uniform rod, of weight W, can turn freely about a hinge at the end, and rests with the other against a rough vertical wall making an angle α with the wall. Shew that this end may rest anywhere on an arc of a circle of angle $2\tan^{-1}[\mu\tan\alpha]$, and that in either of the extreme positions the normal reaction of the wall is $\frac{1}{2}W[\cot^2\alpha+\mu^2]^{-\frac{1}{2}}$, where μ is the coefficient of friction.

4. A thin uniform rod AB, of length $2a$, rests in an oblique position with one end A on a rough horizontal table and the other against a rough vertical wall, the coefficients of friction at the table and wall being μ_1 and μ_2 and the distance of the foot of the rod from the wall being k; shew that the rod is on the point of slipping at the lower end if the vertical plane in which it lies makes an angle θ with the wall given by

$$k\mu_1(\mu_2{}^2\sin^2\theta-\cos^2\theta)^{\frac{1}{2}}=k-2\mu_1(4a^2\sin^2\theta-k^2)^{\frac{1}{2}},$$

and that the inclination of the tangential action at the upper end to the horizon is then $\sec^{-1}(\mu_2\tan\theta)$.

[The resultant reaction at A making an angle λ_1 with the vertical, and the resultant reaction at B making an angle ψ with the vertical must meet in a point D vertically over the centre of gravity G. Hence, if BL be perpendicular to the table and $\angle LAB = a$, we have, by Art. 55,

$$2\tan a = \cot\lambda_1 - \cot\psi = \frac{1}{\mu_1} - \cot\psi.$$

The only forces being in the plane $ABDL$, the end A must be on the point of slipping in the direction LA.

By projecting BD on the normal to the wall at B, we have

$$\sin\psi\sin\theta = \cos\lambda_2 = \frac{1}{\sqrt{1+\mu_2^2}}.$$

Also $k = 2a\sin\theta\cos a$. Hence the first given result.

Also, if R be the reaction at B and $\mu_2 R$ the friction there at an angle χ to the horizontal, then, since we have seen that their resultant lies in the plane ABL, we have, by resolving in a direction perpendicular to this plane, $\mu_2 R\cos\chi\sin\theta = R\cos\theta$.]

5. A hemisphere, whose surface is rough and whose centre is O, is fixed with its base on a horizontal plane. One end of a straight uniform rod is freely jointed to a fixed point A in the plane and the other rests on the surface of the hemisphere at P so that the rod is just on the point of slipping. Shew that the plane through O and the rod makes with the vertical plane through OA an angle $\tan^{-1}(\mu\cos a\operatorname{cosec}\beta)$, where μ is the coefficient of friction, a is the angle OAP, and β is the angle OPA.

6 A heavy circular cylinder rests with its plane base upon a rough horizontal table; if its weight be W and the normal pressure be supposed to be uniformly distributed over the base, shew that the moment of the couple about its axis which would just twist it is $\frac{2}{3}\mu Wa$, where μ is the coefficient of friction and a is the radius of its base.

7. A right circular cone, of weight W and vertical angle $2a$, is placed with its vertex downwards and supported by a circular hole cut in a horizontal table. If μ be the coefficient of friction and b the radius of the hole, shew that the moment of the least couple that will move the cone is $\mu Wb\operatorname{cosec} a$.

8. A smooth pyramidal plug is made to fit symmetrically into an equilateral triangular hole whose side is a and whose plane is horizontal. Prove that to retain it in the hole with its axis vertical, so that its section by the plane of the hole is an equilateral triangle of side c, a couple must be applied of amount $Wh\sqrt{\frac{4c^2}{a^2}-1}$, where W is the weight of the plug and h is the depth of the vertex in this position.

[Let the plug touch the sides BC, CA, AB of the hole in the points A', B', C'. Then easily

$$AC' = \frac{1}{2}\left[a + \sqrt{\frac{4c^2-a^2}{3}}\right] \text{ and } \cos AC'B' = \frac{a+\sqrt{3(4c^2-a^2)}}{4c}.$$

Hence, if O be the centre of the equilateral triangle $A'B'C'$,

$$\sin\theta = \sin OC'A = \sin[AC'B' + 30°] = \frac{a}{2c}.$$

Also, since by symmetry the vertex V of the pyramid is vertically below O, its inclination α to the vertical is given by $\tan\alpha = \frac{OC'}{h} = \frac{c}{h\sqrt{3}}$.

If $C'A$, a perpendicular to $C'A$ on the horizontal plane, and the vertical be taken as the axes of x, y, z, the direction cosines of $C'A$ are (1, 0, 0), and those of the edge are $(-\sin\alpha\cos\theta, -\sin\alpha\sin\theta, \cos\alpha)$, *i.e.* they are proportional to $(-\sqrt{4c^2-a^2}, -a, 2h\sqrt{3})$. Also if the direction of the resultant reaction R at C' make an $\angle\psi$ with the vertical, its direction cosines are $(0, \sin\psi, \cos\psi)$, since it is perpendicular to $C'A$. Since it is also perpendicular to the edge,

$$\therefore \sin\psi(-a) + \cos\psi(2h\sqrt{3}) = 0, \text{ i.e. } \tan\psi = \frac{2h\sqrt{3}}{a}.$$

The horizontal component R_1 of this reaction

$$= R\sin\psi = \frac{W}{3}\tan\psi = \frac{2h\sqrt{3}}{3a}W,$$

since the vertical components of the three equal reactions R balance W.

Also the direction of R_1 is perpendicular to BA. Hence the moment of the required couple = moment about O of the three components R_1 at $A', B', C' = 3R_1 \cdot \frac{2}{3}\frac{c\sqrt{3}}{2}\cos\theta$ = the given result.]

9. A uniform triangular table ABC has three equal legs at A, B, and C which rest on a rough horizontal plane. Find the least couple that will cause the table to move.

[Assume that the table is on the point of turning about a vertical axis meeting the horizontal plane in O. Then the frictions at A, B, C are perpendicular to OA, OB, OC in the same sense and are each equal to $\frac{1}{3}\mu W$ since the thrust of each leg is clearly $\frac{W}{3}$.

If these frictions form a triangle $A'B'C'$ then since they are equivalent to a couple only the resultant force must vanish, and each friction must be proportional to the sine of the angle between the other two. Hence the angles A', B', C' must be equal and hence also the angles AOB, BOC, COA. Hence O must be inside the triangle and be such that

$$\angle BOC = \angle COA = \angle AOB = 120°.$$

Hence O always exists provided no angle of the triangle is greater than 120°, and is easily found as the intersections of circular arcs on BC, CA as bases each containing 120°.

The moment of the couple required to move the table then

$$= \frac{\mu W}{3}[OA + OB + OC].$$

The table might also begin to turn about A. If so the frictions at B and C are each $\frac{\mu W}{3}$, at right angles to AB and AC, so that their resultant is $\frac{2}{3}\mu W \cos\frac{A}{2}$. The resultant force of this resultant and the friction at A cannot be zero if $\frac{2}{3}\mu W \cos\frac{A}{2} > \frac{\mu W}{3}$, *i.e.* if $A < 120°$. Thus the table will only turn about the angle A, if $A \geqq 120°$ and then the moment of the required couple $= \frac{1}{3}\mu W[AB + AC]$.]

175. Principle of Virtual Work. Let a system of forces $P_1, P_2, P_3, \ldots$ act at given points of a material system, and let a small displacement be given to the system consistent with its geometrical conditions. If δp_1 be the displacement of the point of application of P_1 along its line of action, and $\delta p_2, \delta p_3, \ldots$ the displacements similarly of the points of applications of $P_2, P_3, \ldots$, then, if the forces $P_1, P_2, \ldots$ are in equilibrium,

$$P_1\delta p_1 + P_2\delta p_2 + P_3\delta p_3 + \ldots = 0 \quad \ldots\ldots\ldots\ldots(1),$$

when small quantities of the second order are neglected.

Conversely, if $P_1\delta p_1 + P_2\delta p_2 + P_3\delta p_3 + \ldots = 0$, the system of forces is in equilibrium.

We have shewn in Art. 96 that the work done by a force during any displacement is equal to the work done by its component forces.

Hence if X_1, Y_1, Z_1 be the components of P_1 parallel to the axes, and $\delta x_1, \delta y_1, \delta z_1$ the components in the same direction of δp_1, then

$$P_1 . \delta p_1 = X_1 . \delta x_1 + Y_1 . \delta y_1 + Z_1 . \delta z_1,$$

with a similar notation for the other forces.

The relation (1) is thus equivalent to

$$(X_1 . \delta x_1 + Y_1 . \delta y_1 + Z_1 . \delta z_1) + (X_2 . \delta x_2 + Y_2 . \delta y_2 + Z_2 . \delta z_2) + \ldots + \ldots = 0,$$

i.e. to
$$\Sigma X . \delta x + \Sigma Y . \delta y + \Sigma Z . \delta z = 0.$$

We shall assume that any rigid body may be moved from any position to any other position by a motion of translation of any point O' to some other point O and by a rotation of the whole body about some axis passing through O.

This rotation may be resolved into three component rotations about the axes of coordinates and we shall assume that these can be made in any order provided they are small.

[For a proof of these assumptions the reader may refer to *Dynamics of a Particle and of Rigid Bodies*, Arts. 215—218.]

Let us first consider the effect on the coordinates of any point A_1 of these three component rotations, assumed to be through small angles θ_1, θ_2, and θ_3 about the axes.

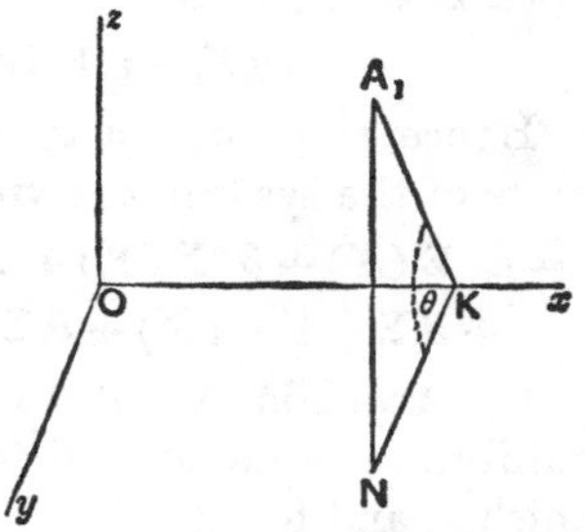

Draw A_1K perpendicular to Ox, A_1N perpendicular to the plane xOy, and let $\angle A_1KN = \theta$.

The y-coordinate of A_1

$$= KN = KA_1 \cos\theta.$$

The rotation about Ox will change this into $KA_1 \cos(\theta + \theta_1)$. Hence the change in the y-coordinate

$$= KA_1 [\cos(\theta + \theta_1) - \cos\theta]$$

$$= KA_1 (-\theta_1 \sin\theta), \text{ if squares of } \theta_1 \text{ are neglected, } = -\theta_1 . z_1.$$

Similarly the change in the z-coordinate

$$= KA_1 [\sin(\theta + \theta_1) - \sin\theta] = KA_1 [\cos\theta . \theta_1] = \theta_1 . y_1.$$

In the same way it may be shewn that a rotation of θ_2 about Oy will produce changes equal to $-\theta_2 . x_1$ and $\theta_2 . z_1$ in the z- and x-coordinates, and that a rotation of θ_3 about Oz will produce changes of $-\theta_3 . y_1$ and $\theta_3 . x_1$ in the x- and y-coordinates. Hence, neglecting squares of small quantities, we see that the resultant changes in the coordinates due to the three rotations

are $\theta_2 . z_1 - \theta_3 . y_1$ parallel to Ox,

$\theta_3 . x_1 - \theta_1 . z_1$ parallel to Oy,

and $\theta_1 . y_1 - \theta_2 . x_1$ parallel to Oz.

If in addition to these rotations the body have been moved through small distances a, b, c parallel to the axes, we have, to the first order of small quantities,

$$\delta x_1 = a + \theta_2 . z_1 - \theta_3 . y_1,$$

$$\delta y_1 = b + \theta_3 . x_1 - \theta_1 . z_1,$$

and

$$\delta z_1 = c + \theta_1 . y_1 - \theta_2 . x_1.$$

The virtual work done by the force P_1 acting at A_1 during the small displacement therefore

$$= X_1 . \delta x_1 + Y_1 . \delta y_1 + Z_1 . \delta z_1$$
$$= aX_1 + bY_1 + cZ_1$$
$$+ \theta_1 (y_1 Z_1 - z_1 Y_1) + \theta_2 (z_1 X_1 - x_1 Z_1) + \theta_3 (x_1 Y_1 - y_1 X_1).$$

Hence, since a, b, c, θ_1, θ_2, and θ_3 are the same for all the points of the system, the virtual work done by all the forces

$$= a . \Sigma (X) + b . \Sigma (Y) + c . \Sigma (Z)$$
$$+ \theta_1 \Sigma (yZ - zY) + \theta_2 \Sigma (zX - xZ) + \theta_3 \Sigma (xY - yX) \ldots (2).$$

By Art. 165 it follows that, if the system of forces is in equilibrium, each of the terms of this expression separately vanishes, and hence

$$P_1 . \delta p_1 + P_2 . \delta p_2 + P_3 . \delta p_3 + \ldots = \text{virtual work of the system} = 0.$$

176. Conversely, let the Virtual Work of the system be zero for *all* displacements.

Choose a simple displacement parallel to the axis of x, so that

$$b = c = \theta_1 = \theta_2 = \theta_3 = 0,$$

but a is not zero. The result (2) of the last article then gives $\Sigma (X) = 0$. Similarly $\Sigma (Y) = 0$, and $\Sigma (Z) = 0$.

Next choose a simple rotation about the axis of x, so that

$$a = b = c = \theta_2 = \theta_3 = 0,$$

whilst θ_1 is not zero. The result (2) then gives $\Sigma (yZ - zY) = 0$, *i.e.* the sum of the moments of the forces of the system about the axis of x is zero, *i.e.* $L = 0$. Similarly, $M = 0$ and $N = 0$. Hence, if the equation of Virtual Work holds, all the conditions of equilibrium of Art. 165 are satisfied, and the system of forces is in equilibrium.

177. Ex. *A regular tetrahedron formed of six light rods, each of length a, rests on a smooth horizontal plane. A ring, of weight W and radius b, is supported by the slant sides. Shew that the stress in any one of the horizontal sides is* $\frac{W}{\sqrt{6}}\left[\frac{1}{3} - \sqrt{3}\frac{b}{a}\right]$.

Give to the system a displacement such that the three slant sides are unaltered in length and the vertex descends in a vertical line. When the slant sides are inclined at θ to the vertical, let the lengths of the sides in contact with the plane be x, so that

$$\frac{2}{3} . \frac{x\sqrt{3}}{2} = a \sin \theta, \text{ and thus } x = a\sqrt{3} \sin \theta.$$

If y be the height of the ring above the plane, then

$$y = a\cos\theta - b\cot\theta.$$

If T be the required stress, reckoned positively as a tension, the equation of virtual work gives

$$-W\delta y - 3T\delta x = 0.$$

$$\therefore \frac{3T}{W} = -\frac{dy}{dx} = \frac{a\sin\theta - \dfrac{b}{\sin^2\theta}}{a\sqrt{3}\cos\theta}.$$

Now in the position of equilibrium $x = a$ and hence $\sin\theta = \dfrac{1}{\sqrt{3}}$.

$$\therefore \frac{T}{W} = \frac{\dfrac{a}{\sqrt{3}} - 3b}{3a\sqrt{2}} = \frac{1}{\sqrt{6}}\left[\frac{1}{3} - \sqrt{3}\frac{b}{a}\right].$$

EXAMPLES

1. A regular octahedron formed of twelve equal rods, each of weight w, freely jointed together, is suspended from one corner. Shew that the thrust in each horizontal rod is $\frac{3}{2}w\sqrt{2}$.

2. A tripod consists of three equal uniform bars, each of length a and weight w, which are freely jointed at one extremity, their middle points being joined by strings, each of length b. The tripod is placed with its free ends in contact with a smooth horizontal plane and a weight W is attached to the common joint; shew that the tension of each string is

$$\tfrac{2}{3}(2W + 3w)\frac{b}{\sqrt{9a^2 - 12b^2}}.$$

3. Twelve similar and uniform rods are jointed at their ends to form an octahedron and are suspended from one of the vertices, being supported by a string joining this vertex to the opposite vertex; the string is elastic and such that the total weight of the rods would stretch it to twice its natural length, the latter being equal to the length of either rod. In the position of equilibrium, shew that the slant rods are inclined to the vertical at an angle $\cos^{-1}\frac{3}{4}$.

4. A parallelopiped is formed of twelve weightless rods which are freely jointed at their ends, and is in equilibrium under the action of four stretched elastic strings which connect the four pairs of opposite vertices. Shew that the tensions of the strings and the actions in the rods are proportional to their lengths.

[If T_1, T_2, T_3, T_4 be the tensions of the strings whose lengths are x, y, z, u, the equation of virtual work gives

$$T_1\delta x + T_2\delta y + T_3\delta z + T_4\delta u = 0.$$

But it is easy to shew that $x^2 + y^2 + z^2 + u^2 =$ four times the sum of the squares of the lengths of the rods, so that $x\delta x + y\delta y + z\delta z + u\delta u = 0$.

Since these two equations are true for all values of the quantities δx, δy, δz, δu, we have

$$\frac{T_1}{x}=\frac{T_2}{y}=\frac{T_3}{z}=\frac{T_4}{u}.$$

The rest follows by considering the forces acting at any corner.]

5. A conical tent resting on a smooth floor is made up of an indefinitely great number of equal isosceles triangular elements, hinged at the vertex and kept in shape by a heavy circular ring placed on it like a necklace. Shew that in equilibrium the semi-vertical angle of the cone is $\sin^{-1}\left(\frac{r}{h}\cdot\frac{3W'}{W+3W'}\right)^{\frac{1}{3}}$, where W, W' are respectively the weights of the cone and the ring, and r, h are respectively the radius of the ring and the slant side of the cone.

6. Three particles, of equal weight w, are in equilibrium on the outer surface of a smooth fixed sphere of radius r; the particles rest symmetrically on the surface of the sphere, being connected by equal strings of length l. Shew, by means of the principle of virtual work, that the tension of each string is $w\left\{\dfrac{\sin^2 a \sin\frac{a}{2}}{3\sin\frac{3a}{2}}\right\}^{\frac{1}{2}}$, where a is an angle of circular measure $\dfrac{l}{r}$.

7. A heavy elastic string, whose natural length is $2\pi a$, is placed round a smooth cone whose axis is vertical and whose semi-vertical angle is a. If W be the weight and λ the modulus of elasticity of the string, prove that it will be in equilibrium when in the form of a circle whose radius is $a\left(1+\dfrac{W}{2\pi\lambda}\cot a\right)$.

8. A smooth paraboloid of revolution is fixed with its axis vertical and vertex upwards; on it is placed a heavy elastic string of unstretched length $2\pi c$; when the string is in equilibrium shew that it rests in the form of a circle of radius $\dfrac{4\pi ac\lambda}{4\pi a\lambda - cW}$, where W is the weight of the string, λ its modulus of elasticity, and $4a$ the latus rectum of the generating parabola.

9. Two equal particles are connected by two given weightless strings, which are placed like a necklace on a smooth cone whose axis is vertical and whose vertex is uppermost; shew that the tension of each string is $\dfrac{W}{\pi}\cot a$, where W is the weight of each particle and $2a$ the vertical angle of the cone.

178. Suppose that at any point (x, y, z) of the path of a particle the component forces on it are X, Y, Z. Then, as in Art. 96, if it undergo a displacement given by δx, δy, δz the work done by the forces on it $= X\delta x + Y\delta y + Z\delta z$.

The work done on it as it moves from some standard position at (x_0, y_0, z_0) to the position (x_1, y_1, z_1)

= the sum of such works as those done in the elementary displacement

$$= \int_{(x_0, y_0, z_0)}^{(x_1, y_1, z_1)} (X\,dx + Y\,dy + Z\,dz) \text{............} (1).$$

This quantity is called the Work Function and is often denoted by W.

If X, Y, Z are such that they are the differential coefficients of some quantity V with respect to x, y, z, then

$$W = \int_{(x_0, y_0, z_0)}^{(x_1, y_1, z_1)} \left(\frac{dV}{dx}\,dx + \frac{dV}{dy}\,dy + \frac{dV}{dz}\,dz\right)$$

$$= \Big[V \Big]_{(x_0, y_0, z_0)}^{(x_1, y_1, z_1)} = V_1 - V_0 \text{ ....} (2),$$

where V_1 and V_0 denote the values of V at the points (x_1, y_1, z_1) and (x_0, y_0, z_0) respectively.

The quantity (2) clearly depends only on the values of V for the initial and final positions of the particle, and not at all on the path by which it passed from the first to the final position.

Such a system of forces, in which the x, y, and z components at any point are the differential coefficients with respect to x, y, and z of some function, so that $X\,\delta x + Y\,\delta y + Z\,\delta z$ is a perfect differential, is called a *Conservative System*.

The quantity V is called the *Potential* of the system.

179. The Potential Energy of the particle due to the given system of forces is the work that the forces would do on it as it moved from any position to the standard position. Thus when it is at the point (x_1, y_1, z_1) the potential energy K

$$= \int_{(x_1, y_1, z_1)}^{(x_0, y_0, z_0)} (X\,dx + Y\,dy + Z\,dz)$$

$$= V_0 - V_1.$$

180. *Coordinates of a system.* A body, which is free to move in two dimensions, has its position known when we are given the coordinates of some point of it and also the angle that a straight line fixed with respect to it makes with the axis of x.

These three quantities $\bar{x}$, $\bar{y}$, and θ may be called the coordinates of the body, and the coordinates of any other point of the body must clearly be expressible in terms of them.

In three dimensions we shall know the position of the body if the coordinates of any three given points of it are known. But the nine coordinates of these points are connected by three relations expressing the invariable lengths of the lines joining them. Thus three of these coordinates may be determined in terms of the rest. The six remaining independent coordinates fix the position of the body and may be called its coordinates.

Or, again, the position of this body, free to move in space, will be given if we know the three coordinates of any point G of it, and also the position with respect to the axes of two known lines AB and CD of it. The direction cosines (l_1, m_1, n_1) and (l_2, m_2, n_2) of these two lines are connected by the relations $l_1^2 + m_1^2 + n_1^2 = 1$, $l_2^2 + m_2^2 + n_2^2 = 1$, and $l_1l_2 + m_1m_2 + n_1n_2 =$ the cosine of the known angle between AB and CD, and hence reduce to three independent quantities.

Hence, again, six independent quantities will fix the position of a body in space and may be called the coordinates of the body.

Thus any independent quantities which, when given, determine the position of a body, are called its Coordinates.

181. *Work function of a body.* If X_1, Y_1, Z_1 be the components of the forces acting at any point (x_1, y_1, z_1) of the body, with similar notations for other particles of the body, the total work done during any elementary displacement of the body

$$= (X_1\delta x_1 + Y_1\delta y_1 + Z_1\delta z_1) + (X_2\delta x_2 + Y_2\delta y_2 + Z_2\delta z_2) + \dots$$

so that

$$\delta W = \Sigma\,(X\delta x + Y\delta y + Z\delta z)\dots\dots\dots\dots\dots(1).$$

Let the independent coordinates of the body be denoted by $\xi, \eta, \zeta, \dots$ so that the coordinates (x_1, y_1, z_1) of each point of the body and the component forces X_1, Y_1, Z_1 can be expressed in terms of $\xi, \eta, \zeta, \dots$. Then (1) can be expressed in the form

$$\delta W = \Xi\,d\xi + \mathrm{H}\,d\eta + \mathrm{Z}\,d\zeta + \dots.$$

The whole work done as the body moves from some standard configuration given by $(\xi_0, \eta_0, \zeta_0, \dots)$ to the configuration given by $(\xi_1, \eta_1, \zeta_1, \dots)$ is, as in the case of a single particle,

$$W = \int_{(\xi_0, \eta_0, \dots)}^{(\xi_1, \eta_1, \dots)} (\Xi\,d\xi + \mathrm{H}\,d\eta + \mathrm{Z}\,d\zeta + \dots).$$

If as before, these quantities Ξ, H, Z, ... can be expressed as the differential coefficients with respect to $\xi, \eta, \zeta, \ldots$ of some quantity V, this gives $W = V_1 - V_0$.

Similarly, if the potential energy K be defined to be the work the system can do as it passes from the position given by $(\xi_1, \eta_1, \zeta_1, \ldots)$ to the standard position we have, as before, $K = V_0 - V_1$.

182. *Position of equilibrium of the system.* The position of equilibrium is, by the Principle of Virtual Work, given by equating δW to zero for every virtual displacement. In other words we find the position of the system for which W is either a maximum or a minimum or stationary.

The quantities $\xi, \eta, \zeta, \ldots$ being independent, we therefore find the position of equilibrium by equating to zero Ξ, H, Z, ... which are the differential coefficients of W with respect to $\xi, \eta, \zeta, \ldots$.

Suppose that the position of equilibrium which is thus found is one in which W is a true maximum, and let the body be slightly displaced into a neighbouring position, and momentarily be at rest. Then the body must (by the principle of Dynamics that the kinetic energy generated is equal to the work done) move so that the work done by the forces is positive, *i.e.* it must move so that W is increased and hence must return towards the position of equilibrium just found, and so this position of equilibrium is stable.

Similarly, if in the position of equilibrium found as above W is a true minimum, the body on being displaced will move so that W is increased and will therefore move further away from the position of equilibrium and this position will be unstable.

Finally if W be neither a true maximum nor a true minimum, *i.e.* if it be a maximum for some displacements and a minimum for others, the equilibrium will be stable for some displacements and unstable for other displacements. The position thus found is then on the whole not a stable one.

To sum up. If the work function W be formed for a body, or system of bodies, and expressed in terms of independent coordinates $\xi, \eta, \zeta, \ldots$, the positions of equilibrium are the positions for which W is a maximum, a minimum or stationary, and those positions only are stable which are such that the corresponding values of W are true maxima.

Since $\delta K = -\delta W$, the opposite will be the case if we consider the Potential Energy. Only those positions are stable which are such that the corresponding values of the Potential Energy K are true minima.

183. Ex. *Three equal spheres rest on a smooth table and are kept in position by a smooth elastic band in the plane of the centres, the band being unstretched when the spheres are in contact. A fourth equal sphere is placed above them. Prove that, if in a position of equilibrium the line joining the centre of the upper sphere to the centre of either of the lower spheres is inclined at an angle θ to the vertical, the equilibrium is stable for symmetrical displacements if $\sin^3\theta < \frac{1}{\sqrt{3}}$.*

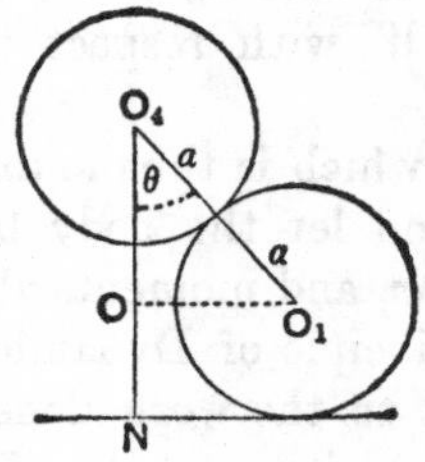

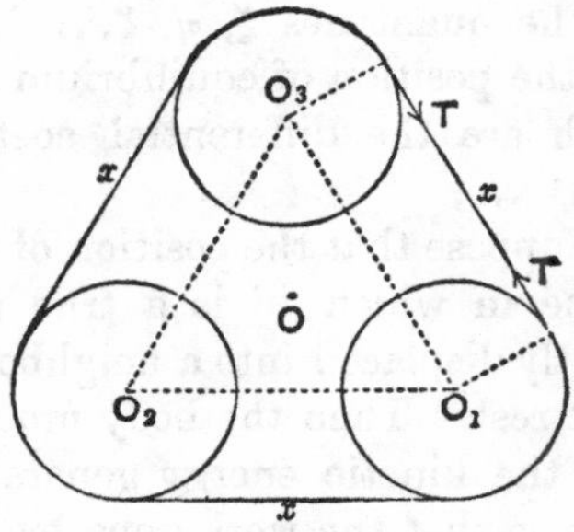

Let θ be the inclination to the vertical of the line joining the centre of the upper sphere to that of one of the lower spheres, when the centres of the latter are at a distance x apart. Then

$$\sin\theta = \frac{OO_1}{2a} = \frac{\frac{2}{3}x\sin 60^\circ}{2a} = \frac{x}{2a\sqrt{3}}.$$

If λ be the coefficient of elasticity, the tension T of the band

$$= \lambda\,\frac{3x + 2\pi a - (6a + 2\pi a)}{6a + 2\pi a} = \frac{3\lambda}{2(\pi+3)a}\cdot(x - 2a)$$

If W_1 be the weight of either sphere, the element δW of the work function for a symmetrical displacement is given by

$$\delta W = -W_1\delta(a + 2a\cos\theta) - 3T\delta x.$$

$$\therefore\ \frac{dW}{d\theta} = W_1\ 2a\sin\theta - \frac{18\sqrt{3}\lambda a}{\pi+3}[\sqrt{3}\sin\theta - 1]\cos\theta.$$

$$\therefore\ \frac{d^2W}{d\theta^2} = W_1.\,2a\cos\theta - \frac{18\sqrt{3}\lambda a}{\pi+3}[\sqrt{3}(\cos^2\theta - \sin^2\theta) + \sin\theta].$$

The position of equilibrium is given by $\frac{dW}{d\theta} = 0$, *i.e.* by

$$W_1\sin\theta = \frac{9\sqrt{3}\lambda}{\pi+3}[\sqrt{3}\sin\theta\cos\theta - \cos\theta].$$

For this value of θ,

$$\frac{d^2W}{d\theta^2}=\frac{18\sqrt{3}\lambda a}{\pi+3}[(\sqrt{3}\sin\theta\cos\theta-\cos\theta)\cot\theta-\sqrt{3}(\cos^2\theta-\sin^2\theta)-\sin\theta]$$

$$=\frac{18\sqrt{3}\lambda a}{\pi+3}\frac{\sqrt{3}\sin^3\theta-1}{\sin\theta}.$$

If $\sin^3\theta<\frac{1}{\sqrt{3}}$, then $\frac{d^2W}{d\theta^2}$ is negative, the corresponding value of W is a maximum, and the equilibrium is stable.

EXAMPLES

1. A solid oblate spheroid is loaded with a weight equal to n times its own weight at one extremity of its axis. Find in what different positions it can be in equilibrium resting on a smooth horizontal plane, and in which of these the equilibrium is stable. Shew that if $e^2<\frac{n}{2n+1}$, there are only two possible positions of equilibrium.

2. The axis of z being vertically upwards, and the origin being A, a uniform square board $ABCD$, of weight W and side $2a$, is mounted so that it can turn freely about AB which is fixed in the direction whose cosines are $(\sin\theta,\ 0,\ \cos\theta)$. A weightless string, fastened to the board at C, passes through a smooth fixed ring at $(0,\ 2a,\ 0)$, and carries a hanging weight w at its other end. Prove that there is equilibrium when the angle ϕ which the board makes with the plane of xz satisfies the equation

$$\frac{\cot\phi}{\sqrt{(3-2\sin\phi)}}=\frac{W}{2w}\sin\theta.$$

Investigate the stability of the equilibrium.

3. A smooth solid circular cone, of height h and vertical angle 2α, is at rest with its axis vertical in a horizontal circular hole of radius a. Shew that if $16a>3h\sin 2\alpha$ the equilibrium is stable, and there are two other positions of unstable equilibrium; and that if $16a<3h\sin 2\alpha$ the equilibrium is unstable, and the position in which the axis is vertical is the only position of equilibrium.

If a weight w be hung on at the vertex of the cone, whose weight is W. prove that the corresponding condition is $16a(w+W)>3Wh\sin 2\alpha$.

4. A uniform right circular cone of height h and vertical angle 2α rests with its vertex downwards and its axis vertical, between two smooth parallel rails at a distance d apart in a horizontal plane. Prove that the equilibrium is stable for angular displacements in which the axis remains in a vertical plane parallel to the rails, if $h<\frac{4}{3}d\operatorname{cosec}2\alpha$.

5. A right cone, whose vertical angle is a right angle, is placed vertex downwards through a square hole in a horizontal plane so as to touch each side. Shew that, if the height h of the cone exceed twice the side a of the square, a position of equilibrium is possible with the axis inclined to the horizon at an angle $\sin^{-1}\sqrt{\frac{2a}{3h-4a}}$. Prove also that this position is stable.

CHAPTER XI

FORCES IN THREE DIMENSIONS (*continued*)

Poinsot's Central Axis. The Cylindroid. Null Lines.

184. *To shew that any system of forces acting on a rigid body can be reduced to a single force together with a couple whose axis is along the direction of the force.*

It has been shewn in Art. 162 that any system can be reduced to a force R acting at any point O and a couple of moment G about a line through O.

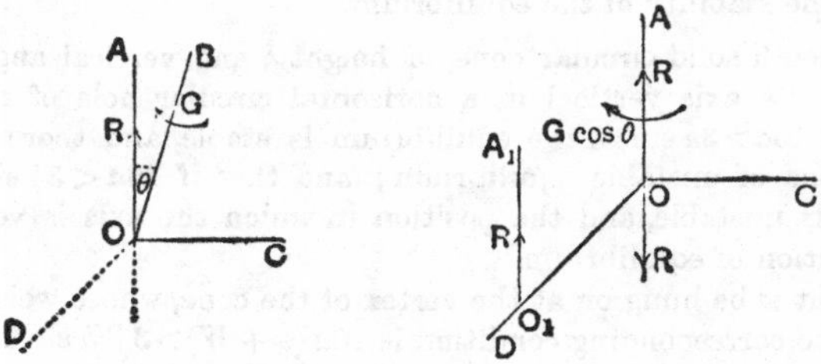

Let OA be the direction of R, OB the axis of the couple G, and let $\angle AOB = \theta$.

In the plane AOB draw OC perpendicular to OA, and draw OD perpendicular to the plane AOC.

By Art. 49 the couple G about OB as axis is equivalent to a couple $G \cos\theta$ about OA as axis and a couple $G \sin\theta$ about OC as axis. This latter couple acts in the plane AOD, and may therefore be replaced by any two equal unlike parallel forces of moment $G \sin\theta$.

Choose for one of these two forces a force R at O in the direction opposite to OA. Then the other force must be equal to R acting parallel to OA at a point O_1 in OD, such that

$$R . OO_1 = G \sin\theta, \textit{ i.e. } OO_1 = \frac{G\sin\theta}{R}.$$

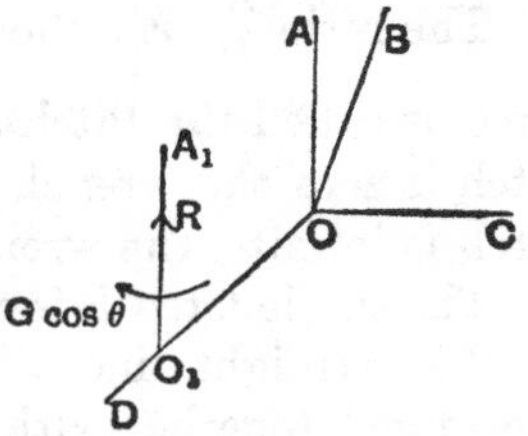

The forces at O now balance; also the axis of the couple $G\cos\theta$ may be transferred from OA to O_1A_1.

We thus have finally a force R along O_1A_1 and a couple of moment $G\cos\theta$ about O_1A_1 as axis.

This axis, O_1A_1, thus obtained is called Poinsot's Central Axis.

It is easily seen that the Central Axis thus determined is unique. For, if possible, let the given system be equivalent to a force along, and a couple about, a line O_1A_1 and also to a force along, and a couple about, another line O_2A_2. By Art. 162, the resultant force is the same in magnitude and direction whatever base point, or origin, is taken. Hence O_2A_2 is parallel to O_1A_1 and the resultant force R is the same for each.

Hence the system $[R; G]$ about O_1A_1 is the same as the system $[R; G']$ about a parallel line O_2A_2. If p be the distance between O_1A_1 and O_2A_2, then R along O_2A_2 is equivalent to R along O_1A_1 and a couple $R.p$ about an axis perpendicular to O_1A_1 [Art. 59]. Hence the second system is equivalent to a force R along O_1A_1, a couple G' about O_1A_1 and a couple $R.p$ about an axis perpendicular to O_1A_1, *i.e.* it is equivalent to a force R along O_1A_1, and a couple about an axis which is not O_1A_1, *i.e.* it is not equal to the system $[R; G]$ with O_1A_1 as axis.

Hence our original supposition is incorrect, and we cannot find two central axes O_1A_1 and O_2A_2, *i.e.* the central axis found in the preceding article is unique.

185. For any origin O the resultant force is the same and equal to that along the central axis. But the resultant couple is not the same. This latter is clearly a minimum for the Central Axis. For if G be the couple for any origin O (not on the central axis), and if its axis be inclined at θ to the resultant force, then the couple for the Central Axis was shewn in the last article to be equal to $G\cos\theta$, which is always less than G.

Hence *the moment of the resultant couple about the Central Axis is less than the moment of the resultant couple corresponding to any point O which is not on the Central Axis.*

186. A single force R together with a couple K whose axis coincides with the direction of the force are, taken together, called a Wrench.

The ratio $\frac{K}{R}$, *viz.* the moment of the couple divided by the force, is called the Pitch and is a linear magnitude. When the pitch is zero the wrench reduces to a single force. When the pitch is infinite, the wrench becomes a couple only.

The single force R is often called the intensity of the wrench.

The straight line along which the single force acts when considered together with the pitch is called a Screw, so that a Screw is a definite straight line associated with a definite pitch

Five quantities are required to determine a Screw. Four are required to give the position of the axis; for example the point in which it cuts one of the coordinate planes and its inclinations to the axes of coordinates. A fifth is required to determine its pitch.

To completely determine a Wrench on a Screw a sixth quantity must be given, *viz.* the Intensity of the Wrench.

187. *Right-handed and left-handed Screws.*

It is clear that combined with the same translation there may be a rotation in either of two opposite directions. When the rotation is the same as in the case of a screw-driver when screwing in a screw, or as in the case of a corkscrew, the screw is said to be right-handed; when the rotation is in the opposite direction the screw is left-handed.

The general definition is as follows; let an observer stand with his body along the axis of the screw, so that the positive direction of the translation is from his feet up through his head; let him also observe a watch whose plane is in the plane of the rotation and whose face is towards him. Then the screw is right or left-handed according as the rotation is opposite to or in the same sense as that of the hands of the watch.

Thus in the figures of this book which are drawn according to the usual conventions of Solid Geometry we have taken the left-handed screw as the positive and standard case. This is clear if we apply the above definition to a screw whose axis is the axis of x; for we have assumed (Art. 47) that a positive couple would tend to rotate the body from Oy to Oz.

188. *Condition that a given system of forces should compound into a single force.*

By Art. 162 the forces are equivalent to a single force R acting at an arbitrary origin O and a single couple G. If θ be the angle between R and the axis of G, then R is equivalent to a force $R\cos\theta$ along the axis OB of the couple, and a force $R\sin\theta$ in the plane of the couple. This force $R\sin\theta$, together with the parallel forces of the couple, are, by Art. 51, equivalent to a parallel force $R\sin\theta$ which does not pass through O and therefore cannot, in general, compound with $R\cos\theta$ into a single force.

But, if $R\cos\theta = 0$, *i.e.* if $\cos\theta = 0$, then we are left with a single force $R\sin\theta$.

Hence θ must be 90°, *i.e.* the angle between the straight lines whose direction cosines are $\left(\frac{X}{R}, \frac{Y}{R}, \frac{Z}{R}\right)$ and $\left(\frac{L}{G}, \frac{M}{G}, \frac{N}{G}\right)$ must be a right angle.

$$\therefore \quad \frac{X}{R}\cdot\frac{L}{G} + \frac{Y}{R}\cdot\frac{M}{G} + \frac{Z}{R}\cdot\frac{N}{G} = \cos 90^\circ = 0.$$

$$\therefore \quad XL + YM + ZN = 0$$

is the required condition.

189. *Invariants.* Whatever origin, or base point, and axes are chosen, for any given system of forces the quantities

$$X^2 + Y^2 + Z^2 \text{ and } LX + MY + NZ$$

are invariable, where $X = \Sigma(X_1)$, etc. and $L = \Sigma(y_1 Z_1 - z_1 Y_1)$, etc.

For, by Arts. 162 and 184, $X^2 + Y^2 + Z^2$ is the square of the resultant force R corresponding to the Central Axis and is therefore invariable.

Again, if, in Art. 162, (l, m, n) are the direction cosines of the resultant force and (l_1, m_1, n_1) those of the axis of the resultant couple, then

$$\frac{X}{R}\cdot\frac{L}{G} + \frac{Y}{R}\cdot\frac{M}{G} + \frac{Z}{R}\cdot\frac{N}{G} = ll_1 + mm_1 + nn_1$$

= the cosine of the angle between the resultant force and the axis of the resultant couple

$= \cos\theta$ (Art. 184).

$$\therefore\ LX + MY + NZ = R\,.\,G\cos\theta = R\,.\,K,$$

where K is the moment of the couple about the Central Axis.

Hence $I \equiv LX + MY + NZ$ is an invariant.

It follows that if K be zero, that is, if the given system reduces to a single force, then $LX + MY + NZ = 0$.

This second invariant will be zero also when the resultant force R is zero. In this case the first invariant is zero also.

The pitch, p, of the resultant Wrench of the system $= \dfrac{K}{R}$

= the invariant I of the system divided by the square of the invariant R.

190. *To find the equation of the Central Axis of any given system of forces.*

With the notation of Art. 162, let (f, g, h) be the coordinates referred to the axes Ox, Oy, Oz of any point Q.

The moment about a line through Q parallel to Ox is clearly obtained by putting $x_1 - f$, $y_1 - g$, $z_1 - h$ instead of x_1, y_1, z_1 in the results of that article.

Hence this moment

$$= \Sigma\,[(y_1 - g)\,Z_1 - (z_1 - h)\,Y_1]$$
$$= \Sigma\,(y_1Z_1 - z_1Y_1) - g\Sigma\,(Z_1) + h\Sigma\,(Y_1)$$
$$= L - gZ + hY,$$

with the notation of that article.

So the moments about lines through Q parallel to the other axes are

$$M - hX + fZ \text{ and } N - fY + gX.$$

Also the components of the resultant force are the same for all points such as Q, and are thus X, Y and Z.

If Q be a point on the central axis, the direction cosines of the axis of the couple corresponding to it are proportional to those of the resultant force. Hence

$$\frac{L - gZ + hY}{X} = \frac{M - hX + fZ}{Y} = \frac{N - fY + gX}{Z}$$

$$= \frac{LX + MY + NZ}{X^2 + Y^2 + Z^2} = \frac{K}{R},$$

by Art. 189.

Hence the equation of the locus of the point (f, g, h), *i.e.* the required equation of the central axis, is

$$\frac{L - yZ + zY}{X} = \frac{M - zX + xZ}{Y} = \frac{N - xY + yX}{Z}$$

$$= \frac{K}{R} = \text{the pitch } p \text{ of the wrench.}$$

191. Ex. 1. *Three forces, each equal to P, act on a body; one at the point* $(a, 0, 0)$ *parallel to Oy, the second at the point* $(0, b, 0)$ *parallel to Oz, and the third at the point* $(0, 0, c)$ *parallel to Ox; the axes being rectangular, find the resultant wrench in magnitude and position.*

Here
$$X = Y = Z = P,$$
$$L = Pb; \quad M = Pc; \quad N = Pa.$$

Hence, if R be the force and K the couple of the wrench, then

$$R = \sqrt{X^2 + Y^2 + Z^2} = P\sqrt{3},$$

and
$$KR = LX + MY + NZ = P^2(a + b + c),$$

so that
$$K = \frac{\sqrt{3}}{3} P(a + b + c).$$

By Art. 190, the equations to the central axis are

$$b - y + z = c - z + x = a - x + y,$$

i.e.
$$x + \frac{a + 2b + 3c}{3} = y + \frac{b + 2c + 3a}{3} = z + \frac{c + 2a + 3b}{3},$$

so that the central axis is a straight line through the point

$$\left(-\frac{a + 2b + 3c}{3}, \; -\frac{b + 2c + 3a}{3}, \; -\frac{c + 2a + 3b}{3}\right),$$

inclined at equal angles to the three axes.

Ex. 2. *Two equal forces act one along each of the straight lines*

$$\frac{x \mp a\cos\theta}{a\sin\theta} = \frac{y - b\sin\theta}{\mp b\cos\theta} = \frac{z}{c};$$

shew that their central axis must, for all values of θ, lie on the surface

$$y\left(\frac{x}{z} + \frac{z}{x}\right) = b\left(\frac{a}{c} + \frac{c}{a}\right).$$

P being the force, then at the point $(a\cos\theta, b\sin\theta, 0)$ we have a force whose components are proportional to $a\sin\theta \,.\, P$, $-b\cos\theta P$, cP, and at the point $(-a\cos\theta, b\sin\theta, 0)$ a force whose components are proportional to $a\sin\theta \,.\, P$, $b\cos\theta P$, cP.

Hence
$$X = \Sigma(X_1) \propto 2a\sin\theta \,.\, P,$$
$$Y = 0, \text{ and } Z \propto 2cP,$$
$$L = \Sigma(y_1 Z_1 - z_1 Y_1) \propto 2bc\sin\theta \,.\, P,$$
$$M = 0, \text{ and } N \propto -2abP.$$

The equations of Art. 190 then become

$$\frac{bc\sin\theta - yc}{a\sin\theta} = \frac{-ab + ya\sin\theta}{c}, \text{ and } za\sin\theta = x.a.$$

Substituting the value of $\sin\theta$ from the second of these equations in the first we have, as the locus of the central axis,

$$y\left(\frac{x}{z}+\frac{z}{x}\right)=b\left(\frac{a}{c}+\frac{c}{a}\right).$$

EXAMPLES

1. Equal forces act along two perpendicular diagonals of opposite faces of a cube of side a; shew that they are equivalent to a single force R acting along a line through the centre of the cube, and a couple $\frac{1}{2}aR$ with the same line for axis.

2. $OBDC$ is a rectangle such that $OB=b$ and $OC=c$; also OA is a perpendicular to its plane; along OA, CD and BD act forces X, Y and Z respectively. Shew that the component force R and couple K of the resultant wrench are $\sqrt{X^2+Y^2+Z^2}$ and $X(Zb-Yc)\div\sqrt{X^2+Y^2+Z^2}$. Shew also that with OA, OB and OC as axes of x, y, and z the equation to the central axis is

$$\frac{x}{X}=\frac{y}{Y}-\frac{ZK}{XYR}=\frac{z}{Z}+\frac{KY}{XZR}.$$

3. OA, OB, OC are the edges of a rectangular parallelopiped of lengths 6, 15 and 8 inches respectively, and OO', AA', BB' and CC' are its diagonals. Calculate the wrench of the forces—130 from B' to O, 68 from A to O', 50 from C' to A' and 68 from B to C.

[$R=-108$, $K=1080$, the central axis being a straight line parallel to OA through the middle point of BC.]

4. OA, OB, OC are three co-terminous edges of a cube and AA', BB', CC', OO' are diagonals; along BC', CA', AB' and OO' act forces equal to X, Y, Z and R respectively; shew that they are equivalent to a single resultant if $(YZ+ZX+XY)\sqrt{3}+R(X+Y+Z)=0$.

5. Forces P, Q, R act along three non-intersecting edges of a cube; find the central axis.

6. Two forces P and Q act along the straight lines whose equations are $y=x\tan a$, $z=c$ and $y=-x\tan a$, $z=-c$ respectively. Shew that their central axis lies on the straight line

$$y=x.\frac{P-Q}{P+Q}\tan a \text{ and } \frac{z}{c}=\frac{P^2-Q^2}{P^2+2PQ\cos 2a+Q^2}.$$

For all values of P and Q, prove that this line is a generator of the surface

$$(x^2+y^2)\,z\sin 2a=2cxy.$$

7. Equal forces act along the axes and along the straight line

$$\frac{x-a}{l}=\frac{y-\beta}{m}=\frac{z-\gamma}{n};$$

find the equations of the central axis of the system.

8. Three forces act along the straight lines

$$x=0,\ y-z=a;\quad y=0,\ z-x=a;\quad z=0,\ x-y=a.$$

Shew that they cannot reduce to a couple.

Prove also that if the system reduces to a single force its line of action must lie on the surface $x^2+y^2+z^2-2yz-2zx-2xy=a^2$.

9. Forces X, Y, Z act along the three straight lines

$$y=b,\ z=-c;\quad z=c,\ x=-a;\quad \text{and}\quad x=a,\ y=-b,$$

respectively; shew that they will have a single resultant if

$$\frac{a}{X}+\frac{b}{Y}+\frac{c}{Z}=0;$$

and that the equations of its line of action are any two of the three

$$\frac{y}{Y}-\frac{z}{Z}-\frac{a}{X}=0,\quad \frac{z}{Z}-\frac{x}{X}-\frac{b}{Y}=0,\quad \frac{x}{X}-\frac{y}{Y}-\frac{c}{Z}=0.$$

10. A single force is equivalent to component forces X, Y and Z along the axes of coordinates and to couples L, M, N about these axes; shew that the magnitude of the single force is $\sqrt{X^2+Y^2+Z^2}$ and that the equation to its line of action is

$$\frac{yZ-zY}{L}=\frac{zX-xZ}{M}=\frac{xY-yX}{N}=1.$$

11. If a force (X, Y, Z) act along a generator of the hyperbolic paraboloid $\frac{x^2}{a^2}-\frac{y^2}{b^2}=2\frac{z}{c}$, and be equivalent to an equal force (X, Y, Z) at the origin together with a couple (L, M, N), shew that

$$aL\pm bM=0,\quad bX\pm aY=0\quad \text{and}\quad cN\pm abZ=0.$$

12. A force P acts along the axis of x and another force nP along a generator of the cylinder $x^2+y^2=a^2$; shew that the central axis lies on the cylinder

$$n^2(nx-z)^2+(1+n^2)^2y^2=n^4a^2.$$

13. Two forces act, one along the line $y=0$, $z=0$ and the other along the line $x=0$, $z=c$. As the forces vary, shew that the surface generated by the axis of their equivalent wrench is $(x^2+y^2)z=cy^2$.

14. A force parallel to the axis of z acts at the point $(a, 0, 0)$ and an equal force perpendicular to the axis of z acts at the point $(-a, 0, 0)$. Shew that the central axis of the system lies on the surface

$$z^2(x^2+y^2)=(x^2+y^2-ax)^2.$$

15. A force F acts along the axis of z, and a force mF along a straight line, intersecting the axis of x at a distance c from the origin and parallel to the plane of yz. Shew that as this straight line turns round the axis of x, the central axis of the forces generates the surface

$$\{m^2z^2+(m^2-1)y^2\}\{c-x\}^2=x^2z^2.$$

16. Along the normal at every point of an octant of an ellipsoid cut off by the principal planes acts a force proportional to the element of surface at P. Shew that these forces are equivalent to a single force acting along the line

$$a\left(x-\frac{4a}{3\pi}\right)=b\left(y-\frac{4b}{3\pi}\right)=c\left(z-\frac{4c}{3\pi}\right),$$

where $2a$, $2b$, $2c$ are the axes of the ellipsoid.

17. Any number of wrenches of the same pitch p act along generators of the same system of the hyperboloid $\frac{x^2}{a^2}+\frac{y^2}{b^2}-\frac{z^2}{c^2}=1$ Shew that they will reduce to a single resultant provided their central axis is parallel to a generator of the cone

$$\left(p+\frac{bc}{a}\right)x^2+\left(p+\frac{ca}{b}\right)y^2+\left(p-\frac{ab}{c}\right)z^2=0.$$

192. *To shew that a given system of forces can be replaced by two forces, equivalent to the given system, in an infinite number of ways and that the tetrahedron formed by the two forces is of constant volume.*

Let the given system have Oz as its Central Axis and let R, K be the resultant force along Oz and the resultant couple about Oz.

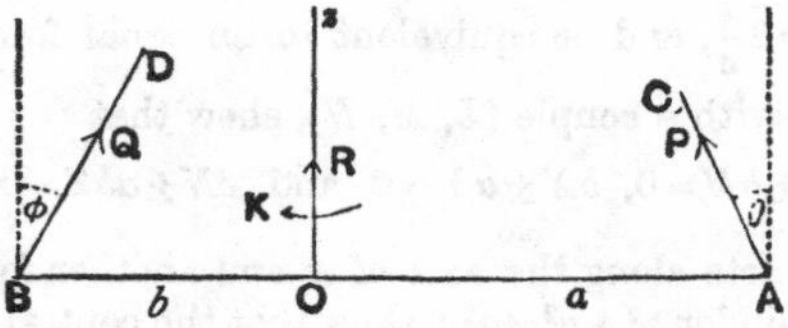

On a line through O perpendicular to Oz take any two points A, B on opposite sides of O such that $OA=a$ and $OB=b$.

Assume that the given system is equivalent to a force P acting through A in a plane perpendicular to OA at an angle θ with Oz, and a force Q acting through B in a plane perpendicular to OB at an angle ϕ with Oz, and let θ and ϕ be measured positively in opposite directions.

Then the resultants of P and Q along Oz and perpendicular to it must be R and zero respectively, and similarly the resultant couples about these two lines must be K and zero.

Hence $$R = P\cos\theta + Q\cos\phi \qquad\text{.....................(1)},$$

and $$0 = P\sin\theta - Q\sin\phi \qquad\text{.....................(2)}.$$

Also $$K = P\sin\theta\,.\,a + Q\sin\phi\,.\,b \qquad\text{..............(3)},$$

and $$0 = P\cos\theta\,.\,a - Q\cos\phi\,.\,b \qquad\text{..............(4)}.$$

(1) and (4) give $$\frac{P\cos\theta}{b} = \frac{Q\cos\phi}{a} = \frac{R}{a+b} \qquad\text{..............(5)}.$$

(2) and (3) give $$P\sin\theta = Q\sin\phi = \frac{K}{a+b} \qquad\text{..............(6)}.$$

Hence $$P^2 = \frac{K^2 + R^2b^2}{(a+b)^2}, \quad Q^2 = \frac{K^2 + R^2a^2}{(a+b)^2},$$

$$\tan\theta = \frac{K}{Rb}, \text{ and } \tan\phi = \frac{K}{Ra}.$$

Whatever be the values of a and b, we thus obtain real values for P, Q, θ, and ϕ, so that our assumption is correct.

Again (5) and (6) give

$$KRa = (a+b)^2 PQ\sin\theta\cos\phi, \text{ and } KRb = (a+b)^2 PQ\cos\theta\sin\phi.$$

Hence, by addition, $KR = PQ(a+b)\sin(\theta+\phi)$ ......(7).

Let AC, BD represent P, Q in magnitude.

The volume of the tetrahedron $ACBD$

$= \frac{1}{3}$ area of the $\Delta ABC \times$ perpendicular from D upon the ΔABC

$= \frac{1}{3} \times \frac{1}{2} AB\,.\,AC \times BD\sin(\theta+\phi)$

$= \frac{1}{6}\,.\,PQ(a+b)\sin(\theta+\phi) = \frac{1}{6}K\ R$, by equation (7),

and it is therefore constant.

Symmetrical Case. If the forces P and Q are equal and at equal distances from the Central Axis, then by equations (1), (2), (3), (4) $\theta = \phi$, so that they are equally inclined to the Central Axis, and

$$R = 2P\cos\theta \text{ and } K = 2Pa\sin\theta.$$

$$\therefore P = \tfrac{1}{2}\sqrt{R^2 + \frac{K^2}{a^2}} \text{ and } \tan\theta = \frac{K}{Ra}.$$

If, in addition, we make $a = \dfrac{K}{\sqrt{3}R}$, then $P = R$ and $\theta = 60°$.

193. *To find the resultant wrench of two given wrenches*

Let AC be the axis of one wrench (R_1, K_1) and BD the axis of the other wrench (R_2, K_2). Let $AB\,(=c)$ be the shortest distance between these two axes, and α the angle between them.

Assume that the required central axis Oz is perpendicular to AB, divides it into parts x and $c-x$, and is inclined at θ to AC and hence $\alpha-\theta$ to BD. Let R and K be the force and moment of the assumed Central Axis.

The conditions of the question then give

$$R = R_1\cos\theta + R_2\cos(\alpha-\theta) \quad\text{..............(1)},$$

$$0 = R_1\sin\theta - R_2\sin(\alpha-\theta) \quad\text{..............(2)},$$

$$K = K_1\cos\theta + K_2\cos(\alpha-\theta) + R_1\sin\theta\,.\,x + R_2(c-x)\sin(\alpha-\theta),$$

and

$$0 = K_1\sin\theta - K_2\sin(\alpha-\theta) - R_1\cos\theta\,.\,x + R_2(c-x)\cos(\alpha-\theta)$$

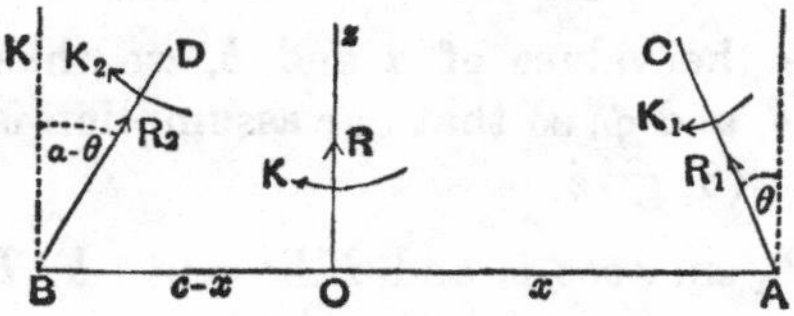

By (1) and (2) these latter equations are

$$K = K_1\cos\theta + K_2\cos(\alpha-\theta) + R_2c\sin(\alpha-\theta) \quad\text{........(3)},$$

and $\quad 0 = K_1\sin\theta - K_2\sin(\alpha-\theta) + R_2c\cos(\alpha-\theta) - Rx\text{...(4)}.$

(1) and (2) give

$$R^2 = R_1^2 + R_2^2 + 2R_1R_2\cos\alpha \quad\text{..............(5)}.$$

Also (2) gives

$$\frac{\sin\theta}{R_2\sin\alpha} = \frac{\cos\theta}{R_1 + R_2\cos\alpha} = \frac{1}{R} \quad\text{..............(6)}.$$

Again, (3) and (6) give

$$RK = (K_1 + K_2\cos\alpha + R_2c\sin\alpha)(R_1 + R_2\cos\alpha)$$
$$+ (K_2\sin\alpha - R_2c\cos\alpha)\,.\,R_2\sin\alpha$$
$$= R_1K_1 + R_2K_2 + (R_1K_2 + R_2K_1)\cos\alpha + R_1R_2c\sin\alpha\text{...(7)}.$$

This gives the value of the Invariant for the given system of forces.

Also (4) with the help of (6) gives

$$R^2x = (K_1 + K_2 \cos\alpha + R_2 c \sin\alpha) \,.\, R_2 \sin\alpha - (K_2 \sin\alpha - R_2 c \cos\alpha)(R_1 + R_2 \cos\alpha)$$

$$= (R_2K_1 - R_1K_2)\sin\alpha + R_2 c\,(R_2 + R_1\cos\alpha) \quad \text{.........(8).}$$

(5), (6), (7) and (8) give R, θ, K and x, and hence the required wrench and the position of its axis.

From these equations we easily obtain

$$\frac{x}{c-x} = \frac{R_2 c\,(R_1\cos\alpha + R_2) - (R_1K_2 - R_2K_1)\sin\alpha}{R_1 c\,(R_1 + R_2\cos\alpha) + (R_1K_2 - R_2K_1)\sin\alpha},$$

and

$$\frac{\cos\theta}{\cos(\alpha-\theta)} = \frac{R_1 + R_2\cos\alpha}{R_1\cos\alpha + R_2}.$$

194. *Resultant wrench of two given forces R_1 and R_2 inclined at a given angle α.*

This is a particular case of the preceding article where $K_1 = K_2 = 0$. We thus have

$$R = \sqrt{R_1^2 + R_2^2 + 2R_1R_2\cos\alpha},$$

$$K \,.\, R = R_1R_2 c\sin\alpha,$$

$$\frac{x}{c-x} = \frac{R_2(R_1\cos\alpha + R_2)}{R_1(R_1 + R_2\cos\alpha)},$$

and

$$\frac{\sin\theta}{R_2\sin\alpha} = \frac{\cos\theta}{R_1 + R_2\cos\alpha} = \frac{1}{\sqrt{R_1^2 + R_2^2 + 2R_1R_2\cos\alpha}},$$

so that

$$\frac{\cos\theta}{\cos(\alpha-\theta)} = \frac{R_1 + R_2\cos\alpha}{R_1\cos\alpha + R_2}.$$

195. *Geometrical construction for the central axis of two forces R_1 and R_2 acting at points A and B in directions AC and BD.*

Let BK be the direction of the resultant R of forces R_2 along BD and R_1 at B parallel to AC, and let it make angles θ and $\alpha - \theta$ with R_1 and R_2, so that BK is parallel to the central axis as in the figure of Art. 193. At B and A, in the plane KBA,

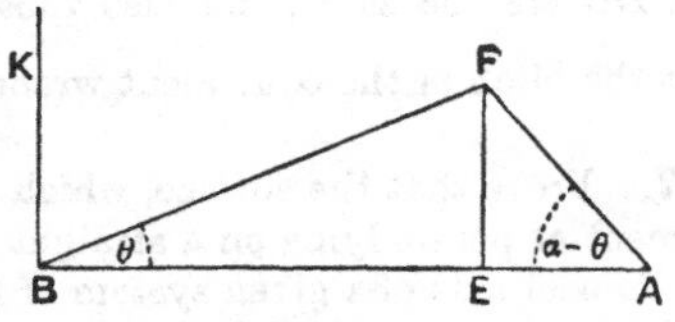

make $\angle ABF = \theta$, and $\angle BAF = \alpha - \theta$; then FE the perpendicular on AB is the central axis, and the couple of the resultant wrench $= K \,.\, EF$.

For, from the equations of the last article, we easily obtain

$$\frac{x}{c-x}=\frac{\sin\theta\cos(\alpha-\theta)}{\cos\theta\sin(\alpha-\theta)}=\frac{AF}{BF}\frac{\cos(\alpha-\theta)}{\cos\theta}=\frac{AE}{BE},$$

so that EF is the central axis.

Also, by the last article,

$$\frac{K}{R}=\frac{R_1R_2c\sin\alpha}{R^2}=\frac{R_1c\sin^2\theta}{R_2\sin\alpha}=\frac{c\sin(\alpha-\theta)\sin\theta}{\sin\alpha}=EF,$$

so that $K=R.EF$.

EXAMPLES

1. If P and Q be two non-intersecting forces whose directions are perpendicular, shew that the distances of the central axis from their lines of action are as Q^2 to P^2.

2. Shew that a system of forces can be replaced in an infinite number of ways by a pair of equal forces whose directions make any given angle with one another, and find the distance between these forces when the angle is known.

3. Two forces P and Q are such that their central axis is given in position and the line of action of P is given. Shew that the locus of the line of action of Q is a conicoid.

4. Shew that the minimum distance between two forces, which are equivalent to a given system $(K; R)$ and which are inclined at a given angle $2a$, is $\frac{2K}{R}\cot a$, and that the forces are then each equal to $\frac{1}{2}R\sec a$.

5. Forces act along the edges of a regular tetrahedron, *viz.* P along BC and DA, Q along CA and DB, and R along AB and DC. Shew that the pitch of the equivalent wrench is $\frac{1}{2\sqrt{2}}$ of the edge of the tetrahedron.

[The three shortest distances between pairs of opposite sides of a regular tetrahedron are mutually perpendicular and meet in a point.]

6. Wrenches of the same pitch p act along the edges of a regular tetrahedron $ABCD$ of side a. If the intensities of the wrenches along AB, DC are the same, and also those along BC, DA and DB, CA, shew that the pitch of the equivalent wrench is $p+\frac{a}{2\sqrt{2}}$.

7. Prove that the surface, which is traced out by the axis of principal moment at points lying on a straight line which intersects at right angles the Poinsot axis of a given system of forces, is a hyperbolic paraboloid.

[Take the given straight line as the axis of x, the Poinsot axis of the system $(R; K)$ as the axis of z, and a perpendicular to both as Oy. Then at the point $(x, 0, 0)$ the force is R parallel to Oz, and the couple is G at an angle θ to R and perpendicular to Ox such that $K=G\cos\theta$ and

$G \sin\theta = Rx$. Hence, if (x, y, z) be any point on the axis of G, we have $\frac{y}{z} = \tan\theta = \frac{Rx}{K}$, so that the locus is the hyperbolic paraboloid

$$Rxz = Ky.]$$

8. At all points on a given straight line are drawn the axes of principal moment corresponding to any given system of forces; shew that these axes lie on a hyperbolic paraboloid and that their ends lie on another given straight line.

[Take the given straight line as Ox, and the common perpendicular to it and the Poinsot axis as Oz. Then clearly there is no component force along Oz or component couple about Oz. The component forces along the axes are thus $(X, Y, 0)$ and the component couples $(L, M, 0)$, so that the component couples at the point $(\xi, 0, 0)$ are L, M, and $-\xi Y$ [Art. 190].

Hence the equation to the axis of principal moment there is

$$\frac{x-\xi}{L} = \frac{y}{M} = \frac{z}{-\xi Y}.$$

Eliminate ξ, and we have the hyperbolic paraboloid $\frac{L}{M} y^2 = xy + \frac{M}{Y} z$.

Also the coordinates (x, y, z) of the end of this axis are given by

$$x = \xi + \lambda L, \quad y = \lambda M, \quad z = -\lambda \xi Y,$$

where λ is some constant.

Eliminating ξ, we have a straight line as the locus of (x, y, z).]

9. Two wrenches of pitches p, p' have axes at a distance $2a$ from one another. If the resultant wrench is of pitch ϖ and its axis is equidistant from the axes of the component wrenches, shew that the angle between them is

$$\tan^{-1} \frac{a(2\varpi - p - p')}{a^2 - (\varpi - p)(\varpi - p')}.$$

10. On three given screws, whose axes are mutually perpendicular and concurrent, there act wrenches of pitches p_1, p_2, and p_3 whose resultant is on a screw of given pitch p. Shew that the locus of this latter screw is the hyperboloid

$$(p-p_1)x^2 + (p-p_2)y^2 + (p-p_3)z^2 + (p-p_1)(p-p_2)(p-p_3) = 0,$$

the axes of coordinates being the axes of the given screws.

11. Shew that a wrench, of which the force is R and the pitch is ϖc may be replaced by forces inclined at an angle 2θ to each other, the shortest distance between them being $2c$, and that their magnitudes are

$$\frac{R}{2}\left[\sqrt{1 + \varpi \tan\theta} \pm \sqrt{1 - \varpi \cot\theta}\right].$$

196. *The axes of two given wrenches intersect at right angles; their intensities are X and Y and their pitches are p_x and p_y; if the pitches are given, to find the locus of the central axis.*

Let OA be the direction of the resultant R of X and Y making an angle θ with Ox, the axis of the first wrench, so that

$$\frac{\cos\theta}{X}=\frac{\sin\theta}{Y}=\frac{1}{R}.$$

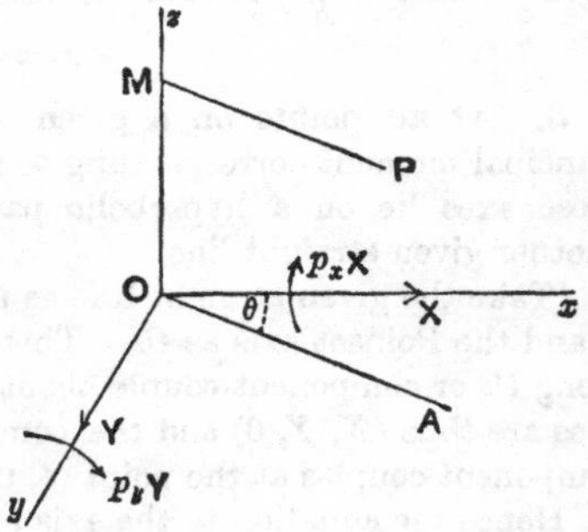

The resultant couple about OA

$$=p_x\,.\,X\cos\theta+p_y\,.\,Y\sin\theta$$
$$=(p_x\cos^2\theta+p_y\sin^2\theta)\,.\,R\ldots(1).$$

The resultant couple about a perpendicular to OA in the plane xOy

$$=-p_x\,.\,X\sin\theta+p_y\,.\,Y\cos\theta=(p_y-p_x)\,R\sin\theta\cos\theta.$$

The latter couple is equivalent to two parallel forces in the plane zOA and, by Art. 51, these and the force R along OA are equivalent to a force R along MP, which is parallel to OA and is such that

$$OM=(p_y-p_x)\sin\theta\cos\theta\ldots\ldots\ldots\ldots\ldots\ldots(2).$$

Transferring the axis of the couple (1) to MP, we then have a wrench, whose axis is MP and whose moment and force are respectively $(p_x\cos^2\theta+p_y\sin^2\theta)\,R$ and R, and whose pitch is therefore given by

$$p=p_x\cos^2\theta+p_y\sin^2\theta.$$

From (2), if (x, y, z) is any point on MP, then

$$z=(p_y-p_x)\frac{xy}{x^2+y^2},$$

i.e. MP lies on the surface

$$(x^2+y^2)\,z=(p_y-p_x)\,xy.$$

This surface is known as the Cylindroid. Also p_x and p_y are called its Principal Pitches.

The equation to this surface follows at once from Art. 190. For the equation to the central axis of the system is

$$\frac{p_x\,.\,X+zY}{X}=\frac{p_y\,.\,Y-zX}{Y}=\frac{-xY+yX}{0},$$

giving $$z\,(X^2+Y^2)=(p_y-p_x)\,XY,$$

and $$xY=yX.$$

Eliminating X, Y we have, as the locus of the central axis,

$$(x^2+y^2)z=(p_y-p_x)xy.$$

197. *Geometrical construction for the cylindroid*

$$(x^2+y^2)z=2axy.$$

In the plane of xy draw a circle of radius a touching the axis of x at the origin O.

Take OA, a radius vector of this circle, at any angle θ to Ox and draw AN perpendicular to Oy. Then

$$AN=a\sin 2\theta.$$

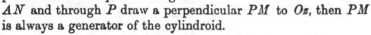

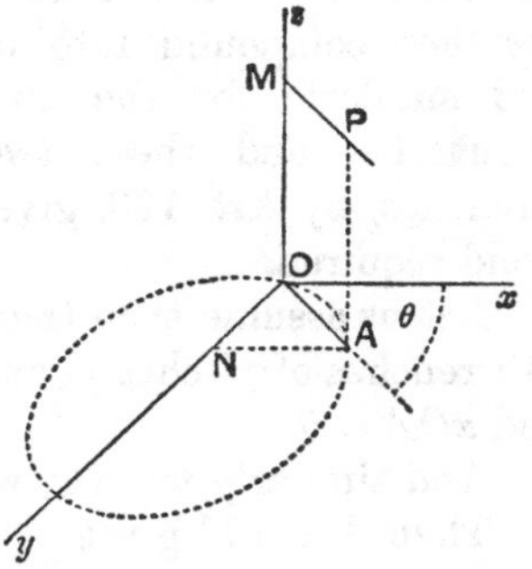

The plane $y=x\tan\theta$, through OA perpendicular to the plane xOy, cuts the cylindroid where

$$z=a\sin 2\theta=AN.$$

Hence, if at A we erect a perpendicular AP equal in length to AN and through P draw a perpendicular PM to Oz, then PM is always a generator of the cylindroid.

198. ***Any wrench may be resolved into two wrenches, whose axes intersect at right angles, in an infinite number of ways.***

For, starting with any wrench (R; pR) about MP as axis (Fig. Art. 196), on any line MO perpendicular to MP take any origin O and any two lines Ox and Oy perpendicular to OM.

Suppose $MO=a$ and $xOA=\theta$ to be given.

Then, as in Art. 196, wrenches (X; p_xX) about Ox and (Y; p_yY) about Oy are equivalent to (R; pR), where

$$p=p_x\cos^2\theta+p_y\sin^2\theta.$$

and $$a=(p_y-p_x)\cos\theta\sin\theta.$$

These give $p_x=p-a\tan\theta$, and $p_y=p+a\cot\theta$.

Also $X=R\cos\theta$ and $Y=R\sin\theta$.

Hence, R and p being given, we have the values of X, p_x and Y, p_y for any assumed values of a and θ.

199. ***Any two wrenches on given screws determine one, and only one, cylindroid.***

Let MP and NQ be the axes of the two wrenches, p and q their pitches, and α the angle and h the shortest distance, MN, between them.

Then if on MN we can find a point O and perpendicular lines Ox, Oy such that when we resolve the wrenches about Ox and Oy, the two component wrenches about Ox have the same pitch and also the two component wrenches about Oy, we have proved the theorem. For the two wrenches about Ox then compound into one wrench, and similarly for the two wrenches about Oy, and these two resultant wrenches, by Art. 196, give the cylindroid required.

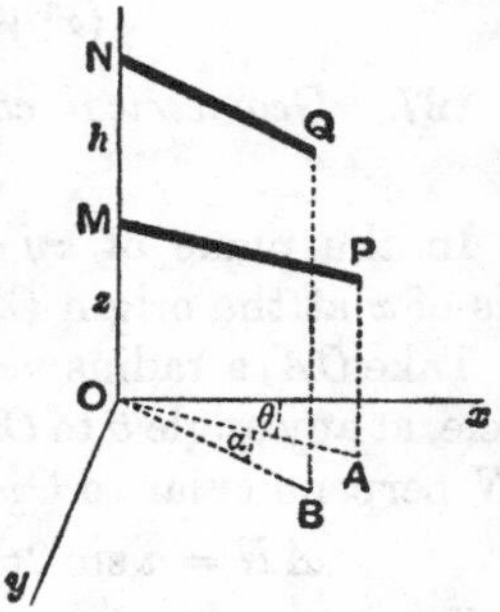

Let us assume then that the wrench about MP is equivalent to wrenches of pitches p_x and p_y about Ox and Oy, where $OM = z$ and $xOA = \theta$.

And similarly for NQ, where $xOB = \theta + \alpha$.

Then Art. 196 gives

$$p = p_x \cos^2\theta + p_y \sin^2\theta = \frac{p_x + p_y}{2} + \frac{p_x - p_y}{2}\cos 2\theta \quad \ldots(1),$$

$$z = (p_y - p_x)\cos\theta\sin\theta = \frac{p_y - p_x}{2}\sin 2\theta \quad \ldots\ldots(2),$$

$$q = p_x\cos^2(\theta+\alpha) + p_y\sin^2(\theta+\alpha)$$

$$= \frac{p_x + p_y}{2} + \frac{p_x - p_y}{2}\cos(2\theta + 2\alpha) \quad \ldots(3),$$

and $z + h = (p_y - p_x)\cos(\theta+\alpha)\sin(\theta+\alpha)$

$$= \frac{p_y - p_x}{2}\sin(2\theta + 2\alpha) \quad \ldots\ldots(4).$$

Hence we have

$$p + q = p_x + p_y + (p_x - p_y)\cos\alpha\cos(2\theta+\alpha) \quad \ldots\ldots(5),$$

$$p - q = \qquad (p_x - p_y)\sin\alpha\sin(2\theta+\alpha) \quad \ldots\ldots(6),$$

and $$h = \frac{p_y - p_x}{2}[\sin(2\theta+2\alpha) - \sin 2\theta]$$

$$= (p_y - p_x)\cos(2\theta+\alpha)\sin\alpha \quad \ldots\ldots(7).$$

(6) and (7) give

$$\tan(2\theta+\alpha) = \frac{q-p}{h} \quad \ldots\ldots(8).$$

From (5) and (6) we have

$$p_x + p_y = p + q - (p - q)\cot\alpha\cot(2\theta+\alpha) = p + q + h\cot\alpha,$$

and $p_y - p_x = (q - p)\operatorname{cosec}\alpha\operatorname{cosec}(2\theta+\alpha) = \sqrt{h^2 + (q-p)^2}\operatorname{cosec}\alpha.$

These equations determine p_x and p_y.

Also (2) gives

$$z = \frac{1}{2}\frac{q-p}{\sin\alpha\sin(2\theta+\alpha)}\sin 2\theta$$

$$= \frac{1}{2}\frac{q-p}{\sin\alpha}[\cos\alpha - \sin\alpha\cot(2\theta+\alpha)] = \frac{1}{2}[(q-p)\cot\alpha - h].$$

The values thus found uniquely determine the positions of Ox and Oy and the corresponding wrenches about them.

200. The quantities having been determined as in the previous article, if R and R' are the intensities of the wrenches about MP and NQ, these two wrenches are equivalent to forces

$$R\cos\theta + R'\cos(\theta+\alpha) \text{ along } Ox,$$
$$R\sin\theta + R'\sin(\theta+\alpha) \text{ along } Oy,$$

and to couples

$$p_x[R\cos\theta + R'\cos(\theta+\alpha)] \text{ about } Ox,$$

and

$$p_y[R\sin\theta + R'\sin(\theta+\alpha)] \text{ about } Oy.$$

These, by Art. 196, clearly compound into a wrench about a screw of the cylindroid

$$z(x^2+y^2) = (p_y - p_x)xy = xy\operatorname{cosec}\alpha\sqrt{h^2+(q-p)^2}.$$

Ex. A screw of pitch 1 in the line $y=2$, $z=\sqrt{3}$, and a screw of pitch 4 in the line $(y-2)\sqrt{3} = x-3$, $z=2\sqrt{3}$, determine a cylindroid; shew that its principal pitches are $4-2\sqrt{3}$ and $4+2\sqrt{3}$ and that the equations of its principal axes are $y-2=(\sqrt{3}\mp 2)(x-3)$.

201. *Resultant wrench of wrenches on screws of the same cylindroid.*

The principal pitches of the cylindroid being p_x and p_y, a wrench of intensity R and the proper pitch for an inclination θ to Ox is, by Art. 196, equivalent to

a wrench $(R\cos\theta;\ p_x R\cos\theta)$ about Ox,

and a wrench $(R\sin\theta;\ p_y R\sin\theta)$ about Oy.

Hence, if each of the given wrenches is replaced by wrenches about Ox and Oy, they are equivalent to a system of forces and couples given by

$$X = \Sigma(R\cos\theta),$$
$$Y = \Sigma(R\sin\theta),$$
$$L = \Sigma(p_x R\cos\theta) = p_x X,$$

and

$$M = \Sigma(p_y R\sin\theta) = p_y Y,$$

i.e. to a wrench $(X; p_x X)$ about Ox and a wrench $(Y; p_y Y)$ about Oy.

These are wrenches of the same pitch as the principal pitches of the cylindroid, and thus equivalent to a wrench $[S; qS]$ on a screw of the cylindroid inclined at ϕ to Ox, where

$$\tan\phi = \frac{Y}{X} = \frac{\Sigma(R\sin\theta)}{\Sigma(R\cos\theta)},$$

$$S = \sqrt{X^2+Y^2} = \sqrt{\{\Sigma(R\cos\theta)\}^2 + \{\Sigma(R\sin\theta)\}^2},$$

and $$q = p_x\cos^2\phi + p_y\sin^2\phi.$$

202. It follows from the previous work that wrenches on screws of the same cylindroid are in equilibrium if their forces, when transferred to the same point parallel to their original directions, are in equilibrium; for then X and Y are both zero.

In particular, three wrenches on screws of the same cylindroid are in equilibrium if the intensity of each is proportional to the sine of the angle between the other two.

203. *To shew that the work done by a wrench, of intensity R and pitch p about a given screw, when the body is given a small twist $\delta\omega$ about another screw of pitch p_1, is*

$$R\,.\,\delta\omega\,\{(p+p_1)\cos\theta - h\sin\theta\},$$

where θ is the angle between the axes of the two screws and h is the shortest distance between them.

Let Ox be the axis of the screw p_1, MP the axis of the wrench, OM the shortest distance h between them; let MK be parallel to Ox and ML be perpendicular to MK and OM.

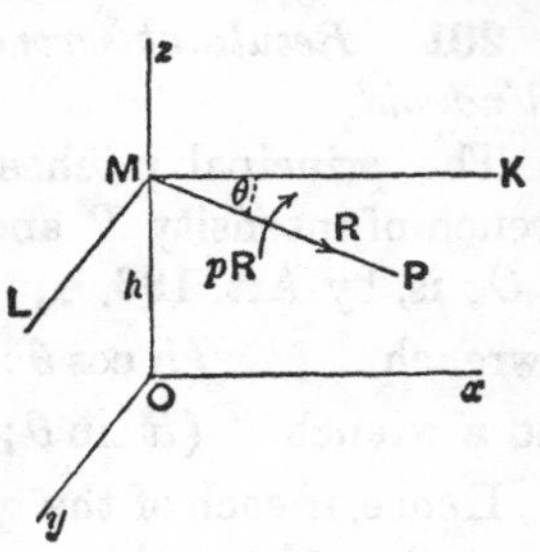

The force R is equivalent to $R\cos\theta$ along MK and $R\sin\theta$ along ML.

The component $R\cos\theta$ is equivalent to $R\cos\theta$ along Ox and a couple $R\cos\theta\,.\,h$ about Oy.

The component $R\sin\theta$ is equivalent to $R\sin\theta$ along Oy and a couple $-R\sin\theta\,.\,h$ about Ox.

The couple pR about MP is equivalent to couples $pR\cos\theta$ and $pR\sin\theta$ about MK and ML, and the axes of these may be removed to Ox and Oy.

The given wrench is thus equivalent to

a force $R \cos \theta$ along Ox,

a force $R \sin \theta$ along Oy,

a couple $R(p \cos \theta - h \sin \theta)$ about Ox,

and a couple $R(p \sin \theta + h \cos \theta)$ about Oy.

Now the displacement of the body consists of an angular displacement $\delta\omega$ about Ox, and a linear displacement $p_1 . \delta\omega$ along Ox.

Due to the angular displacement the work done by the couples is $R(p \cos \theta - h \sin \theta) . \delta\omega$ (Art. 97) and that by the forces is zero.

Due to the linear displacement the work done by the couples is zero and that done by the forces is $R \cos \theta . p_1 \delta\omega$.

Hence the total work done in the small displacement

$$= R\delta\omega \{(p + p_1) \cos \theta - h \sin \theta\}.$$

This work is not altered if p and p_1 be interchanged, so that the work done is the same as it would be if we had a wrench $(R ; p_1 R)$ about a screw Ox, and the body were displaced through the small angle $\delta\omega$ about a screw of pitch p on MP as axis.

204. From the principle of Virtual Work it follows that a body free to move only on a screw of axis Ox, and acted upon by a wrench on the screw MP, will be in equilibrium if

$$(p + p_1) \cos \theta - h \sin \theta = 0.$$

Screws which satisfy this condition are called Reciprocal Screws; they are therefore such that a body acted on by a wrench, of any intensity and the proper pitch, about either of them will be at rest if free to move only about the other.

It follows from the above condition that two screws whose axes intersect, so that $h = 0$, are reciprocal either if they are at right angles or if their pitches are equal and opposite.

205. *If a screw γ is reciprocal to each of two screws α and β, it is reciprocal to any screw δ on the cylindroid determined by α and β.*

For, by Art. 201, α, β, and δ being three screws on the same cylindroid, a wrench about δ is equivalent to two wrenches about α and β if the intensity of the wrench δ is equivalent to the two component intensities of the wrenches α and β. Replace the wrench δ by these two component wrenches about α and β.

Since the screws γ and α are reciprocal, the virtual work of the wrench about α for a displacement about γ is zero. Similarly for the component wrench about β.

Hence the total virtual work of the wrench δ for a displacement about γ is zero. Hence δ and γ are reciprocal.

206. NUL LINES AND PLANES. Suppose that, corresponding to any origin or base point O', the resultant force is R and the resultant couple is G. Take any line through O' perpendicular to the axis of G; then the sum of the moments of the forces of the system about this line is zero; for the axis of G has no component along it and R meets it.

For this reason the line is called a nul line, and its locus, which is the line perpendicular to the axis of G, is called the nul plane of O'. Also the point O' is called the nul point of the plane.

207. *To find the equation to the nul plane of a given point (f, g, h) referred to any axes Ox, Oy, Oz.*

Let X, Y, Z be the component forces along Ox, Oy, Oz, and L, M, N the component couples about them.

By Art. 190 the component couples about lines parallel to the axes through (f, g, h) are

$$L - gZ + hY, \quad M - hX + fZ, \quad N - fY + gX,$$

and these are proportional to the direction cosines of the axis of the resultant couple at (f, g, h), which is the normal to the nul plane there.

Hence the equation to the nul plane is

$$(x-f)(L-gZ+hY)+(y-g)(M-hX+fZ) \\ +(z-h)(N-fY+gX)=0,$$

i.e. $$x(L-gZ+hY)+y(M-hX+fZ) \\ +z(N-fY+gX)=fL+gM+hN \quad \ldots(1).$$

Conversely, if we want the nul point of the plane

$$lx+my+nz=1 \quad \ldots\ldots\ldots\ldots\ldots\ldots\ldots(2),$$

then on comparing it with (1) we have

$$\frac{L-gZ+hY}{l}=\frac{M-hX+fZ}{m}=\frac{N-fY+gX}{n}=fL+gM+hN.$$

Since the point (f, g, h) also must lie on the plane (2), we have, on solving,

$$\frac{f}{X-nM+mN}=\frac{g}{Y-lN+nL}=\frac{h}{Z-mL+lM}=\frac{1}{lX+mY+nZ},$$

giving the nul point of (2).

208. *Condition that the straight line*

$$\frac{x-f}{l}=\frac{y-g}{m}=\frac{z-h}{n}$$

may be a nul line for the same system of forces.

The component couples about lines through (f, g, h) parallel to the axes are

$$L-gZ+hY,\quad M-hX+fZ,\text{ and } N-fY+gX.$$

Hence the moment of the couple about the given line

$$=l(L-gZ+hY)+m(M-hX+fZ)+n(N-fY+gX),$$

and hence is zero if

$$X(mh-ng)+Y(nf-lh)+Z(lg-mf)=Ll+Mm+Nn,$$

i.e. if

$$\begin{vmatrix} X, & Y, & Z \\ l, & m, & n \\ f, & g, & h \end{vmatrix}=Ll+Mm+Nn.$$

This is therefore the condition that the given line may be a nul line of the system.

Ex. *Shew that among the nul lines of any system of forces four are generators of any hyperboloid, two belonging to one system of generators and two to the other system.*

Let the hyperboloid be $\frac{x^2}{a^2}+\frac{y^2}{b^2}-\frac{z^2}{c^2}=1$, and referred to its centre and axes let the system be given by $(X, Y, Z; L, M, N)$. Any generator is

$$\frac{x-a\cos\theta}{a\sin\theta}=\frac{y-b\sin\theta}{-b\cos\theta}=\frac{z}{c}.$$

By the previous article this is a nul line of the system if

$$X(-bc\sin\theta)+Yca\cos\theta+Zab=La\sin\theta-Mb\cos\theta+Nc,$$

i.e. if $\sin\theta\left(\frac{X}{a}+\frac{L}{bc}\right)-\cos\theta\left(\frac{Y}{b}+\frac{M}{ca}\right)=\frac{Z}{c}-\frac{N}{ab}$, which clearly gives two values of θ, in general. Hence two generators belonging to one system are nul lines. Similarly for the other system.

209. *To shew that a given system of forces may be replaced by two forces, one of which acts along a given line OA.*

With O as origin, or base-point, let R and G be the resultant force and couple. Through OA and R let a plane be drawn, and let it cut the plane of the resultant couple (*i.e.* the plane COD perpendicular to the axis of G) in OB. Resolve R into two forces, one, P_1, along OA, and the other, P_2, along OB.

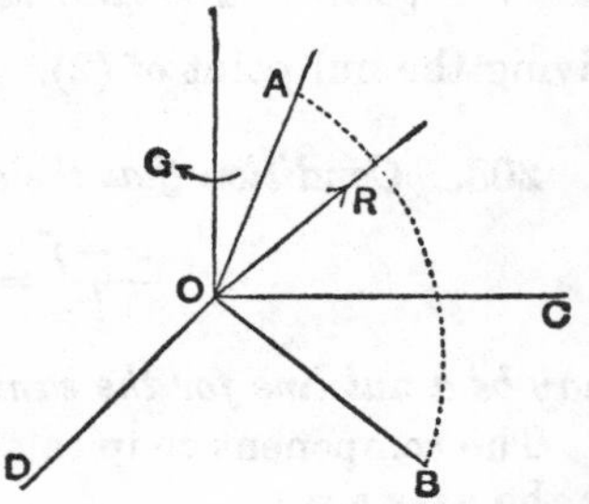

The force P_2 along OB, when compounded with the two forces in the plane BOC which form the resultant couple, will give a force P_2 in this plane which is parallel to OB.

It follows that the given system of forces is equivalent to some force P_1 acting along the given straight line OA, together with a second force P_2 which acts somewhere in the nul plane of O.

Such forces as P_1 and P_2 are called conjugate forces, and their lines of action are called conjugate lines.

Whatever point O we take on OA the force P_2 will still lie in its nul plane, so that, as O moves along OA, its null plane continually turns round so that it always passes through the line conjugate to OA. Hence the conjugate line of OA may be determined by taking any two convenient points on it, and obtaining the equations to their nul planes by Art. 207. The conjugate line is then the intersection of these two planes.

Thus suppose we require the conjugate line, with respect to the system of forces given by $(X, Y, Z;\ L, M, N)$, of the line

$$\frac{x-f}{l}=\frac{y-g}{m}=\frac{z-h}{n} \qquad \text{.....................(1)}.$$

By Art. 207 the nul plane of (f, g, h) is

$$x(L-gZ+hY)+y(M-hX+fZ)$$
$$+z(N-fY+gX)=Lf+Mg+Nh \text{...(2)}.$$

Another point on (1) is $\left(f-\frac{hl}{n},\ g-\frac{mh}{n},\ 0\right)$.

Its nul plane is

$$x\left[L-gZ+\frac{mhZ}{n}\right]+y\left[M+fZ-\frac{hlZ}{n}\right]$$
$$+z\left[N-fY+gX+\frac{hlY}{n}-\frac{mhX}{n}\right]$$
$$=L\left(f-\frac{hl}{n}\right)+M\left(g-\frac{mh}{n}\right)\ldots\ldots(3).$$

(2) and (3) give the required conjugate line.

Subtracting (2) from (3), we have

$$x[nY-mZ]+y[lZ-nX]+z[mX-lY]$$
$$=Ll+Mm+Nn\ldots(4).$$

(2) and (4) give the conjugate line more easily.

It is easily seen that (4) is the nul plane of the point at infinity which lies on (1). For the coordinates of this point are $(l\rho, m\rho, n\rho)$, where ρ is infinite.

EXAMPLES

1. Examine the case in which the straight line OA of Art. 209 is itself a nul line of the system.

2. A straight line is given by the equations $Ax+By+Cz=D$, $A'x+B'y+C'z=D'$; shew that its conjugate is given by equating to zero any two of the determinants

$$\left\|\begin{matrix} L', & M', & N', & Lx+My+Nz \\ A, & B, & C, & D \\ A', & B', & C', & D' \end{matrix}\right\|,$$

where L', M' N' are the component couples at the point (x, y, z) and L, M, N those at the origin.

3. A system of forces given by $(X, Y, Z; L, M, N)$ is replaced by two forces, one acting along the axis of x, and another force. Shew that the magnitudes of the forces are

$$\frac{LX+MY+NZ}{L} \quad\text{and}\quad \frac{[(MY+NZ)^2+L^2(Y^2+Z^2)]^{\frac{1}{2}}}{L}.$$

[Let P be the force along the axis of x; then the components of the other force must be $X-P$, Y, Z; let it act at the point $(f, g, 0)$. Since these two forces are equal to the given system we have, by Art. 162,

$$L=gZ,\quad M=-fZ\quad\text{and}\quad N=fY-g(X-P).$$

Hence the given results, and also the equation to the line of action of the second force.]

4. Shew that the wrench $(X, Y, Z; L, M, N)$ is equivalent to two forces, one along the line $x=y=z$, and the other along the line given by

$$Lx+My+Nz=0,$$
$$x(Y-Z)+y(Z-X)+z(X-Y)=L+M+N,$$

and find the magnitudes of the two forces.

5. Shew that in general two systems of forces have only one pair of conjugate lines in common.

6. Forces act along generators of the same system of a hyperboloid. Shew that two generators of the same system are nul lines of the system of forces.

7. Shew that the nul planes of a series of points which lie in a straight line AB pass through a second straight line CD; and that if the series of lines AB be generators of a hyperboloid, the lines CD will also be generators of a hyperboloid.

8. A system of forces is reduced to two forces one of which acts along an assigned line. Shew (i) that the four lines of action of two such pairs of forces are generators of the same system of a hyperboloid of one sheet; (ii) that lines meeting two such forces and the central axis generate a hyperbolic paraboloid, one set of whose generators is perpendicular to the central axis.

9. Referred to the same origin and axes of coordinates two systems of forces are given by $(X, Y, Z; L, M, N)$ and $(X', Y', Z'; L', M', N')$. Shew that, in general, a unique pair of lines can be found such that forces along them may be equivalent to either system; and prove that the shortest distance between them is

$$\sqrt{\Pi^2 - 4II'}\,(R^2R'^2 - K^2)^{-\frac{1}{2}},$$

where
$$\Pi = LX' + MY' + NZ' + L'X + M'Y + N'Z,$$
$$K = XX' + YY' + ZZ',$$

and I, I', R, R' have their usual significance.

[Using the equations of Art. 190 for the central axes of the two systems, we obtain for the shortest distance ξ, and the angle a, between these central axes, $\cos a = \frac{K}{RR'}$,

and $$\xi = \left(\Pi - \frac{IK}{R^2} - \frac{I'K}{R'^2}\right) \div \sqrt{R^2R'^2 - K^2} = \left[\frac{\Pi}{K} - \frac{I}{R^2} - \frac{I'}{R'^2}\right]\cot a.$$

Also, by Art. 192, if AC and BD be the pair of lines to be found, we see that AB must be perpendicular to each central axis, and hence must lie along the shortest distance ξ between them. The equations for θ and ϕ of Art. 192 become

$$\tan\theta = \frac{I}{R^2b};\ \tan\phi = \frac{I}{R^2a};\ \tan(\theta - a) = \frac{I'}{R'^2(b+\xi)};$$

and $$\tan(\phi + a) = \frac{I'}{R'^2(a-\xi)}.$$

Eliminating θ and ϕ from these equations, we see that a and $-b$ are the roots of the equation

$$y^2 - y\left[\frac{\Pi}{K} - \frac{2I}{R^2}\right]\cot a + \left[\frac{I^2I'^2}{R^2R'^2} - \frac{I\xi\cot a}{R^2}\right] = 0.$$

Hence the required distance = the difference of the roots of this equation = the given answer, on reduction.]

10. Shew that the nul lines which are common to three given systems of forces are generators of the same system of a hyperboloid of one sheet.

CHAPTER XII

MACHINES

210. IN the present chapter we shall explain and discuss the equilibrium of some of the simpler machines.

We shall suppose the different portions of these machines to be smooth and rigid, that all cords or strings used are perfectly flexible, and that the forces acting on the machines always balance, so that they are at rest. In actual practice these conditions are not even approximately satisfied in the cases of many machines.

A machine is always used in practice to overcome some resistance; the force we exert on the machine is the Power or Effort; the resistance to be overcome, in whatever form it may appear, is called the Weight or Resistance.

211. Mechanical Advantage. If in any machine an effort P balance a resistance W, the ratio $\frac{W}{P}$, *i.e.* $\frac{\text{Resistance}}{\text{Effort}}$, is called the Mechanical Advantage of the machine, so that

$$\text{Resistance} = \text{Effort} \times \text{Mechanical Advantage}.$$

The term Force-Ratio is sometimes used instead of Mechanical Advantage. Almost all machines are constructed so that the mechanical advantage is a ratio greater than unity.

Velocity Ratio. The velocity ratio of any machine is the ratio of the distance through which the point of application of the effort or "power" moves to the distance through which the point of application of the resistance, or "weight," moves in the same time; so that

$$\text{Velocity Ratio} = \frac{\text{Distance through which } P \text{ moves}}{\text{Distance through which } W \text{ moves}}.$$

If the machine be such that no work has to be done in lifting its component parts, and if it be perfectly smooth throughout, it will be found that the Mechanical Advantage and the Velocity Ratio are equal, so that in this case

$$\frac{W}{P} = \frac{\text{Distance through which } P \text{ moves}}{\text{Distance through which } W \text{ moves}},$$

and then $P \times$ distance through which P moves

$$= W \times \text{distance through which } W \text{ moves},$$

i.e. work done by $P =$ work done against W.

212. The following we shall thus find to be a universal principle, known as the **Principle of Work,** *viz.*, *Whatever be the machine we use, provided that there be no friction and that the weight of the machine be neglected, the work done by the effort is always equivalent to the work done against the weight, or resistance.*

It may be regarded as an extension of the Principle of Virtual Work, in which, instead of virtual displacements, we have actual finite displacements which are consistent with the geometrical relations of the machine.

Assuming that the machine we are using gives mechanical advantage, so that the effort is less than the weight, the distance moved through by the effort is therefore greater than the distance moved through by the weight in the same proportion. This is sometimes expressed in popular language in the form: *What is gained in power is lost in speed.*

More accurate is the statement that mechanical advantage is always gained at a proportionate diminution of speed. No work is ever gained by the use of a machine though mechanical advantage is generally obtained. Some work, in practice, is always lost by the use of any machine.

The uses of a machine are

(1) to enable a man to lift weights or overcome resistances much greater than he could deal with unaided, *e.g.* by the use of a system of pulleys, or a wheel and axle, or a screw-jack, etc.,

(2) to cause a motion imparted to one point to be changed into a more rapid motion at some other point, *e.g.* in the case of a bicycle,

(3) to enable a force to be applied at a more convenient point or in a more convenient manner, *e.g.* in the use of a poker to stir the fire, or in the lifting of a bucket of mortar by means

of a long rope passing over a pulley at the top of a building, the other end being pulled by a man standing on the ground.

213. The Lever. The Lever consists essentially of a rigid bar, straight or bent, which has one point fixed about which the rest of the lever can turn. This fixed point is called the Fulcrum, and the perpendicular distances between the fulcrum and the lines of action of the effort and the weight are called the arms of the lever.

CLASS I. Here the effort P and the weight W act on opposite sides of the fulcrum C.

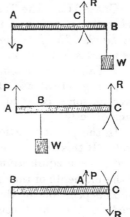

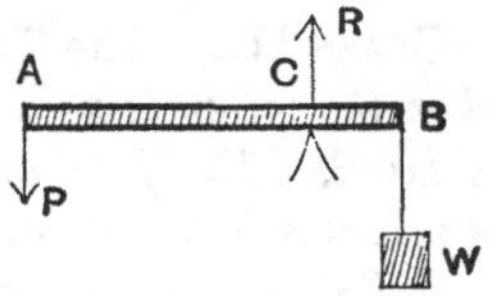

CLASS II. Here the effort P and the weight W act on the same side of the fulcrum C, but the former acts at a greater distance than the latter from the fulcrum.

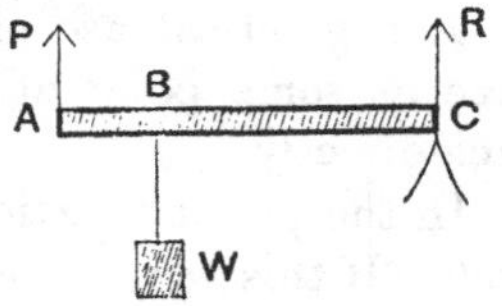

CLASS III. Here the effort P and the weight W act on the same side of the fulcrum C, but the former acts at a less distance than the latter from the fulcrum.

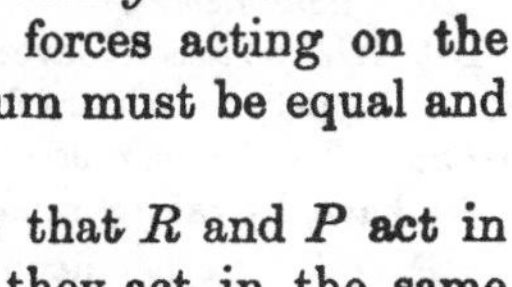

214. ***Conditions of equilibrium of a straight lever.***

In each case we have three parallel forces acting on the body, so that the reaction, R, at the fulcrum must be equal and opposite to the resultant of P and W.

In the first and third classes we see that R and P act in opposite directions; in the second class they act in the same direction. In all three classes, since the resultant of P and W passes through C, we have, as in Art. 31, $P \,.\, AC = W \,.\, BC$.

Since $\dfrac{W}{P} = \dfrac{AC}{BC} = \dfrac{\text{arm of } P}{\text{arm of } W}$, we observe that generally in Class I, and always in Class II, there is mechanical advantage, but that in Class III there is mechanical disadvantage.

215. Examples of the different classes of levers are:

CLASS I. A Poker (*when used to stir the fire, the bar of the grate being the fulcrum*); A Claw-hammer (*when used to extract nails*); A Crowbar (*when used with a point in it resting on a fixed support*); A Pair of Scales; The Brake of a Pump.

Double levers of this class are: A Pair of Scissors, A Pair of Pincers.

CLASS II. A Wheelbarrow; A Cork Squeezer; A Crowbar (*with one end in contact with the ground*); An Oar (*assuming the end of the oar in contact with the water to be at rest*).

A Pair of Nutcrackers is a double lever of this class.

CLASS III. The Treadle of a Lathe; The Human Forearm (*when the latter is used to support a weight placed on the palm of the hand. The fulcrum is the elbow, and the tension exerted by the muscles is the effort*).

A Pair of Sugar-tongs is a double lever of this class.

The practical use of levers of the latter class is to apply a force at some point at which it is not convenient to apply the force directly.

In the previous article we neglected the weight of the lever itself. If this weight be taken into consideration, we obtain the conditions of equilibrium by equating to zero the algebraic sum of the moments of the forces about the fulcrum C.

The principle of the lever was known to Archimedes who lived in the third century B.C.; until the discovery of the Parallelogram of Forces in the sixteenth century it was the fundamental principle of Statics.

216. *Bent Levers.*

Let AOB be a bent lever, of which O is the fulcrum, and let OL and OM be the perpendiculars from O upon the lines of action AC and BC of the effort P and resistance W.

We have, by taking moments about O,

$$\frac{P}{W}=\frac{OM}{OL}=\frac{\text{perpendicular from fulcrum on direction of resistance}}{\text{perpendicular from fulcrum on direction of effort}}.$$

To obtain the reaction at O let the directions of P and W meet in C. Since there are only three forces acting on the body, the direction of the reaction R at O must pass through C, and then, by Lami's Theorem, we have

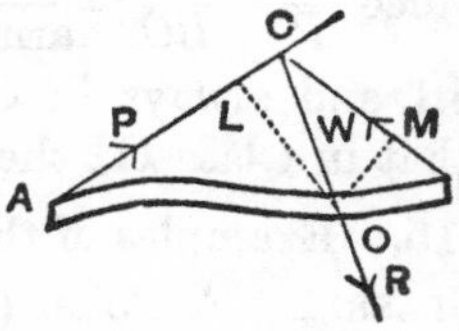

$$\frac{R}{\sin ACB}=\frac{P}{\sin BCO}=\frac{W}{\sin ACO}.$$

The reaction may also be obtained by resolving the forces R, P, and W in two directions at right angles.

In the above articles we have neglected any friction at the fulcrum; we have also assumed the forces acting on the lever to be in a plane which is perpendicular to the axis about which it can turn. If the forces act in any other directions the consideration of the equilibrium would be an example of Chap. X.

217. Pulleys. A pulley is composed of a wheel of wood, or metal, grooved along its circumference to receive a string or rope; it can turn freely about an axle passing through its centre perpendicular to its plane, the ends of this axle being supported by a frame of wood called the block. A pulley is said to be movable or fixed according as its block is movable or fixed.

The weight of the pulley is often so small, compared with the weights which it supports, that it may be neglected; such a pulley is called a weightless pulley. We shall always neglect the weight of the string or rope which passes round the pulley.

We shall also consider the pulley to be perfectly smooth, so that the tension of a string which passes round a pulley is constant throughout its length.

218. We shall discuss three systems of pulleys and shall follow the usual order; there is no particular reason for this order, but it is convenient to retain it for purposes of reference.

FIRST SYSTEM OF PULLEYS. *Each string attached to the supporting beam. To find the relation between the effort or "power" and the weight.*

In this system of pulleys the weight is attached to the lowest pulley, and the string passing round it has one end attached to the fixed beam, and the other end attached to the next highest pulley; the string passing round the latter pulley has one end attached to the fixed beam, and the other to the next pulley, and so on; the effort is applied to the free end of the last string.

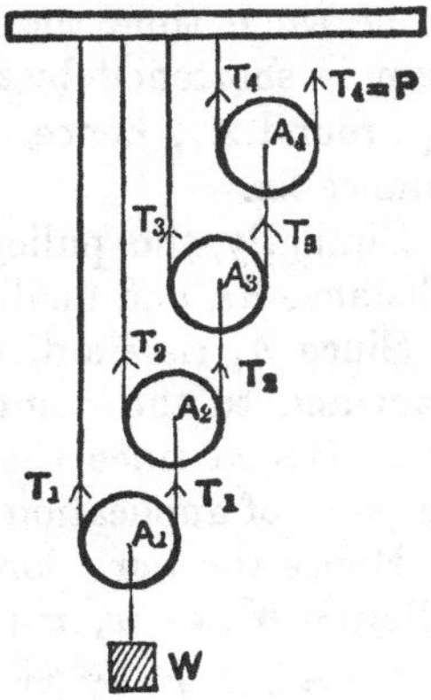

Often there is an additional fixed pulley over which the free end of the last string passes; the effort may then be applied as a downward force.

Let A_1, A_2, ... be the pulleys, beginning from the lowest, and let the tensions of the strings passing round them be T_1, T_2, ... Let W be the weight and P the effort.

For the equilibrium of the pulleys A_1, A_2, ... we have, if w_1, w_2, ... be their weights,

$$T_1 = \frac{W}{2} + \frac{w_1}{2}, \quad T_2 = \tfrac{1}{2}(T_1 + w_2) = \frac{W}{2^2} + \frac{w_1}{2^2} + \frac{w_2}{2},$$

$$T_3 = \tfrac{1}{2}(T_2 + w_3) = \frac{W}{2^3} + \frac{w_1}{2^3} + \frac{w_2}{2^2} + \frac{w_3}{2},$$

....

If there were n movable pulleys, we should have finally

$$P = T_n = \frac{W}{2^n} + \frac{w_1}{2^n} + \frac{w_2}{2^{n-1}} + \ldots + \frac{w_n}{2}.$$

$$\therefore \quad 2^n P = W + w_1 + 2w_2 + 2^2 w_3 + \ldots + 2^{n-1} w_n.$$

If the weight of each pulley is w, then

$$2^n P = W + w(1 + 2 + 2^2 + \ldots + 2^{n-1}) = W + w(2^n - 1).$$

It follows that the mechanical advantage, $\frac{W}{P}$, depends on the weight of the pulleys.

Let R be the stress on the beam. Since R, together with the force P, supports the system of pulleys, together with the weight W, we have

$$R + P = W + w_1 + w_2 + \ldots + w_n.$$

219. Verification of the Principle of Work. If the weight W be raised through a distance x, the pulley A_2 would, if the distance A_1A_2 remained unchanged, rise a distance x; but, at the same time, the length of the string joining A_1 to the beam is shortened by x, and a portion x of the string therefore slips round A_1; hence, altogether, the pulley A_2 rises through a distance $2x$.

Similarly, the pulley A_3 rises a distance $4x$, the pulley A_4 a distance $8x$, and finally the pulley A_n a distance $2^{n-1}x$.

Since A_n rises a distance $2^{n-1}x$, the strings joining it to the beam and to the point at which P is applied both shorten by 2^{n-1}. Hence, since the slack string runs round the pulley A_n, the point of application of P rises through $2^n x$.

Hence the work done on the weight and the weights of the pulleys $= W . x + w_1 . x + w_2 . 2x + w_3 . 4x + w_4 . 8x + \ldots + w_n . 2^{n-1}x$

$= 2^n x . P$, by the last article, $=$ work done by the effort.

220. Second system of pulleys. *The same string passing round all the pulleys. To find the relation between the effort and the weight.*

In this system there are two blocks, each containing pulleys the upper block being fixed and the lower block movable. The same string passes round all the pulleys as in the figures and is fastened to either the upper or the lower block.

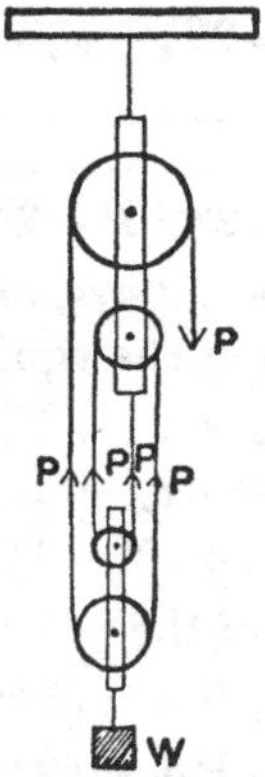

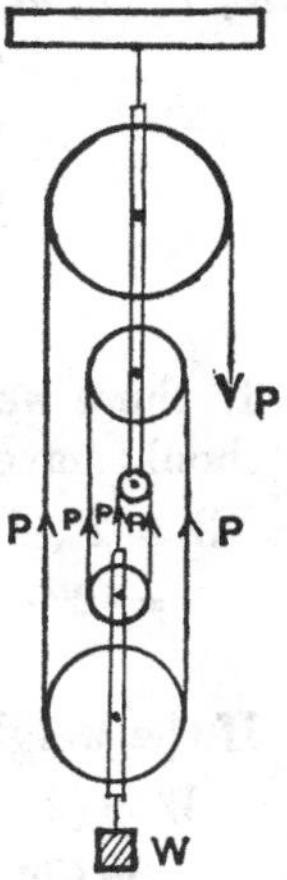

In either case, let n be the number of portions of string at the lower block. Since we have only one string passing over smooth pulleys, the tension of each of these portions is P, so that $nP = W + w$, where W is the weight supported and w is the weight of the lower block.

In practice the pulleys of each block are often placed parallel to one another, so that the strings are not mathematically parallel; they are, however, very approximately parallel, so that the above relation is still very approximately true.

221. Third system of pulleys. *All the strings attached to the weight. To find the relation between the effort and the weight.*

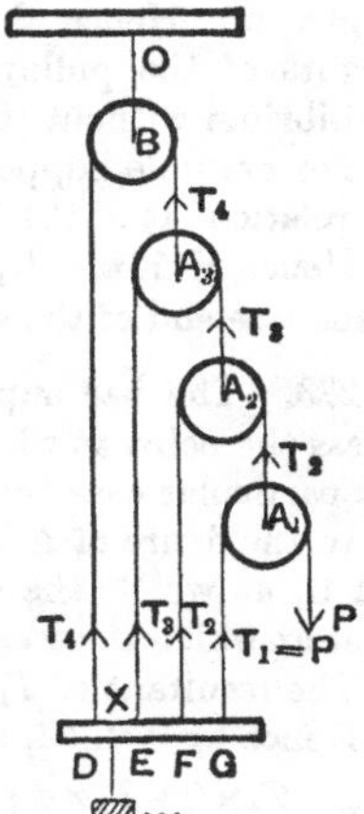

In this system the string passing round any pulley is attached at one end to a bar, from which the weight is suspended, and at the other end to the next lower pulley; the string round the lowest pulley is attached at one end to the bar, whilst at the other end of this string the power is applied. In this system the upper pulley is fixed.

Let A_1, A_2, A_3, ... be the movable pulleys, beginning from the lowest, and let the tensions of the strings passing round these pulleys respectively be T_1, T_2, T_3, ...

If the power be P, we have clearly $T_1 = P$.

Considering the equilibrium of the pulleys in order, we have, if w_1, w_2, ... be their weights,

$$T_2 = 2T_1 + w_1 = 2P + w_1,$$
$$T_3 = 2T_2 + w_2 = 2^2P + 2w_1 + w_2,$$
$$T_4 = 2T_3 + w_3 = 2^3P + 2^2w_1 + 2w_2 + w_3,$$
$$\ldots\ldots\ldots\ldots\ldots\ldots\ldots\ldots\ldots\ldots$$
$$T_n = 2^{n-1}P + 2^{n-2}w_1 + 2^{n-3}w_2 + \ldots + w_{n-1}.$$

If there were n pulleys, of which $(n-1)$ would be movable, we should have, from the equilibrium of the bar,

$$W = T_n + T_{n-1} + \ldots + T_2 + T_1$$
$$= (2^n - 1)P + (2^{n-1} - 1)w_1 + (2^{n-2} - 1)w_2 + \ldots$$
$$+ (2^2 - 1)w_{n-2} + (2 - 1)w_{n-1} \ldots\ldots\ldots(1).$$

If the weights of the pulleys are all equal to w, this gives

$$W = (2^n - 1)P + w[2^{n-1} + 2^{n-2} + \ldots + 2 - (n-1)]$$
$$= (2^n - 1)P + w[2^n - n - 1].$$

Stress on the supporting beam. This stress balances the effort, the weight, and the weight of the pulleys, and therefore equals

$$P + W + w_1 + w_2 + \ldots + w_n.$$

222. In this system we observe that, the greater the weight of each pulley, the less is P required to be in order that it may support a given weight W. Hence the weights of the pulleys assist the effort. If the weights of the pulleys be properly chosen, the system will remain in equilibrium without the application of any effort whatever.

For example, suppose we have three movable pulleys, each of weight w, the relation (1) of the last article will become $W = 15P + 11w$.

Hence, if $11w = W$, we have P zero, so that no power need be applied at the free end of the string to preserve equilibrium.

223. The bar supporting the weight W will not remain horizontal, unless the point at which the weight is attached be properly chosen. In any particular case the proper point of attachment can be easily found.

In the figure of Art. 221 let the distances between the points D, E, F, and G, at which the strings are attached, be successively a, and let the point at which the weight is attached be X.

The resultant of T_1, T_2, T_3, and T_4 must pass through X.

Hence by Art. 34, if the weights of the pulleys are neglected,

$$DX = \frac{T_4 \times 0 + T_3 \times a + T_2 \times 2a + T_1 \times 3a}{T_4 + T_3 + T_2 + T_1} = \frac{4P \cdot a + 2P \cdot 2a + P \cdot 3a}{8P + 4P + 2P + P} = \frac{11}{15}DE.$$

224. This system of pulleys was not however designed in order to lift weights. If it be used for that purpose it is soon found to be unworkable. Its use is to give a short strong pull. For example it is used on board a yacht to set up the back stay. In the figure of Art. 221, $DEFG$ is the deck of the yacht to which the strings are attached and there is no W. The strings to the pulleys A_1, A_2, A_3, A_4 are inclined to the vertical and the point O is at the top of the mast which is to be kept erect. The resistance in this case is the force at O necessary to keep the mast up, and the effort is applied as in the figure.

225. VERIFICATION OF THE PRINCIPLE OF WORK. Suppose the weight W to ascend through a space x. The string joining B to the bar shortens by x, and hence the pulley A_3 descends a distance x. Since the pulley A_3 descends x and the bar rises x, the string joining A_3 to the bar shortens by $2x$, and this portion slides over A_3; hence the pulley A_2 descends a distance equal to $2x$ together with the distance through which A_3 descends, *i.e.* A_2 descends a distance $2x + x$, or $3x$. Hence the string A_2F shortens by $4x$, which slips over the pulley A_2, so that the pulley A_1 descends a distance $4x$ together with the distance through which A_2 descends, *i.e.* $4x + 3x$, or $7x$. The first, second, third, ... $(n-1)$th movable pulleys therefore descend $x, 3x, 7x, \ldots (2^{n-1}-1)x$ and the point of application of P descends $(2^n - 1)x$, so that the velocity ratio is $2^n - 1$.

The work done by the effort and the weights of the pulleys [which in this case assist the power]

$$= P.(2^n - 1)x + w_1(2^{n-1} - 1)x + w_2(2^{n-2} - 1)x + \ldots$$
$$+ 3w_{n-2}x + w_{n-1}x$$

$$= x.W, \text{ by Art. 221}, = \text{work done on the weight } W.$$

226. The Wheel and Axle. This machine consists of a strong circular cylinder, or axle, terminating in two pivots, A and B, which can turn freely on fixed supports. To the cylinder is rigidly attached

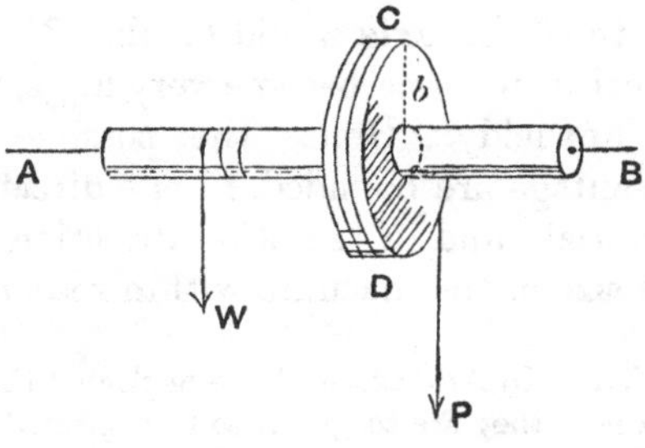

a wheel, CD, the plane of the wheel being perpendicular to the axle.

Round the axle is coiled a rope, one end of which is firmly attached to the axle, and the other end of which is attached to the weight. Round the circumference of the wheel, in a direction opposite to that of the first rope, is coiled a second rope, having one end firmly attached to the wheel, and having the "power," or effort, applied at its other end. The circumference of the wheel is grooved to prevent the rope from slipping off.

If a be the radius of the axle, and b be the radius of the wheel, the condition of equilibrium is, by taking moments about the fixed axis, $P \cdot b = W \cdot a$ ....(1).

Hence the mechanical advantage $= \dfrac{W}{P} = \dfrac{b}{a} = \dfrac{\text{radius of the wheel}}{\text{radius of the axle}}$.

VERIFICATION OF THE PRINCIPLE OF WORK. Let the machine turn through four right angles. A portion of string whose length is $2\pi b$ becomes unwound from the wheel, and hence P descends through this distance. At the same time a portion equal to $2\pi a$ becomes wound upon the axle, so that W rises through this distance. The work done by P is therefore $P \times 2\pi b$ and that done against W is $W \times 2\pi a$. These are equal by the relation (1). Also the velocity ratio (Art. 211)

$$= \frac{2\pi b}{2\pi a} = \frac{b}{a} = \text{the mechanical advantage.}$$

Theoretically, by making the quantity $\dfrac{b}{a}$ very large, we can make the mechanical advantage as great as we please; practically however there are limits. Since the pressure of the fixed supports on the axle must balance P and W, it follows that the thickness of the axle, *i.e.* $2a$, must not be reduced unduly, for then the axle would break. Neither can the radius of the wheel in practice become very large, for then the machine would be unwieldy. Hence the possible values of the mechanical advantage are bounded, in one direction by the strength of our materials, and in the other direction by the necessity of keeping the size of the machine within reasonable limits.

227. In Art. 226 we have neglected the thicknesses of the ropes. If, however, they are too great to be neglected, compared with the radii of the wheel and axle, we may take them into consideration by supposing the tensions of the ropes to act along their middle threads.

Suppose the radii of the ropes which pass round the axle and wheel to be x and y respectively; the distances from the line joining the pivots at which the tensions now act are $(a+x)$ and $(b+y)$ respectively. Hence the condition of equilibrium is $P(b+y)=W(a+x)$, so that

$$\frac{P}{W}=\frac{\text{sum of the radii of the axle and its rope}}{\text{sum of the radii of the wheel and its rope}}.$$

228. Other forms of the Wheel and Axle are the Windlass, used for drawing water from a well, and Capstan, used on board ship. In these machines the effort instead of being applied, as in Art. 226, by means of a rope passing round a cylinder, is applied at the ends of a spoke, or spokes, which are inserted in a plane perpendicular to the axle.

In the Windlass the axle is horizontal, and in the Capstan it is vertical. In the latter case the resistance consists of the tension T of the rope round the axle, and the effort consists of the forces applied at the ends of bars inserted into sockets at the point A of the axle. The advantage of pairs of arms is that the strain on the bearings of the capstan is thereby much diminished or destroyed. The condition of equilibrium may be obtained as in Art. 226.

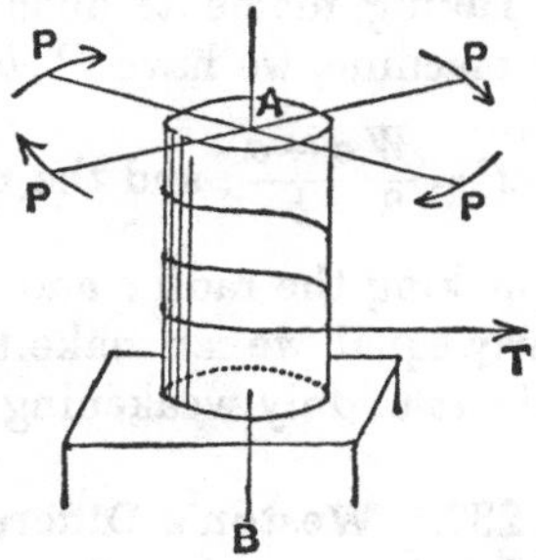

229. Differential Wheel and Axle. A slightly modified form of the ordinary wheel and axle is the differential wheel and axle. In this machine the axle consists of two cylinders, having a common axis, joined at their ends, the radii of the two cylinders being different. One end of the rope is wound round one of these cylinders, and its other end is wound in a contrary direction round the other cylinder. Upon the slack portion of the rope is slung a pulley to which the weight is attached. The part of the rope which passes round the smaller cylinder tends to turn the machine in the same direction as the effort.

As before, let b be the radius of the wheel and let a and c be the radii of the portions AC and CB of the axle, a being the smaller. Since the pulley is smooth, the tension T of the string round it is the same throughout its length, and hence, for the equilibrium of the weight, we have $T = \frac{1}{2}W$.

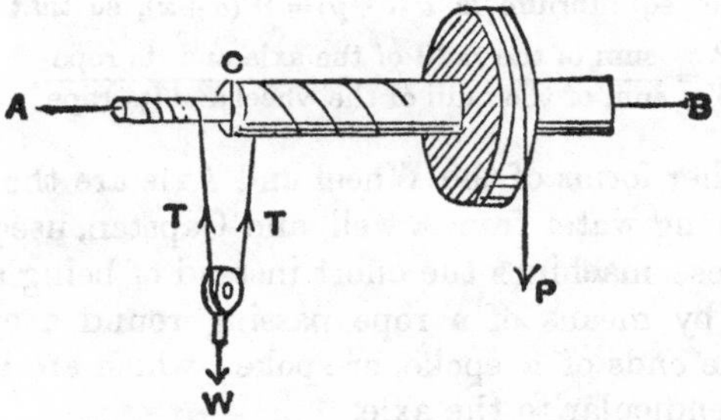

Taking moments about the line AB for the equilibrium of the machine, we have $P \cdot b + T \cdot a = T \cdot c$.

$$\therefore P = \frac{W}{2}\frac{c-a}{b}, \text{ and the mechanical advantage} = \frac{W}{P} = \frac{2b}{c-a}.$$

By making the radii c and a of the two portions of the axle very nearly equal, we can make the mechanical advantage very great without unduly weakening the machine.

230. Weston's Differential Pulley.

In this machine there are two blocks; the upper contains two pulleys of nearly the same size which turn together as one pulley; the lower consists of one pulley to which the weight W is attached.

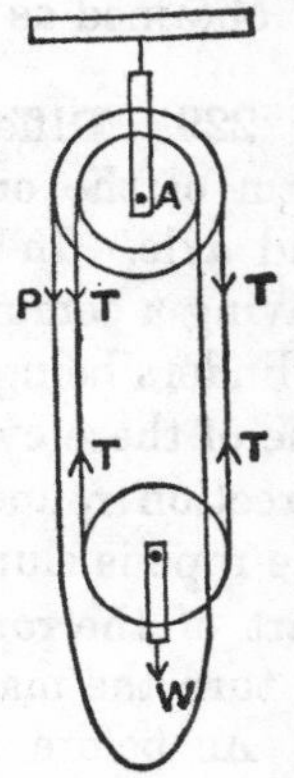

The figure represents a section of the machine. An endless chain passes round the larger of the upper pulleys, then round the lower pulley and the smaller of the upper pulleys; the remainder of the chain hangs slack and is joined on to the first portion of the chain. The effort P is applied as in the figure. The chain is prevented from slipping by small projections on the surfaces of the upper pulleys, or by depressions in the pulleys into which the links of the chain fit.

If T be the tension of the portions of the chain which support the weight W, we have, since these portions are approximately vertical, on neglecting the weight of the chain and the lower pulley, $2T = W$ ....(1).

If R and r be the radii of the larger and smaller pulleys of the upper block, we have, by taking moments about the centre A of the upper block,

$$P . R + T . r = T . R.$$

Hence

$$P = \frac{W}{2}\frac{R-r}{R}, \text{ and the mechanical advantage} = \frac{W}{P} = \frac{2R}{R-r}.$$

Since R and r are nearly equal, this mechanical advantage is therefore very great.

The differential pulley-block avoids one great disadvantage of the differential wheel and axle. In the latter machine a very great amount of rope is required in order to raise the weight through an appreciable distance.

231. *Wheel and Axle with the pivot resting on rough bearings.*

Let the central circle represent the pivots A or B of Fig., Art. 226 (much magnified) when looked at endways.

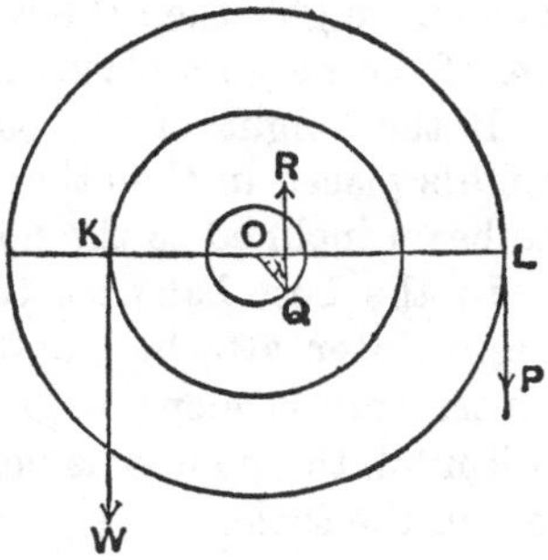

The resultant action between these pivots and the bearings on which they rest must be vertical, since it balances P and W. Also it must make an angle λ, the angle of friction, with the normal at the point of contact Q, if we assume that P is just on the point of overcoming W. Hence Q cannot be at the lowest point of the pivot, but must be as denoted in the figure, where OQ makes an angle λ with the vertical. The resultant reaction at Q is thus vertical.

Since R balances P and W, $\therefore R = P + W$ ....(1).

Also, by taking moments about O, we have

$$P . b - R . c \sin\lambda = W . a \text{.................(2)},$$

where c is the radius of the pivot and b, a the radii of the wheel and the axle (as in Art. 226).

Hence $$P = W\frac{a + c\sin\lambda}{b - c\sin\lambda}.$$

If P be only just sufficient to support W, *i.e.* if the machine be on the point of motion in the direction ↻, then, by changing the sign of λ, we have $P_1 = W\dfrac{a - c\sin\lambda}{b + c\sin\lambda}$.

In this case the point of contact Q is on the left of the vertical through O.

232. The Common Balance. The Common Balance consists of a rigid beam AB, carrying a scale-pan suspended from each end, which can turn freely about a fulcrum O outside the beam. The fulcrum and the beam are rigidly connected and, if the balance be well constructed, at the point O is a hard steel wedge, whose edge is turned downward and which rests on a small plate of agate.

The body to be weighed is placed in one scale-pan and in the other are placed weights, whose magnitudes are known; these weights are adjusted until the beam of the balance rests in a horizontal position. If OH be perpendicular to the beam, and the arms HA and HB be of equal length, and if the centre of gravity of the beam lie in the line OH, and the scale-pans be of equal weight, then the weight of the body is the same as the sum of the weights placed in the other scale-pan.

If the weight of the body be not equal to the sum of the weights placed in the other scale-pan, the balance will rest with the beam inclined to the horizon.

In the best balances the beam is usually provided with a long pointer attached to the beam at H. The end of this pointer travels along a graduated scale and, when the beam is horizontal, the pointer is vertical and points to the zero graduation on the scale.

233. *To find the position of equilibrium of a balance when the weights placed in the scale-pans are not equal.*

Let the weights placed in the scale-pans be P and W, the former being the greater; let S be the weight of each scale-pan, and let the weight of the beam (and the parts rigidly connected with it) be W', acting at a point K on OH.

[*The figure is drawn out of proportion so that the points may be distinctly marked; K is actually very near the beam.*]

When in equilibrium let the beam be inclined at an angle θ to the horizontal and let $OH = h$, $OK = k$ and $AH = HB = a$. Let the horizontal through O meet the verticals through A, K and B in L, G and M.

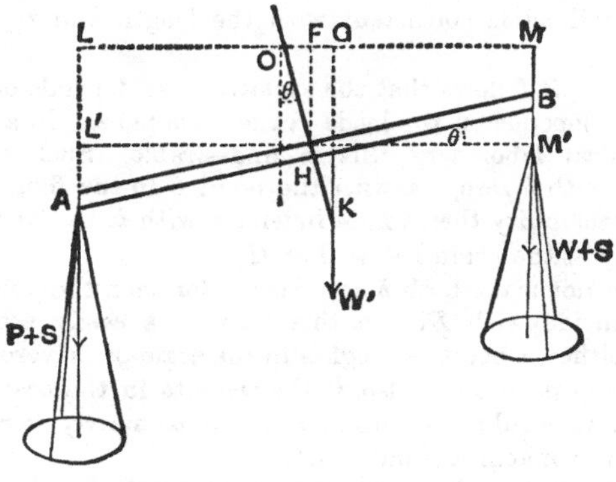

Taking moments about O, we have

$$(P+S)\,OL = (W+S)\,OM + W'.OG,$$

i.e. $(P+S)(a\cos\theta - h\sin\theta) = (W+S)(a\cos\theta + h\sin\theta) + W'.k\sin\theta.$

$$\therefore \tan\theta = \frac{(P-W)\,a}{W'k + (P+W+2S)\,h}.$$

234. *Requisites of a good balance.*

(1) The balance must be true. This will be the case if the arms of the balance be equal, if the weights of the scale-pans be equal, and if the centre of gravity of the beam be on the line through the fulcrum perpendicular to the beam; for the beam will now be horizontal when equal weights are placed in the scale-pans.

To test whether the balance is true, first see if the beam is horizontal when the scale-pan is empty, and then make the beam horizontal by putting sufficient weights in one scale-pan to balance the weight of a body placed in the other; now interchange the body and the weights; if they still balance one another, the balance must be true; if in the second case the beam assumes any position inclined to the vertical, the balance is not true.

(2) The balance must be sensitive, *i.e.* the beam must, for any difference, however small, between the weights in the scale-pans, be inclined at an appreciable angle to the horizon.

For a *given* difference between P and W, the greater the inclination of the beam to the horizon the more sensitive is the balance; also the less the difference, $P-W$, between the weights required to produce a given inclination θ, the greater is the sensitiveness of the balance.

The sensitiveness is therefore appropriately measured by

$$\frac{\tan\theta}{P-W},\ \text{i.e. by}\ \frac{a}{W'k+(P+W+2S)h}.\quad \text{(Art 233.)}$$

Thus the sensitiveness of a balance will be great if the arm a be fairly long in comparison with the distances h and k and the weight W' of the beam be as small as is consistent with the length and rigidity of the machine.

If h is not zero, it follows that the sensitiveness depends on the values of P and W, *i.e.* depends on the loads in the scale-pans. In a balance for use in a chemical laboratory this is undesirable. Such balances are therefore made with h zero, *i.e.* with the point O in the figure coinciding with H. The sensibility then varies inversely with k, the distance of the centre of gravity of the beam below O or H.

But we must not make both h and k zero; for then the points O and K would both coincide with H. In this case it is easily seen that the balance would either, when the weights in the scale-pans were equal, be in equilibrium in any position or else, if the weights in the scale-pans were not equal, that it would take up a position as nearly vertical as the mechanism of the machine would allow.

(3) The balance must be stable and must quickly take up its position of equilibrium.

The determination of the time taken by the machine to take up its position of equilibrium is essentially a dynamical question. We may however assume that this condition is best satisfied when the moment of the forces about the fulcrum O is greatest. When the weights in the scale-pans are each P, the moment of the forces tending to restore equilibrium

$$=(P+S)(a\cos\theta+h\sin\theta)-(P+S)(a\cos\theta-h\sin\theta)+W'.k\sin\theta$$

$$=[2(P+S)h+W'.k]\sin\theta.$$

This expression is greatest when h and k are greatest.

Since the balance is most sensitive when h and k are small, and most stable when these quantities are large, we see that in any balance great sensitiveness and quick weighing are to a certain extent incompatible. In practice this is not very important; for in balances where great sensitiveness is required (such as balances used in a laboratory) we can afford to sacrifice quickness of weighing; the opposite is the case when the balance is used for ordinary commercial purposes. To ensure as much as possible both the qualities of sensitiveness and quick weighing, the balance should be made with fairly light long arms, and at the same time the distance of the fulcrum from the beam should be considerable.

235. By the method of double weighing the weight of a body may be accurately determined even if the balance be not accurate.

Place the body to be weighed in one scale-pan and in the other pan put sand, or other suitable material, sufficient to balance the body. Next remove the body, and in its place put known weights sufficient to again balance the sand. The weight of the body is now clearly equal to the sum of the weights.

This method is used even in the case of extremely good machines when very great accuracy is desired. It is known as Borda's Method.

236. The Steelyards. The Common, or Roman, Steelyard is a machine for weighing bodies and consists of a bar, AB, movable about a fixed fulcrum at a point C.

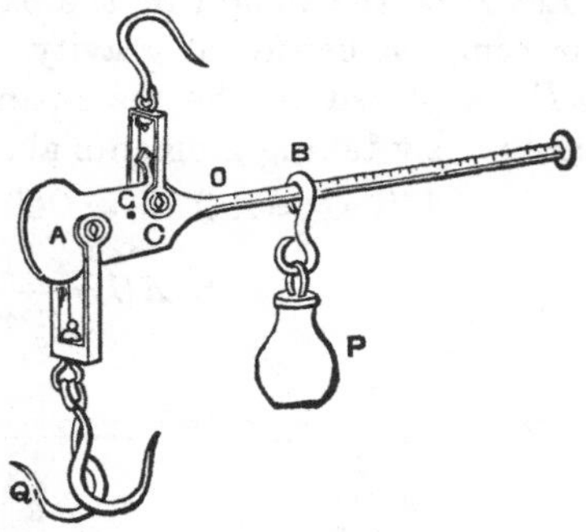

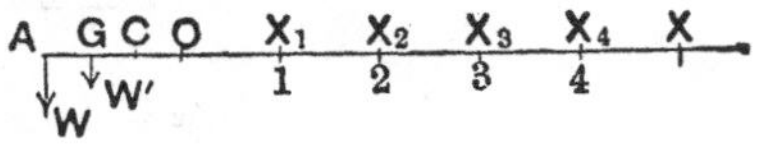

At the point A is attached a hook or scale-pan to carry the body to be weighed, and on the arm CB slides a movable weight P. The point at which P must be placed, in order that the beam may rest in a horizontal position, determines the weight of the body in the scale-pan. The arm CB has numbers engraved on it at different points of its length, so that the graduation at which the weight P rests gives the weight of the body.

Let W' be the weight of the steelyard and the scale-pan, and G be the point of the beam through which W' acts. The beam is usually constructed so that G lies in the shorter arm AC.

When there is no weight in the scale-pan, let O be the point in CB at which the movable weight P must be placed to balance W'. Moments about C give $W'.GC = P.CO$ ....(i).

This condition determines the position of the point O which is the zero of graduation.

When the weight in the scale-pan is $W (= nP)$, let X_n be the point at which P must be placed. Taking moments, we have

$$nP.CA + W'.GC = P.CX_n \text{.............(ii)}.$$

From (i) and (ii), $OX_n = n.CA$.

Hence, to graduate the steelyard, we must mark off from O successive distances CA, $2CA$, $3CA$, ... and at their extremities put the figures 1, 2, 3, ... The intermediate spaces can be subdivided to shew fractions of P lbs.

237. The Danish Steelyard consists of a bar AB, terminating in a heavy knob, or ball, B. At A is attached a hook or scale-pan to carry the body to be weighed.

The weight of the body is determined by observing about what point of the bar the machine balances. This is often done by having a loop of string, which can slide along the bar, and finding where the loop must be to give equilibrium.

Let P be the weight of the bar and scale-pan, and let G be their common centre of gravity. When a body of weight W $(=nP)$ is placed in the scale-pan, let C be the position of the fulcrum. By taking moments about C, we have

$$AC \,.\, nP = AC \,.\, W = CG \,.\, P = (AG - AC) \,.\, P.$$

$$\therefore\ AC = \frac{1}{n+1} \,.\, AG.$$

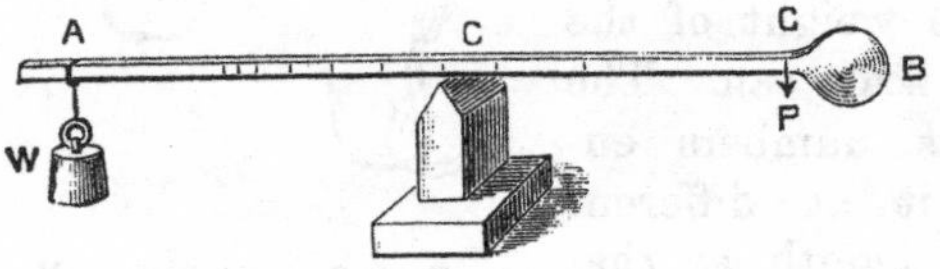

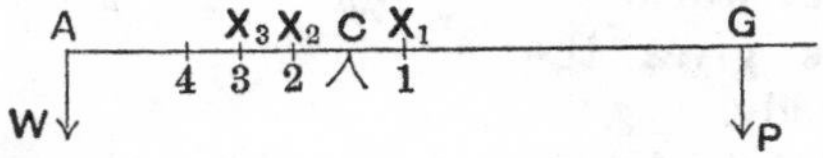

Hence, to graduate this steelyard, we must mark off from A distances equal to $\frac{1}{2}AG$, $\frac{1}{3}AG$, $\frac{1}{4}AG$, ... and at their extremities put the figures 1, 2, 3, ...

Similarly, for fractional weights, take distances equal to $\frac{2}{3}AG$, $\frac{3}{4}AG$, $\frac{4}{5}AG$, ..., and at the ends mark $\frac{1}{2}$, $\frac{1}{3}$, $\frac{1}{4}$, ...

The point G is easily determined since it is the position of the fulcrum when the steelyard balances without any weight in the scale-pan.

238. The Screw. A Screw consists of a cylinder of metal round the outside of which runs a protuberant thread of metal.

Let $ABCD$ be a solid cylinder, and let $EFGH$ be a rectangle, whose base EF is equal to the circumference of the solid cylinder. On EH and FG take points $L, N, Q, \ldots$ and $K, M, P, \ldots$ such that $EL, LN, \ldots FK, KM, MP, \ldots$ are all equal, and join $EK, LM, NP, \ldots$.

Wrap the rectangle round the cylinder, so that the point E coincides with A and EH with the line AD. The point F will coincide with E at A and the lines EK, LM, NP, ... will now become a continuous spiral line on the surface of the cylinder and, if we imagine the metal along this spiral line to become protuberant, we shall have the thread of a screw.

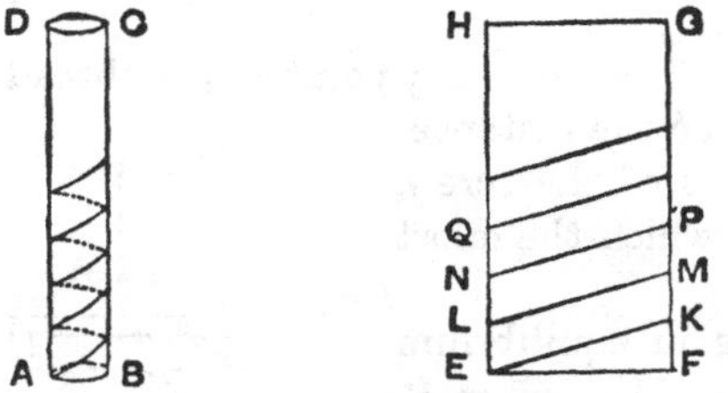

It is evident that the thread is an inclined plane running round the cylinder, and that its inclination to the horizon is the same everywhere and equal to the angle KEF. This angle is often called the angle of the screw, and the distance between two consecutive threads, measured parallel to the axis, is often called the pitch of the screw. The pitch is also defined by other writers as the distance advanced along the axis of the screw for an angular turn equal to unit angle, so that the pitch $= \frac{KF}{2\pi}$ with this definition.

Also tan (angle of screw) $= \frac{FK}{EF}$

$$= \frac{\text{distance between successive threads}}{\text{circumference of a circle whose radius is the distance from the axis of any point of the screw.}}$$

The section of the thread of the screw has, in practice, various shapes. The only kind that we shall consider has the section rectangular.

239. The screw usually works in a fixed support, along the inside of which is cut out a hollow of the same shape as the thread of the screw, and along which the thread slides. The only movement admissible to the screw is to revolve about its axis, and at the same time to move in a direction parallel to its length. If the screw were placed in an upright position, and a weight placed on its top, the screw would, in general, revolve and descend. Hence, if the screw is to remain in equilibrium, some force must act on it; this force is usually applied at one end of a horizontal arm, the other end of which is rigidly attached to the screw.

240. *In a smooth screw, to find the relation between the effort and the weight.*

Let a be the distance of any point on the thread of the screw from its axis, and b the distance, AB, from the axis of the screw, of the point at which the effort is applied.

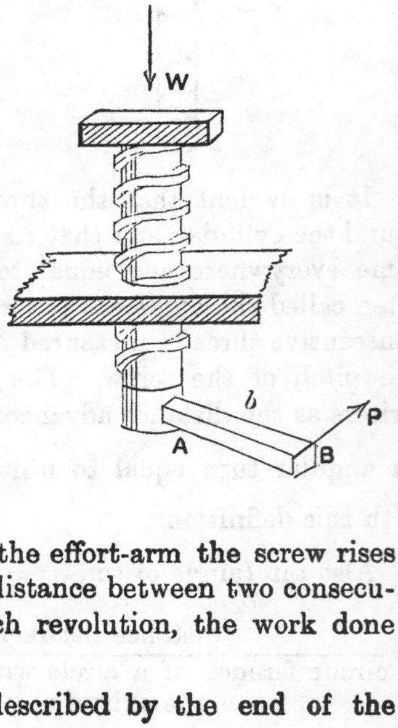

The screw is in equilibrium under the action of the effort P, the weight W, and the reactions at the points in which the fixed block touches the thread of the screw. Let R, S, T, ... denote the reactions of the block at different points of the thread of the screw. These will all be perpendicular to the thread of the screw, since it is smooth, and therefore do no work as the screw revolves.

For each revolution made by the effort-arm the screw rises through a distance equal to the distance between two consecutive threads. Hence, during each revolution, the work done by the effort is

$P \times$ circumference of the circle described by the end of the effort-arm,

and that done against the weight is

$W \times$ distance between two consecutive threads.

These are equal by the Principle of Work, and hence the

$$\text{mechanical advantage} = \frac{W}{P} = \frac{2\pi b}{2\pi a \tan \alpha}$$

$$= \frac{\text{circumference of a circle whose radius is the effort-arm}}{\text{distance between consecutive threads of the screw}}.$$

241. Equilibrium of a rough screw. *To find the relation between the effort and the resistance in the case of a screw, when friction is taken into account.*

With the same notation as in Art. 240, let the screw be on the point of motion *downwards*, so that the friction acts *upwards* along the thread.

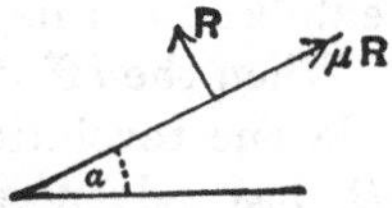

The vertical pressures of the block are $R(\cos\alpha + \mu\sin\alpha)$, $S(\cos\alpha + \mu\sin\alpha)$, ... and the horizontal components of these pressures are $R(\sin\alpha - \mu\cos\alpha)$, $S(\sin\alpha - \mu\cos\alpha)$, ...

By resolving vertically, and taking moments about the axis of the screw, we have

$$W = (R + S + T + \ldots)(\cos\alpha + \mu\sin\alpha) \quad \ldots\ldots\ldots(1),$$

and $$P \, . \, b = a(R + S + T + \ldots)(\sin\alpha - \mu\cos\alpha) \quad \ldots\ldots(2).$$

Hence $$\frac{P \, . \, b}{W} = a\frac{\sin\alpha - \mu\cos\alpha}{\cos\alpha + \mu\sin\alpha} = a\frac{\sin(\alpha - \lambda)}{\cos(\alpha - \lambda)}.$$

$$\therefore \ \frac{P}{W} = \frac{a}{b}\tan(\alpha - \lambda).$$

Similarly, if the screw be on the point of motion upwards, we have, by changing the sign of μ,

$$\frac{P_1}{W} = \frac{a}{b}\frac{\sin\alpha + \mu\cos\alpha}{\cos\alpha - \mu\sin\alpha} = \frac{a}{b}\tan(\alpha + \lambda).$$

If the effort have any value between P and P_1, the screw will be in equilibrium, but the friction will not be limiting friction. It will be noted that if the angle α of the screw be equal to the angle of friction, λ, then the value of the effort P is zero. In this case the screw will just remain in equilibrium supported only by the friction along the thread of the screw. If $\alpha < \lambda$, P will be negative, *i.e.* the screw will not descend unless it is forced down.

242. Theoretically, the mechanical advantage in the case of the screw can be made as large as we please, by decreasing sufficiently the distance between the threads of the screw. In practice, however, this is impossible; for, if we diminish the distance between the threads to too small a quantity, the threads themselves would not be sufficiently strong to bear the strain put upon them.

By means of **Hunter's Differential Screw** this difficulty

may be overcome. In this machine we have a screw AD working in a fixed block. The inside of the screw AD is hollow and is grooved to admit a smaller screw DE. The screw DE is fastened at E to a block, so that it cannot rotate, but can only move in the direction of its length.

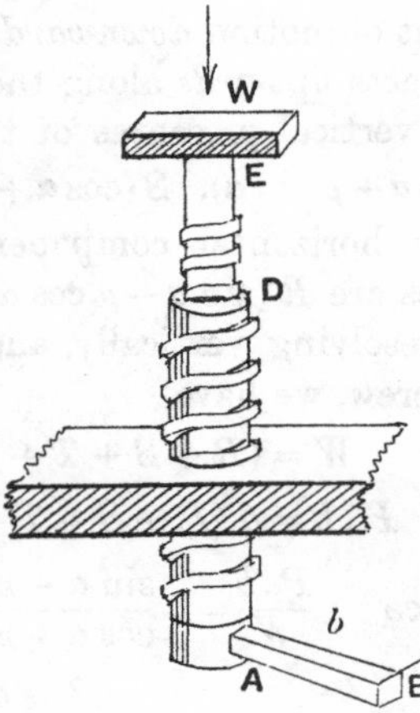

When the effort-arm AB has made one revolution, the screw AD has advanced a distance equal to the distance between two consecutive threads, and at the same time the smaller screw goes into DA a distance equal to the distance between two consecutive threads of the smaller screw. Hence the smaller screw, and therefore also the weight, advances a distance equal to the *difference* of these two distances.

Hence, just as in Art. 240, if the screw be smooth we have by the Principle of Work

$$\frac{W}{P}=\frac{2\pi b}{2\pi a\tan\alpha-2\pi a'\tan\alpha'}$$

$$=\frac{\text{circum. of the circle described by the end of the power-arm}}{\text{difference of the distances between consecutive threads of the two screws.}}$$

By making the distances between consecutive threads of the two screws nearly equal, we can make the mechanical advantage very great without weakening the machine.

243. The Wedge is a piece of iron, or metal, which has two plane faces meeting in a sharp edge. It is used to split wood or other tough substances, its edge being forced in by repeated blows applied by a hammer to its upper surface.

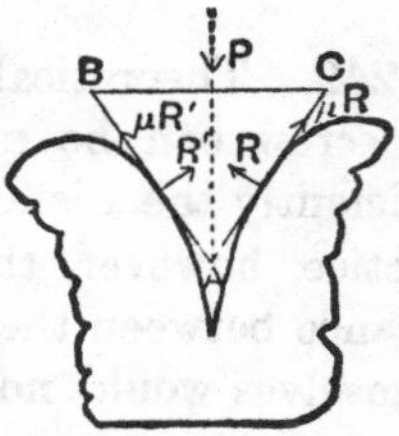

The problem of the action of a wedge is essentially a dynamical one. We shall only consider the statical problem when the wedge is just kept in equilibrium by a steady force applied to its upper surface.

Let ABC be a section of the wedge and let its faces be equally inclined to the base BC. Let the angle CAB be α.

Let P be the force applied to the upper face, R and R' the normal reactions of the wood at the points where the wedge touches the wood, and μR and $\mu R'$ the frictions, it being assumed that the wedge is on the point of being pushed in. Let P be applied at the middle point of BC in a direction perpendicular to BC, and let the weight of the wedge be negligible in comparison with P.

Resolving along and perpendicular to BC, we have

$$\mu R \sin\frac{\alpha}{2} - R\cos\frac{\alpha}{2} = \mu R' \sin\frac{\alpha}{2} - R'\cos\frac{\alpha}{2} \quad \ldots\ldots(1),$$

and
$$P = \mu(R+R')\cos\frac{\alpha}{2} + (R+R')\sin\frac{\alpha}{2} \quad \ldots\ldots\ldots(2).$$

Hence $R = R'$, and $\dfrac{2R}{P} = \dfrac{1}{\mu\cos\frac{\alpha}{2} + \sin\frac{\alpha}{2}} = \dfrac{\cos\lambda}{\sin\left(\frac{\alpha}{2}+\lambda\right)}$,

if λ be the coefficient of friction.

The splitting power of the wedge is measured by R. For a given force P this splitting power is therefore greatest when α is least. Theoretically this will be when α is zero, *i.e.* when the wedge is of infinitesimal strength. Practically the wedge has the greatest splitting power when it is made with as small an angle as is consistent with its strength.

244. If the force of compression exerted by the wood on the wedge be great enough, the force P may not be large enough to make the wedge on the point of motion down; in fact the wedge may be on the point of being forced out. If P_1 be the value of P in this case, its value is found by changing the sign of μ in Art. 243, so that we should have

$$P_1 = 2R\left(\sin\frac{a}{2} - \mu\cos\frac{a}{2}\right) = 2R\sec\lambda\sin\left(\frac{a}{2}-\lambda\right).$$

If $\frac{a}{2}$ be $> \lambda$, the value of P_1 is positive.

If $\frac{a}{2}$ be $< \lambda$, P_1 is negative and the wedge could therefore only be on the point of slipping out if a pull were applied to its upper surface.

If $\frac{a}{2} = \lambda$, the wedge will just stick fast without the application of any force.

245. The Inclined Plane. The Inclined Plane, considered as a mechanical power, is a rigid plane inclined at an angle to the horizon.

It is used to facilitate the raising of heavy bodies.

The equilibrium of a particle on an inclined plane has already been considered in Arts. 78—80.

To find the work done in dragging a body up a rough inclined plane by means of a force parallel to the plane.

As in Art. 78 the force P_1 which would just move the body up the plane is $W(\sin\alpha + \mu\cos\alpha)$.

Hence the work done in dragging it from A to $C = P.AC$

$= W.AC\sin\alpha + \mu W.AC\cos\alpha = W.BC + \mu W.AB$

= work done in dragging the body through the same vertical height without the intervention of the plane

+ the work done in dragging it along a horizontal distance equal to the base of the inclined plane and of the same roughness as the plane.

246. From the preceding article we see that, if our inclined plane be rough, the work done by the power is more than the work done against the weight. This is true for any machine; the principle may be expressed thus:

In any machine, the work done by the power is equal to the work done against the weight, together with the work done against the frictional resistances of the machine, and the work done against the weights of the component parts of the machine.

The ratio of the work done on the weight to the work done by the effort is, for any machine, called the efficiency of the machine, so that, during any small displacement,

$$\text{Efficiency} = \frac{\text{Useful work done by the machine}}{\text{Work supplied to the machine}}.$$

Let P_0 be the effort required if there were no friction, and P the actual effort. Then, by Art. 211,

Work done against the weight

$= P_0 \times$ distance through which its point of application moves,

and work supplied to the machine

$= P \times$ distance through which its point of application moves.

Hence, by division,

$$\text{Efficiency} = \frac{P_0}{P} = \frac{\text{Effort when there is no friction}}{\text{Actual effort}}.$$

We can never get rid entirely of frictional resistances, or make our machine without weight, so that some work must always be lost through these two causes. Hence the efficiency of the machine can never be so great as unity. The more nearly the efficiency approaches to unity, the better is the machine.

There is no machine by whose use we can create work, and in practice, however smooth and perfect the machine may be, we always lose work. The only use of any machine is to multiply the force we apply, whilst at the same time the distance through which the force works is more than proportionately lessened.

247. Friction exerts such an important influence on the practical working of machines that the theoretical investigations are not of much actual use and recourse must for any particular machine be had to experiment. The method is the same for all kinds of machines.

The velocity ratio can be obtained by experiment; for in all machines it equals the distance through which the effort moves divided by the corresponding distance through which the weight, or resistance, moves. Call it n.

Let the weight raised be W. Then the theoretical effort P_0, corresponding to no friction, is $\frac{W}{n}$. Find by experiment the actual value of the effort P which just raises W. The actual mechanical advantage of the machine is $\frac{W}{P}$, and the efficiency of it is, by Art. 246, $\frac{P_0}{P}$.

248. As an example take the case of a class-room model of a differential wheel and axle on which some experiments were performed. The machine was not at all in good condition and was not cleaned before use, and no lubricants were used for the bearings of either it or its pulley.

With the notation of Art. 229 the values of a, b, and c were found to be $1\frac{1}{2}$, $6\frac{3}{4}$, and 3 inches, so that the value of the velocity ratio $n = \frac{2b}{c-a} = 9$. This value was also verified by experiment; for it was found that for every inch that W went up, P went down nine inches.

P was measured by means of weights put into a scale-pan whose weight is included in that of P; similarly for W. The weight of the pulley to which W is attached was also included in the weight of W.

The corresponding values of P and W, in grammes' weight, are given in the following table; the value of P was that which just overcame the weight W The third column gives the corresponding values of P_0, *i.e.* the effort which would have been required had there been no frictional resistances.

W	P	$P_0=\frac{W}{n}$	$E=\frac{P_0}{P}$	$M=\frac{W}{P}$
50	28	5·55	·2	1·79
100	36	11·11	·31	2·78
150	45	16·67	·37	3·3
250	60	27·78	·46	4·17
450	90	50	·56	5
650	119	72·22	·61	5·46
850	147	94·44	·64	5·78
1050	175	116·67	·67	6
1250	203	138·88	·68	6·16
1450	232	161·11	·694	6·25

The fourth column gives the values of E, the corresponding efficiency, and the last column gives the values of M, the mechanical advantage.

On plotting out on squared paper the above results, the points giving P are found to roughly be on a straight line going through the third and last of the above. Hence the relation between P and W is of the form $P=aW+b$, where a and b are constants.

Also $P=45$ when $W=150$, and $P=232$ when $W=1450$.

Hence $a=\cdot 144$ and $b=23\cdot 4$ approximately, so that $P=\cdot 144\,W+23\cdot 4$.

This is called the Law of the Machine.

Also
$$P_0=\tfrac{1}{9}W=\cdot 111\,W.$$

Hence
$$E=\frac{P_0}{P}=\frac{\cdot 111\,W}{\cdot 144\,W+23\cdot 4} \text{ and } M=\frac{W}{P}=\frac{W}{\cdot 144\,W+23\cdot 4}.$$

These give E and M for any weight W.

The values of E and M get bigger as W increases. Assuming the above value of E to be true for all values of W, then its greatest value is when W is infinitely great, and is then about ·77, so that in this machine at least 23 % of the work put into it is lost.

So the greatest value of the mechanical advantage $=\frac{1}{\cdot 144}=$ about 7.

If the machine had been well cleaned and lubricated before the experiment, much better results would of course have been obtained.

249. Just as in the example of the last article, so, with any other machine, the actual efficiency is found to fall considerably short of unity.

There is one practical advantage which, in general, belongs to machines having a comparatively small efficiency. It can be shewn that, in any machine in which the magnitude of the effort applied has no effect on the friction, the load does not run down of its own accord when no effort is applied provided that the efficiency is less than $\frac{1}{2}$. Examples of such machines are a Screw whose angle is small and whose "Power" or effort is applied horizontally as in Art. 241, and an Inclined Plane where the effort acts up the plane.

In machines where the friction does depend on the effort applied no such general rule can be theoretically proved, and each case must be considered separately. But it may be taken as a rough general rule that where the effort has a comparatively small effect on the amount of friction then the load will not run down if the efficiency be less than $\frac{1}{2}$. Such a machine is said not to "reverse" or "overhaul."

Thus in the case of the Differential Pulley (Art. 230), as usually constructed the efficiency is less than $\frac{1}{2}$, and the load W does not run down when no force P is applied, that is, when the machine is left alone and the chain let go. This property of not overhauling compensates, in great measure, for the comparatively small efficiency.

In a wheel and axle the mechanical advantage is usually great and the efficiency usually considerably more than $\frac{1}{2}$; but the fact that it reverses does not always make it a more useful machine than the Differential Pulley.

The student, who desires further information as to the practical working of machines, should consult Sir Robert Ball's *Experimental Mechanics* or other works on Applied Mechanics.

EXAMPLES

1. If the centre of gravity of a wheel and axle be at a distance a from the axis, shew that the wheel can rest with the plane through the axis and the centre of gravity inclined at an angle less than θ to the vertical, where $\sin\theta = \frac{b}{a}\sin\phi$, b being the radius of the axle, and ϕ the angle of friction.

2. The arms of a balance are equal in length but the beam is unjustly loaded; if a body be placed in succession in each scale-pan and weighed, shew that its true weight is the arithmetic mean between its apparent weights.

3. The arms of a balance are of unequal length, but the beam remains in a horizontal position when the scale-pans are not loaded; shew that, if a body be placed successively in each scale-pan, its true weight is the geometrical mean between its apparent weights.

Shew also that if a tradesman appear to weigh out equal quantities of the same substance, using alternately each of the scale-pans, he will defraud himself.

4. If a balance be unjustly weighted, and have unequal arms, and if a tradesman weigh out to a customer a quantity of some substance by weighing equal portions in the two scale-pans, shew that he will defraud himself if the centre of gravity of the beam be in the longer arm.

5. A common steelyard is graduated on the assumption that its weight is Q and the movable weight is W, both which assumptions are incorrect. If two masses whose real weights are P and R appear to weigh $P+X$ and $R+Y$, shew that the weight of the movable weight and the steelyard are less than their assumed values by $\frac{W}{D}(X-Y)$ and $\frac{Q}{D}(X-Y)-\frac{a}{bD}(PY-RX)$, where $D=P-R+X-Y$, and b and a are the distances (both measured in the same direction) from the fulcrum to the centre of gravity of the bar and to the point of attachment of the substance to be weighed.

Shew also that a body whose real weight is S appears to weigh

$$S+\frac{S(X-Y)+PY-RX}{P-R}.$$

6. *Shew that the efficiency of a screw is greatest when its angle is* $45^\circ-\frac{\lambda}{2}$.

The force required to lift the weight W, when there is friction, is $W\frac{a}{b}\tan(\alpha+\lambda)$, and, when there is no friction, it is $W\frac{a}{b}\tan\alpha$.

As in Art. 247 the efficiency, E, = the ratio of these

$$=\frac{\tan\alpha}{\tan(\alpha+\lambda)}=\frac{\sin(2\alpha+\lambda)-\sin\lambda}{\sin(2\alpha+\lambda)+\sin\lambda}=1-\frac{2\sin\lambda}{\sin(2\alpha+\lambda)+\sin\lambda}.$$

$\therefore$ E is greatest when $2\alpha+\lambda=90^\circ$.

7. An ordinary block and tackle has two pulleys in the lower block and two in the upper. What force must be exerted to lift a load of 300 lbs.? If on account of friction a given force will only lift ·45 times as much as if the system were frictionless, find the force required.

[75 lbs.; $166\frac{2}{3}$ lbs.]

8. In a block and tackle the velocity ratio is 8 : 1. The friction is such that only 55 % of the force applied can be usefully employed. Find what force will raise 5 cwt. by its use. [$1\frac{3}{22}$ cwt.]

9. In some experiments with a screw-jack the values of the load W were 150, 180, 210, 240 and 270 lbs. wt. and the corresponding values of the effort P were found to be 20·9, 22·7, 25·75, 28·4 and 31·4 lbs. wt.; assuming that $P=a+bW$, find the approximate values of a and b.

[5·3; ·097.]

10. In some experiments with a model block and tackle (the second system of pulleys), the values of W (including the weight of the lower block) and P expressed in grammes weight were found to be as follows:

$$W = 75, \quad 175, \quad 275, \quad 475, \quad 675, \quad 875, \quad 1075;$$
$$P = 25, \quad 48, \quad 71, \quad 119, \quad 166, \quad 214, \quad 264.$$

Also there were five strings at the lower block. Find an approximate relation between P and W, and the corresponding values for the efficiency and mechanical advantage. Draw the graphs of P, P_0, E, and M.

[$P = 7{\cdot}3 + {\cdot}236\,W$.]

11. The following table gives the load in tons upon a crane, and the corresponding effort in lbs. wt.:

Load	1,	3,	5,	7,	8,	10,	11.
Effort	9,	20,	28,	37,	42,	51,	56.

Find the approximate law of the machine, and calculate the efficiency at the loads 5 and 10 tons, given that the velocity-ratio is 500.

[$P = 4{\cdot}3 + 4{\cdot}7\,W$; ·8 and ·88.]

12. A weight is lifted by a screw-jack, of pitch $\frac{1}{4}$ inch, the force being applied at right angles to a lever of length 15 inches. The values of the weight in tons, and the corresponding force in lbs., are given in the following table:

Weight	1,	2·5,	5,	7,	8,	10.
Force	24,	32,	46,	57,	63,	73.

Find the approximate law of the machine, and calculate its efficiency for the weights 4 and 9 tons. [$P = 18{\cdot}5 + 5{\cdot}5\,W$; ·59 and ·79.]

CHAPTER XIII

EQUILIBRIUM OF STRINGS AND CHAINS

250. A PERFECTLY flexible string is such that the action across any normal section of it is a single force whose direction is along the tangent to the string. This section is supposed to be so small that the string may be treated as a curved line. The string offers no resistance to being bent at any point and hence possesses no rigidity of shape. A chain, whose links are very small and perfectly smooth, may be treated as a flexible string.

In the case of a string which is not perfectly flexible, or of a wire, the actions across any normal section do not reduce to a single tangential force, but to a single force and a couple as we have already seen in Chapter VII.

251. *A uniform heavy inextensible string hangs freely under the action of gravity; to find the equation of the curve which it forms.*

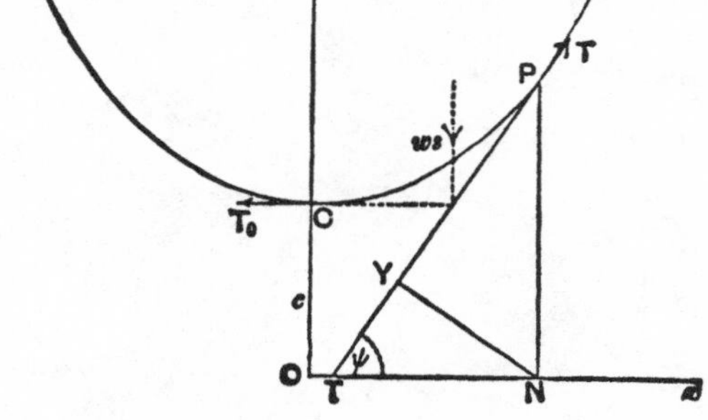

Let C be the lowest point of the curve, P any point of the string such that the arc $CP = s$; let T be the tension at P and T_0 the tension at C.

Then the portion CP of the string is in equilibrium under the action of T, T_0, and its weight, which is ws, where w is the weight of the string per unit of length.

If ψ be the inclination of the tangent at P to the horizontal, we then have

$$T\cos\psi = T_0 \quad\text{.....................(1)},$$

and

$$T\sin\psi = ws \quad\text{.....................(2)}.$$

Let the tension at the lowest point be equal to the weight of a length c of the string, so that $T_0 = wc$, and hence

$$\tan\psi = \frac{ws}{T_0} = \frac{s}{c} \quad\text{...................(3)},$$

i.e. $\dfrac{dy}{dx} = \dfrac{s}{c}$, if the axes of x and y be respectively horizontal and vertical. Differentiating, we have

$$\frac{d^2y}{dx^2} = \frac{1}{c}\frac{ds}{dx} = \frac{1}{c}\sqrt{1+\left(\frac{dy}{dx}\right)^2}.$$

$$\therefore \frac{\dfrac{d^2y}{dx^2}}{\sqrt{1+\left(\dfrac{dy}{dx}\right)^2}} = \frac{1}{c}.$$

On integration, $\log\left[\dfrac{dy}{dx} + \sqrt{1+\left(\dfrac{dy}{dx}\right)^2}\right] = \dfrac{x}{c} + \text{const.}$

If the axis of y be taken to pass through C, we have $\dfrac{dy}{dx} = 0$ when $x = 0$, and hence this constant vanishes.

$$\therefore \frac{dy}{dx} + \sqrt{1+\left(\frac{dy}{dx}\right)^2} = e^{\frac{x}{c}}.$$

Now $$-\frac{dy}{dx} + \sqrt{1+\left(\frac{dy}{dx}\right)^2} = \frac{1}{\dfrac{dy}{dx} + \sqrt{1+\left(\dfrac{dy}{dx}\right)^2}} = e^{-\frac{x}{c}}.$$

Hence, by subtraction,

$$\frac{dy}{dx} = \frac{1}{2}\left[e^{\frac{x}{c}} - e^{-\frac{x}{c}}\right] \quad\text{.....................(4)}.$$

On integration, $y = \frac{c}{2}\left[e^{\frac{x}{c}} + e^{-\frac{x}{c}}\right] + A.$

If the origin O be taken at a depth c below C, we have $y = c$ when $x = 0$, and hence $A = 0$.

The equation to the curve with the axes so chosen is thus

$$y = \frac{c}{2}\left[e^{\frac{x}{c}} + e^{-\frac{x}{c}}\right] = c \cosh \frac{x}{c} \quad \text{..............(5).}$$

This curve is known as the Common Catenary, and Ox is called its Directrix.

Equations (3) and (4) give

$$s = c \tan \psi = c\frac{dy}{dx} = \frac{c}{2}\left[e^{\frac{x}{c}} - e^{-\frac{x}{c}}\right] = c \sinh \frac{x}{c} \quad \text{......(6).}$$

From these two equations we have

$$y^2 = c^2 + s^2 \quad \text{.........................(7).}$$

This, together with equation (3), gives

$$y = c \sec \psi \quad \text{and} \quad s = c \tan \psi \quad \text{...........(8).}$$

If PN be the ordinate at P and NY be drawn perpendicular to PT, then the angle $YNP = \psi$ and hence, from (8), we have

$$NY = c = OC,$$

and $$PY = s = \text{the arc } CP.$$

(1) now gives $\quad T = wc \sec \psi = wy,$

i.e. the tension at any point P of a Catenary is equal to the weight of the portion of the string whose length is the vertical distance between P and the directrix.

The quantity c, which defines the size of the Catenary, is called its Parameter.

252. For all values of c, an approximation to the form of the curve for small values of x is $2c(y-c) = x^2$; this is seen from equation (5) of the last article by expanding the right-hand side in powers of x. Hence, in the neighbourhood of the lowest point of the Catenary, the curve approximates in form to a parabola.

For large values of x compared with c, the value of $e^{-\frac{x}{c}}$ is negligible, and the Catenary then approximates in shape to the Exponential Curve.

253. **Ex. 1.** *A uniform chain, of length $2l$, has its ends attached to two points in the same horizontal line at a distance $2a$ apart. If l is only a little greater than a, shew that the tension of the chain is approximately equal to the weight of a length $\sqrt{\frac{a^3}{6(l-a)}}$ of the chain, and that the "sag," or depression of the lowest point of the chain below its ends, is $\frac{1}{2}\sqrt{6a(l-a)}$ nearly.*

Since l is very little greater than a, the tension of the chain must be very great and hence c must be large.

Equation (6) of Art. 251 then becomes $l=\frac{c}{2}\left[e^{\frac{a}{c}}-e^{-\frac{a}{c}}\right]$.

The Exponential Theorem gives, on expanding,

$$l=\frac{c}{2}\left[2\cdot\frac{a}{c}+2\cdot\frac{a^3}{6c^3}+\dots\right]=a+\frac{a^3}{6c^2}+\frac{a^5}{120c^4}+\dots.$$

A first approximation is then $l-a=\frac{a^3}{6c^2}$, and hence $c=\sqrt{\frac{a^3}{6(l-a)}}$, so that the tension at the lowest point is equal to a weight of this length of the chain.

The ordinate of the end of the chain is, by equation (5) of Art. 251, given by

$$y=\frac{c}{2}\left[e^{\frac{a}{c}}+e^{-\frac{a}{c}}\right]=\frac{c}{2}\left[2+2\cdot\frac{a^2}{2c^2}+2\cdot\frac{a^4}{24c^4}+\dots\right]=c+\frac{a^2}{2c}+\frac{a^4}{24c^3}+\dots.$$

Hence the sag of the lowest point

$$=y-c=\frac{a^2}{2c}\text{ approx.}=\frac{a^2}{2}\sqrt{\frac{6(l-a)}{a^3}}=\tfrac{1}{2}\sqrt{6a(l-a)}.$$

It follows easily that if d be the sag of the lowest point, then the tension there is approximately equal to a length $\frac{a^2}{2d}$ of the chain.

Ex. 2. *A box kite is flying at a height h with a length l of wire paid out, and with the vertex of the catenary on the ground; shew that at the kite the inclination of the wire to the ground is $2\tan^{-1}\frac{h}{l}$, and that its tensions there and at the ground are $w\frac{l^2+h^2}{2h}$ and $w\frac{l^2-h^2}{2h}$, where w is the weight of the wire per unit of length.*

With the figure of Art. 251, P is the kite and C is on the ground, so that $NP=h+c$ and hence, from the triangle NYP,

$$(h+c)^2=NY^2+YP^2=c^2+l^2.$$

$$\therefore\ c=\frac{l^2-h^2}{2h}\text{ and } NP=c+h=\frac{l^2+h^2}{2h}.$$

Also $\cos\psi=\frac{c}{c+h}=\frac{l^2-h^2}{l^2+h^2}$, so that $\tan\frac{\psi}{2}=\frac{h}{l}$.

Also the required tensions at P and C are $w\,.\,PN$ and $w\,.\,c$.

Ex. 3. *A heavy uniform string 90 inches long hangs over two smooth pegs at different heights. The parts which hang vertically are of lengths 30 and 33 inches. Prove that the vertex of the catenary divides the whole string in the ratio 4 : 5, and find the distance between the pegs.*

Let s_1 and s_2 be the arcual distances of the two pegs from the vertex C of the catenary, so that $s_1+s_2=27$. If c be the parameter of the catenary then, by equation (7) of Art. 251, $s_1^2+c^2=30^2$ and $s_2^2+c^2=33^2$; for the ends of the strings must lie on the directrix of the catenary by the last property of Art. 251, since the tension of the string is unaltered by its passing round a smooth peg.

Hence, easily, $s_1=10$, $s_2=17$ and $c=20\sqrt{2}$, so that $\dfrac{s_1+30}{s_2+33}=\dfrac{4}{5}$.

Also, from equation (5) of Art. 251, we have

$$x=c\log_e\frac{y+\sqrt{y^2-c^2}}{c}.$$

Hence, when $y=30$ or 33, $x=10\sqrt{2}\log_e 2$, or $10\sqrt{2}\log_e \frac{25}{8}$.

$$\therefore\ x_1+x_2=20\sqrt{2}\log_e 2\cdot 5=28\cdot 284\times\cdot 9163=25\cdot 92,$$

so that the horizontal and vertical distances between the pegs are 25·92 and 3 inches.

Ex. 4. *A uniform chain, of length 2l and weight W, is suspended from two points, A and B, in the same horizontal line. A load P is now suspended from the middle point D of the string; if AB=2a, find the depth below AB of the final position of D.*

Let C be the lowest point of the catenary of which DA is a part; let its parameter be c, let the arc $CD=s$ and let the ordinate of D be y. Then

$$P=\text{vertical component of the tensions at } D=2T\sin\psi \text{ (Art. 251)}$$

$$=2\cdot\frac{W}{2l}y\cdot\frac{s}{y}=W\frac{s}{l}.$$

Also, by equation (7) of Art. 251,

$$s^2+c^2=y^2,\ \text{and}\ (s+l)^2+c^2=(y+h)^2 \ \ldots\ldots\ldots\ldots\ldots\ldots(1),$$

if h be the required depth.

These equations give $s=\dfrac{lP}{W}$, and $y=\dfrac{l^2}{2h}\dfrac{W+2P}{W}-\dfrac{h}{2}$ ....(2).

Also, if x be the abscissa of D, then by equations (5) and (6) of Art. 251,

we have $$y+s=ce^{\frac{x}{c}},\ \text{and}\ y+h+s+l=ce^{\frac{x+a}{c}}$$

$$\therefore\ e^{\frac{a}{c}}=1+\frac{h+l}{y+s}.$$

On substitution from (1) and (2), we have an equation to give h.

Ex. 5. *A chain, of length 2l, is hung over two small smooth pulleys which are in the same horizontal line at a distance 2a apart; to find the positions of equilibrium and to determine whether they are stable.*

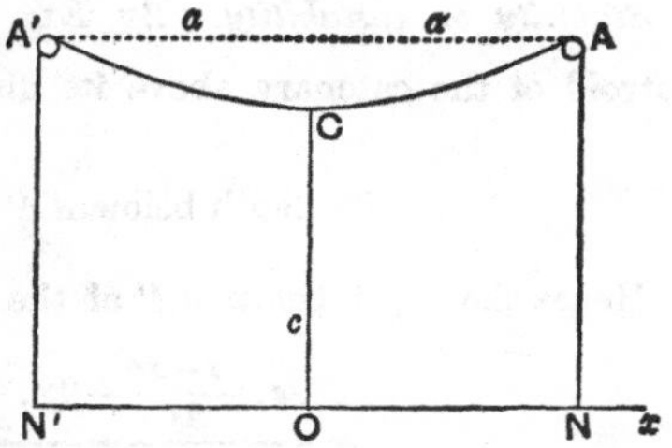

Since the tension of the chain is unaltered by passing over the pulley, and since on one side it is equal to the weight of the free part AN, and on the other it is equal to the weight of the chain that would stretch vertically down to the directrix (Art. 251), it follows that N and also N' lie on the directrix of the catenary.

Hence $\quad l = \text{arc } CA + \text{line } AN = \frac{c}{2}\left[e^{\frac{a}{c}} - e^{-\frac{a}{c}}\right] + \frac{c}{2}\left[e^{\frac{a}{c}} + e^{-\frac{a}{c}}\right] = ce^{\frac{a}{c}}$ ...(1),

where c is the parameter of the catenary.

Equation (1) cannot be solved algebraically, but a graphic solution may be obtained as follows when a and l are given numerically. Put $\frac{a}{c} = X$; then $e^X = \frac{l}{a}X$ ....(2).

Draw the curve $Y = e^X$ and the straight line $Y = \frac{l}{a}X$.

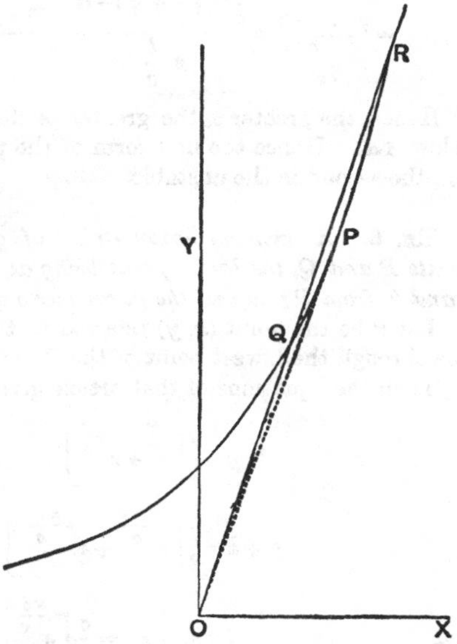

The points, Q and R, in which they cut give, on measurement of their abscissae, approximate solutions for X and hence for c. It is clear that there will be two real, coincident or imaginary solutions according as

$$\frac{l}{a} \gtreqless \tan POX,$$

where OP is the tangent from O to the curve.

Now P is given by

$$\frac{Y}{X} = \tan POX = \frac{dY}{dX} = e^X = Y,$$

and is therefore the point $(1, e)$, so that $\tan POX = e$.

There are therefore two, one, or no possible catenaries according as $l \gtreqless ae$.

One catenary will be somewhat as drawn in the first figure and the other as in the annexed one.

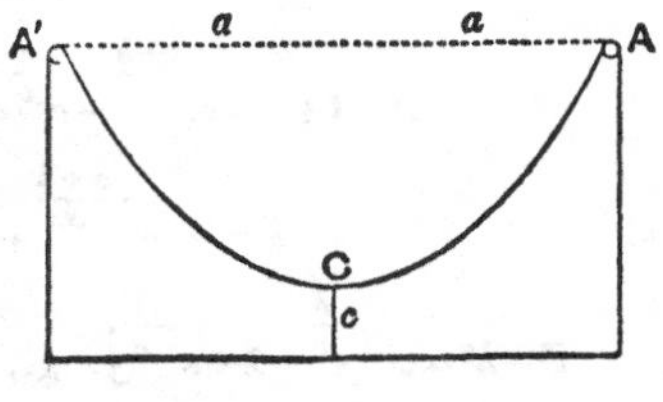

The parameter c of the first case is clearly greater than in the second.

Stability or instability. By Ex. 2 of Art. 144 the height of the centroid of the catenary above its directrix $=\frac{cx+ys}{2s}$.

$\therefore$ its depth below $AA'=y-\frac{cx+ys}{2s}=\frac{ys-xc}{2s}$.

Hence the depth below AA' of the C.G. of the whole chain

$$=\frac{2s.\frac{ys-xc}{2s}+2y.\frac{y}{2}}{2s+2y}=\frac{y^2+ys-xc}{2(y+s)}.$$

Now in the above case $x=a$; $y=\frac{c}{2}\left(e^{\frac{a}{c}}+e^{-\frac{a}{c}}\right)$, $s=\frac{c}{2}\left(e^{\frac{a}{c}}-e^{-\frac{a}{c}}\right)$, and $l=y+s=ce^{\frac{a}{c}}$.

$\therefore$ depth below AA' of the centre of gravity of the whole chain

$$=\frac{y.e^{\frac{a}{c}}-a}{2e^{\frac{a}{c}}}=\frac{\frac{l}{c}.\frac{c}{2}\left[\frac{l}{c}+\frac{c}{l}\right]-a}{2.\frac{l}{c}}=\frac{l^2+c^2-2ac}{4l}=\frac{(l^2-a^2)+(c-a)^2}{4l}.$$

Hence, the greater c, the greater is the depth of the centre of gravity below AA'. Hence the first form of the possible curves is the stable one, and the second is the unstable.

Ex. 6. *A uniform heavy string of given length l is attached to two points P and Q, the latter point being at horizontal and vertical distances, h and k, from P; to find the parameter c of the catenary in which it rests.*

Let P be the point (x, y) referred to the directrix Ox and the vertical line through the lowest point of the Catenary (Art. 251).

Then the equations of that article give

$$y=\frac{c}{2}\left[e^{\frac{x}{c}}+e^{-\frac{x}{c}}\right] \quad\text{...(1)},$$

$$y+k=\frac{c}{2}\left[e^{\frac{x+h}{c}}+e^{-\frac{x+h}{c}}\right] \quad\text{..............................(2)},$$

and $$l=s_Q-s_P=\frac{c}{2}\left[e^{\frac{x+h}{c}}-e^{-\frac{x+h}{c}}\right]-\frac{c}{2}\left[e^{\frac{x}{c}}-e^{-\frac{x}{c}}\right]\text{...(3)}.$$

$$\therefore\ l+k=ce^{\frac{x+h}{c}}-ce^{\frac{x}{c}}=ce^{\frac{x}{c}}\left(e^{\frac{h}{c}}-1\right) \quad\text{..........................(4)},$$

and $$l-k=-ce^{-\frac{x+h}{c}}+ce^{-\frac{x}{c}}=ce^{-\frac{x}{c}}\left(1-e^{-\frac{h}{c}}\right) \quad\text{..............(5)}.$$

$$\therefore\ l^2-k^2=c^2\left[e^{\frac{h}{c}}-2+e^{-\frac{h}{c}}\right],\ \text{giving}\quad \pm\sqrt{l^2-k^2}=c\left[e^{\frac{h}{2c}}-e^{-\frac{h}{2c}}\right]\text{...(6)}.$$

This equation cannot be solved algebraically. A graphic solution may be obtained by putting $\frac{h}{2c}=X$, so that (6) gives

$$\sinh X = \pm\frac{\sqrt{l^2-k^2}}{h}X \quad \text{.........................(7)},$$

and hence X is given as the point in which the straight lines

$$Y=\pm\frac{\sqrt{l^2-k^2}}{h}X$$

meet the curve $Y=\sinh X$.

On drawing the curve we obtain two equal and opposite values for X, and hence two equal and opposite values for c, provided that $\frac{\sqrt{l^2-k^2}}{h}>1$, *i.e.* provided that l is greater than the length PQ.

The value of c being now known to any degree of approximation, equation (4) gives x and then equation (1) gives y. The solution is therefore complete. It is clear that only the positive value of c need be taken. For a negative value of c would make y negative.

Supposing that a value X_1 as an approximate solution of (7) has been obtained by graphic methods, a further approximation may be obtained by analysis. For, taking the upper sign in (6) and putting $\frac{\sqrt{l^2-k^2}}{h}=\lambda$, we have to solve $\sinh X=\lambda X$, where X_1 is an approximate solution.

Putting $X=X_1+\xi$, where ξ is small, we have

$$\sinh(X_1+\xi)=\lambda(X_1+\xi),$$

i.e. $\sinh X_1+\xi\cosh X_1+\ldots=\lambda(X_1+\xi)$, by Taylor's theorem.

Hence, neglecting squares of ξ, we have

$$\xi=\frac{\lambda X_1-\sinh X_1}{\cosh X_1-\lambda}=\frac{\sqrt{l^2-k^2}\,X_1-h\sinh X_1}{h\cosh X_1-\sqrt{l^2-k^2}},$$

so that $X_1+\xi$ is a second approximation.

EXAMPLES

1. Shew that the maximum tension in a wire which weighs ·15 lb. per yard and hangs with a sag of 1 foot in a horizontal span of 100 feet is about $62\frac{1}{2}$ lbs. wt.

2. A telegraph is constructed of No. 8 iron wire which weighs 7·3 lbs. per 100 feet; the distance between the posts is 150 feet and the wire sags 1 foot in the middle. Shew that it is screwed up to a tension of about 205 lbs. wt.

3. A trolley-wire is carried on poles round a curve of 1200 feet radius. The poles are 40 yards apart, and in the middle of each span the wire sags down 6 inches below the points of support. If the wire weighs $1\frac{1}{2}$ lbs. per yard, shew that the resultant horizontal pull on each pole is very nearly 180 lbs. wt.

4. From statical considerations shew that the tangents at any two points P and Q of a Common Catenary intersect on the vertical through the centre of gravity G of the arc PQ.

5. A uniform heavy chain, of length 155 feet, is suspended from two points in a horizontal plane which are 150 feet apart; shew that the tension at the lowest point is nearly 1·08 times the weight of the chain.

6. If a uniform chain be fixed at its ends, and any number of its links be free to move along smooth horizontal wires in the same vertical plane as the chain, shew that the parts of the chain between these successive links are all arcs of the same catenary.

7. If the velocity of the wind be the same at all heights, and its effect on the string attached to a kite be negligible, shew that as the kite ascends the force required to hold it diminishes.

8. A uniform chain, of length $2l$ and weight W, is suspended from two points, A and B, in the same horizontal line. A load P is now suspended from the middle point D of the chain and the depth of this point below AB is found to be h. Shew that each terminal tension is

$$\frac{1}{2}\left[P\frac{l}{h}+W\frac{h^2+l^2}{2hl}\right].$$

9. An inelastic string, of length $2l$ and weight w per unit of length, is suspended from two points at the same level and at a distance $2d$ apart. Find equations to determine the lowest point of the string below the point of suspension.

10. A heavy uniform string, of length l, is suspended from a fixed point A, and its other end B is pulled horizontally by a force equal to the weight of a length a of the string. Shew that the horizontal and vertical distances between A and B are $a\sinh^{-1}\frac{l}{a}$ and $\sqrt{l^2+a^2}-a$.

11. A uniform chain, of length l, is to be suspended from two points, A and B, in the same horizontal line so that either terminal tension is n times that at the lowest point. Shew that the span AB must be

$$\frac{l}{\sqrt{n^2-1}}\log_e[n+\sqrt{n^2-1}].$$

If $l=100$ feet and $n=3$ shew, from the Tables, that the length is about 62·3 feet.

12. A chain reaches vertically to the ground from the bows of a ship about to be launched, and is then laid in a straight line away from the ship for a distance of 80 yards. Its end is attached to a heavy anchor which requires a horizontal pull of 25 tons to move it. The vertical part of the chain is 50 feet long. What must the chain weigh per foot in order that, when its whole length is lifted off the ground by the ship moving horizontally, the tension may be just sufficient to move the anchor? [68·6... lbs.]

13. A barge is tied by a uniform chain, 20 feet long, one end of which is attached to the bow B and the other to the top A of a post, A being 12 feet higher than B; the stream exerts a force of $7\frac{1}{2}$ lbs. weight on the barge and the chain has a mass of $\frac{1}{2}$ lb. per foot; shew that the distance of B from the vertical through A is $30 \log_e \frac{5}{3}$ feet.

14. A wire 140 yards long hangs between two points, 138 yards apart horizontally and 50 feet vertically; shew that the tension at its lowest point is about 495 lbs. wt., the wire weighing half a pound per foot.

15. A string, of length l, hangs between two points (not in the same vertical) and makes angles α, β with the vertical at the points of support. Shew that, if k is the height of one point above the other and the vertex of the catenary does not lie between them, then

$$k \cos \frac{\alpha - \beta}{2} = l \cos \frac{\alpha + \beta}{2}.$$

16. A heavy chain, of length $2l$, has one end tied at A and the other is attached to a small heavy ring which can slide on a rough horizontal rod which passes through A. If the weight of the ring be n times the weight of the chain, shew that its greatest possible distance from A is

$$\frac{2l}{\lambda} \log [\lambda + \sqrt{1+\lambda^2}], \quad \text{where} \quad \frac{1}{\lambda} = \mu (2n+1)$$

and μ is the coefficient of friction.

17. One end of a heavy rough uniform string is fastened to a point P at a height h above a table, and a length $z [< l - h]$ of the string rests on the table in a vertical plane through P. Shew that the string is pulled as far away from P as is possible consistent with equilibrium when z is given by the equation

$$z^2 - 2(l + \mu h) z + l^2 - h^2 = 0,$$

where μ is the coefficient of friction and l is the length of the string.

[The tension at the lowest point is μ times the weight of the part on the table, so that $c = \mu z$. Also $(h+c)^2 = c^2 + (l-z)^2$.]

18. A heavy chain, of total length l, rests with one end on a rough table and the other on a rough floor; it is stretched out as far as possible so that the equilibrium is limiting; if μ be the coefficient of friction both of the table and floor and b be the height of the table, shew that the length on the floor is

$$\frac{\mu}{4}\left[b + \frac{2l}{\mu} - \sqrt{b^2 + \frac{4lb}{\mu}}\right] - \frac{b}{2\mu}, \quad \text{provided that } l > b + \frac{b}{\mu}.$$

19. A chain, of length $2l$ and weight $2W$, hangs with one end A attached to a fixed point in a smooth horizontal wire, and the other end B attached to a smooth ring which slides along the wire. Initially A and B are together. Shew that the work done in drawing the ring along the wire till the chain at A is inclined at an angle of 45° to the vertical is

$$Wl [1 - \sqrt{2} + \log_e (1 + \sqrt{2})].$$

[Use Art. 89 and the result of Ex. 2 of Art. 144.]

20. A uniform string, of length $18a$ feet and weight $3W$ pounds, lies on a horizontal smooth table along with a rod, of weight $2W$ pounds, to the ends of which its ends are attached. When the middle point of the string is raised to a height $7a$, the pressure on the table just vanishes. Shew that the length of the rod is $16a \log_e 2$ feet, and that the work done is

$$\tfrac{1}{8} Wa\,(15 + 64 \log_e 2) \text{ foot-pounds.}$$

21. A telegraph wire is made of a given material, and such a length l is stretched between two posts, distant d apart and of the same height, as will produce the least possible tension at the posts. Shew that $l = \frac{d}{\lambda} \sinh \lambda$, where λ is given by the equation $\lambda \tanh \lambda = 1$.

254. *A string hangs under gravity and it is loaded so that the weight on each element of it is proportional to the horizontal projection of that element; to shew that it will hang in the form of a parabola.*

Let T be the tension at any point P and T_0 that at the lowest point A. Draw the tangent PT and the perpendiculars PN, PM to the horizontal and vertical lines through A.

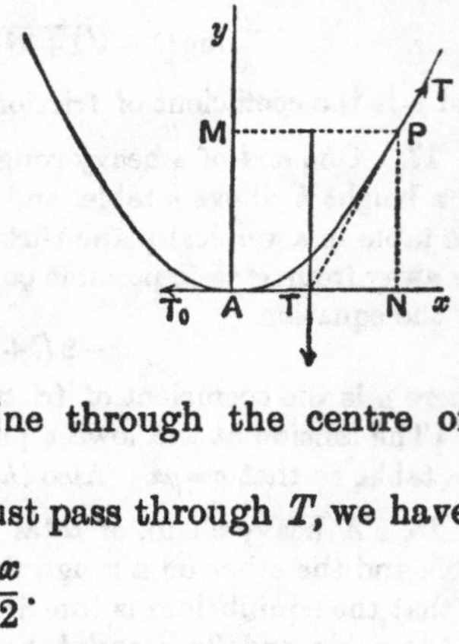

Since the weight of each element of the string AP is proportional to its projection on MP, it is clear that the abscissa of the centre of gravity of AP is the same as that of MP, *i.e.* the vertical line through the centre of gravity bisects MP.

Hence, since this vertical line must pass through T, we have

$$AT = TN = \frac{x}{2}.$$

Now PTN is clearly a triangle of forces for the arc AP.

If then w be the loading per unit of horizontal span of the string, we have $\dfrac{T}{PT} = \dfrac{T_0}{TN} = \dfrac{\text{wt. of } AP}{PN} = \dfrac{wx}{y}$.

If T_0 be equal to the load on a length c of the horizontal span, this gives $\dfrac{wc}{\frac{x}{2}} = \dfrac{wx}{y}$, *i.e.* $x^2 = 2cy$, so that the curve is a parabola of latus rectum $2c$.

Also $$T = w\frac{x}{y}.PT = w\frac{x}{y}\sqrt{y^2+\frac{x^2}{4}}$$
$$= w\sqrt{x^2+\frac{x^4}{4y^2}} = w\sqrt{2cy+c^2}.$$

Since the vertical through T bisects MP it will also bisect the straight line AP. The curve is thus such that the line parallel to its axis through the intersection of two of its tangents bisects the line joining their points of contact. But this is a fundamental property of the parabola; so that it is clear without any analysis that the curve is a parabola.

255. *The Catenary, when very tightly stretched, becomes ultimately a parabola to a first approximation.*

Since c is the length of the string whose weight is equal to the tension at the lowest point, it becomes very large.

The equation to the Catenary, referred to axes through the lowest point C, is

$$y+c=\frac{c}{2}\left[e^{\frac{x}{c}}+e^{-\frac{x}{c}}\right]=\frac{c}{2}\left[2+2.\frac{x^2}{\underline{|2}c^2}+2.\frac{x^4}{\underline{|4}c^4}+\dots\right],$$

by the Exponential Theorem, $= c+\dfrac{x^2}{2c}+\dfrac{x^4}{24c^3}+\dots,$

i.e. $$2cy = x^2+\frac{x^4}{12c^2}+\dots.$$

When c is made very large, a first approximation to the form of the Catenary is therefore the parabola $x^2 = 2cy$. We should expect this result from another point of view. For when the Catenary is very tightly stretched its inclination to the horizontal is very small, so that the load on any element, *viz.* its weight which is proportional to the length of the element, is therefore very nearly proportional to the projection of the element on the horizontal. The case of this article is then very nearly that of the last article.

256. Suspension Bridges. In the case of a suspension bridge we have two chains hung up so as to be parallel, their ends being firmly fixed to supports. From different points of these chains hang supporting chains or rods which carry the roadway of the bridge, these supporting rods being usually at equal horizontal distances from one another.

The weight of the chain itself and the weights of the supporting rods may be neglected in comparison with the weight of the roadway. The weight supported by each of the rods may then be taken to be the weight of equal portions of the roadway, so that we have the case of Art. 254. Thus the figure of the chain of a suspension bridge will approximate very closely to that of a parabola; the closer the supporting bars, and the lighter they and the chain are in comparison with the roadway, the more nearly will this figure approach to a parabola.

Ex. 1. The whole load of a suspension bridge is 200 tons evenly distributed over its horizontal span, which is 150 feet, and its height is 20 feet; shew that the tensions at the lowest point and at the points of support are $187\frac{1}{2}$ and $212\frac{1}{2}$ tons weight respectively.

Ex. 2. Shew that the tensions at the ends of a chain whose span is 30 feet, the depth at its lowest point below the level of the supports 10 feet, and the load, which is uniformly distributed across the span, half a ton per foot run of the span, are each 9·375 tons wt.

Ex. 3. If a chain, of length l and weight W, is stretched tightly between two points A and B which are not in the same horizontal and k is the vertical depth of the middle point C of the chain below the line AB, shew that the tension of the chain is approximately $\frac{Wl}{8k}$.

[Draw ac, cb vertical each to represent $\frac{W}{2}$; through a and b draw aO, bO parallel to the tangents at A and B. Then aO, bO, and cO represent the tensions T, T', and T_1 at A, B, and C.

Since $Oa^2+Ob^2=2Oc^2+2cb^2$, $\therefore \quad T^2+T'^2=2T_1^2+\frac{W^2}{2}$.

Also $T_1=\frac{T+T'}{2}-\frac{W}{l}k$ from the properties of the Catenary.

Eliminating T_1, we have $2T^2+2T'^2=\left(T+T'-\frac{2Wk}{l}\right)^2+W^2$.

But in the limit $T=T'$, so that this equation gives

$$0=-\frac{8Wk}{l}T+\frac{4W^2k^2}{l^2}+W^2,$$

i.e. $$T=\frac{Wl}{8k}+\frac{kW}{2l}=\frac{Wl}{8k} \text{ approximately, since } k \text{ is small.]}$$

257. *General equations of equilibrium of a string which rests under the action of any given forces.*

Suppose the string to be in one plane in which the forces act. Let P be any point (x, y) of the string, whose arcual distance from a fixed point C is s. Let Q be a point very close to P, so that $PQ = \delta s$, and hence Q is the point $(x + \delta x, y + \delta y)$.

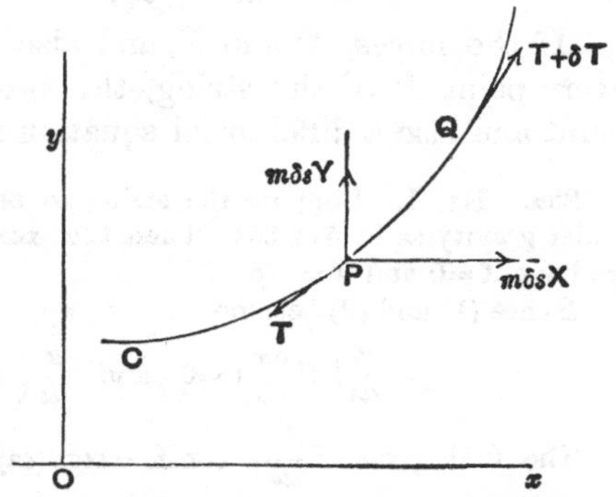

Let T and $T + \delta T$ be the tensions at P and Q.

Let the forces per unit of mass of the string at P be X and Y, so that on the element PQ the components parallel to the axes are $m\delta s . X$ and $m\delta s . Y$, where m is the mass per unit of length.

[Strictly speaking these components do not act at P, but at the various points between P and Q; we are however ultimately taking PQ very small, so that no error is introduced by making them act at P.]

The tension at P resolved parallel to the axis of x is $T\dfrac{dx}{ds}$, and this is clearly equal to some function, $f(s)$, of the arc s, since it depends on the position of P.

The tension at Q, resolved parallel to Ox, similarly

$$= f(s + \delta s) = f(s) + \delta s f'(s) + \dots, \quad \text{by Taylor's theorem,}$$

$$= T\frac{dx}{ds} + \delta s \frac{d}{ds}\left(T\frac{dx}{ds}\right) + \dots.$$

Equating the forces on PQ in the direction of the axis of x, we then have

$$T\frac{dx}{ds} = m\delta s . X + T\frac{dx}{ds} + \delta s \frac{d}{ds}\left(T\frac{dx}{ds}\right)$$

+ terms containing squares and higher powers of δs.

After cancelling and dividing by δs, make δs very small, *i.e.* make Q approach indefinitely close to P, and we have

$$\frac{d}{ds}\left(T\frac{dx}{ds}\right) + mX = 0 \dots\dots\dots\dots\dots\dots\dots(1).$$

Similarly, resolving the forces parallel to the axis of y, we have

$$\frac{d}{ds}\left(T\frac{dy}{ds}\right)+mY=0 \text{ ....(2).}$$

If the forces, X and Y, and also the mass m, be known for every point P of the string, the two equations give T at any point and also a differential equation for the form of the string.

258. Ex. 1. Suppose the string to be uniform and to hang freely under gravity as in Art. 251. Then, the axes being horizontal and vertical, we have $X=0$ and $Y=-g$.

Hence (1) and (2) become

$$\frac{d}{ds}\left(T\frac{dx}{ds}\right)=0 \quad\text{and}\quad \frac{d}{ds}\left(T\frac{dy}{ds}\right)=mg.$$

The first gives $T\dfrac{dx}{ds}=\text{const.}=mgc$ (say), and hence the horizontal tension is constant throughout the string.

On substituting for T in the second equation, we have $\dfrac{d}{ds}\left(c\dfrac{dy}{dx}\right)=1$,

i.e.
$$\frac{d^2y}{dx^2}=\frac{1}{c}\frac{ds}{dx}=\frac{1}{c}\sqrt{1+\left(\frac{dy}{dx}\right)^2},$$

which is the differential equation of Art. 251.

Ex. 2. Let the string be, as in the case of the Suspension Bridge, so loaded that on each element the weight is proportional to the horizontal projection.

Then $\qquad X=0$ and $Y\delta s=-\lambda\delta x$.

Hence (1) and (2) give

$$\frac{d}{ds}\left(T\frac{dx}{ds}\right)=0 \quad\text{and}\quad \frac{d}{ds}\left(T\frac{dy}{ds}\right)=\lambda m\frac{dx}{ds}.$$

$$\therefore\ T\frac{dx}{ds}=\text{const.}=C \quad\text{and}\quad \frac{d}{ds}\left(C\frac{dy}{dx}\right)=\lambda m\frac{dx}{ds},\quad \textit{i.e. } C\frac{d^2y}{dx^2}=\lambda m.$$

On integrating we see that the curve is a parabola, as in Art. 254.

If the mass m of the chain per unit of length vary in any way, we have similarly

$$\frac{d}{ds}\left(C\frac{dy}{dx}\right)=mg.$$

If m be given in terms of the position of the element δs, this equation gives the form of the curve. It also gives the variation of the mass when the form of the curve of the string is given.

259. Catenary of uniform strength. Let us find the equation of the curve in which a string would hang if its mass at each point P were proportional to the tension there, so that the strength of the string is to be everywhere proportional to the force it has to exert.

In this case $X = 0$, $Y = -g$, and $m \propto T$, *i.e.* $m = \lambda T$, where λ is some constant.

The equations of Art. 257 then give

$$T\frac{dx}{ds} = C, \quad \text{and} \quad \frac{d}{ds}\left(T\frac{dy}{ds}\right) = \lambda Tg \quad \text{.........(1).}$$

Substituting for T, we have $\frac{d}{ds}\left(\frac{dy}{dx}\right) = \lambda g\frac{ds}{dx}$,

i.e. $$\frac{d^2y}{dx^2} = \lambda g\left(\frac{ds}{dx}\right)^2 = \lambda g\left[1 + \left(\frac{dy}{dx}\right)^2\right].$$

$$\therefore \quad \frac{\frac{d^2y}{dx^2}}{1 + \left(\frac{dy}{dx}\right)^2} = \lambda g.$$

On integration, this gives $\tan^{-1}\left(\frac{dy}{dx}\right) = \lambda gx + C_1$.

If we choose the lowest point of the string as origin, then $\frac{dy}{dx} = 0$ when $x = 0$, and hence $C_1 = 0$.

$$\therefore \quad \frac{dy}{dx} = \tan(\lambda gx) = \tan\left(\frac{x}{a}\right), \quad \text{if} \quad \frac{1}{a} = \lambda g.$$

Integrating, we have

$$y = -a\log\cos\frac{x}{a} = a\log\sec\frac{x}{a},$$

the constant of integration vanishing since x and y vanish together.

The curve has two vertical asymptotes when $x = \pm\frac{\pi}{2}a$.

Law of variation of the mass of the string.

From (1), $$T = C\frac{ds}{dx} = C\sqrt{1 + \left(\frac{dy}{dx}\right)^2} = C\sec\frac{x}{a}.$$

Hence $s = a\log\tan\left(\frac{\pi}{4} + \frac{x}{2a}\right)$, if s is measured from the lowest point.

$$\therefore \quad e^{\frac{s}{a}} + e^{-\frac{s}{a}} = \tan\left(\frac{\pi}{4} + \frac{x}{2a}\right) + \cot\left(\frac{\pi}{4} + \frac{x}{2a}\right) = 2\sec\frac{x}{a}.$$

$$\therefore \quad T = C \cdot \tfrac{1}{2}\left(e^{\frac{s}{a}} + e^{-\frac{s}{a}}\right).$$

Hence the mass per unit of length at any point distant s from the lowest point varies as $\frac{1}{2}(e^{\frac{s}{a}} + e^{-\frac{s}{a}})$, *i.e.* varies as

$$\cosh\left(\frac{s}{a}\right).$$

260. If the string does not necessarily lie in one plane, and X, Y, Z are the component forces parallel to the axes, the equations of equilibrium are, similarly as in Art. 257,

$$\frac{d}{ds}\left(T\frac{dx}{ds}\right) + mX = 0,$$

$$\frac{d}{ds}\left(T\frac{dy}{ds}\right) + mY = 0,$$

and
$$\frac{d}{ds}\left(T\frac{dz}{ds}\right) + mZ = 0.$$

Hence
$$T\frac{d^2x}{ds^2} + \frac{dT}{ds}\frac{dx}{ds} + mX = 0 \quad\quad (1),$$

$$T\frac{d^2y}{ds^2} + \frac{dT}{ds}\frac{dy}{ds} + mY = 0 \quad\quad (2),$$

and
$$T\frac{d^2z}{ds^2} + \frac{dT}{ds}\frac{dz}{ds} + mZ = 0 \quad\quad (3).$$

Multiply these equations by $\frac{dx}{ds}$, $\frac{dy}{ds}$, $\frac{dz}{ds}$ and add;

then, since
$$\left(\frac{dx}{ds}\right)^2 + \left(\frac{dy}{ds}\right)^2 + \left(\frac{dz}{ds}\right)^2 = 1, \text{ and}$$

$$\frac{dx}{ds}\frac{d^2x}{ds^2} + \frac{dy}{ds}\frac{d^2y}{ds^2} + \frac{dz}{ds}\frac{d^2z}{ds^2} = \frac{1}{2}\frac{d}{ds}\left[\left(\frac{dx}{ds}\right)^2 + \left(\frac{dy}{ds}\right)^2 + \left(\frac{dz}{ds}\right)^2\right] = 0,$$

we have
$$\frac{dT}{ds} + m\left(X\frac{dx}{ds} + Y\frac{dy}{ds} + Z\frac{dz}{ds}\right) = 0,$$

i.e.
$$T = C - \int m\,(Xdx + Ydy + Zdz).$$

Hence, if the external forces have no component along the tangent to the string, the tension is constant.

Again from (1), (2), and (3) eliminating T and $\frac{dT}{ds}$, we have

$$X\left[\frac{dy}{ds}\frac{d^2z}{ds^2}-\frac{dz}{ds}\frac{d^2y}{ds^2}\right]+Y\left[\frac{dz}{ds}\frac{d^2x}{ds^2}-\frac{dx}{ds}\frac{d^2z}{ds^2}\right]$$
$$+Z\left[\frac{dx}{ds}\frac{d^2y}{ds^2}-\frac{dy}{ds}\frac{d^2x}{ds^2}\right]=0,$$

i.e. $$X\lambda+Y\mu+Z\nu=0,$$

where (λ, μ, ν) are the direction cosines of the binormal of the curve in which the string lies.

Hence the resulting external force at any point P of the string is perpendicular to the binormal at P, *i.e.* it always lies in the osculating plane at the point P of the string.

If we multiply (1), (2), and (3) by $\frac{d^2x}{ds^2}$, $\frac{d^2y}{ds^2}$, and $\frac{d^2z}{ds^2}$ and add, we have $\frac{T}{\rho^2}+m\left(X\frac{d^2x}{ds^2}+Y\frac{d^2y}{ds^2}+Z\frac{d^2z}{ds^2}\right)=0,$

i.e. $$\frac{T}{\rho}+mR=0,$$

where ρ is the radius of curvature, and R is the resolved part of the external force along the principal normal whose direction cosines are

$$\rho\frac{d^2x}{ds^2},\quad \rho\frac{d^2y}{ds^2},\quad \rho\frac{d^2z}{ds^2}.$$

Once more, integrating the first three equations of this article, we have

$$T\frac{dx}{ds}+\int mX ds=A,\quad T\frac{dy}{ds}+\int mY ds=B,$$

and $$T\frac{dz}{ds}+\int mZ ds=C.$$

Hence the equation of the curve in which the string lies is given by

$$\frac{A-\int mX ds}{\frac{dx}{ds}}=\frac{B-\int mY ds}{\frac{dy}{ds}}=\frac{C-\int mZ ds}{\frac{dz}{ds}},$$

where A, B, and C are constants.

261. *String on a given smooth surface under given forces.*

Let R be the pressure of the surface at any point of the string and (l_1, m_1, n_1) the direction cosines of the normal drawn inwards. The equations of equilibrium are then

$$\frac{d}{ds}\left(T\frac{dx}{ds}\right) + mX - Rl_1 = 0 \quad \text{..............(1)},$$

$$\frac{d}{ds}\left(T\frac{dy}{ds}\right) + mY - Rm_1 = 0 \quad \text{..............(2)},$$

and

$$\frac{d}{ds}\left(T\frac{dz}{ds}\right) + mZ - Rn_1 = 0 \quad \text{..............(3)}.$$

Also the equation of the surface is known, say

$$f(x, y, z) = 0 \quad \text{......................(4)}.$$

Since $l_1\frac{dx}{ds} + m_1\frac{dy}{ds} + n_1\frac{dz}{ds}$ is proportional to

$$\frac{df}{dx}\frac{dx}{ds} + \frac{df}{dy}\frac{dy}{ds} + \frac{df}{dz}\frac{dz}{ds}$$

and is therefore zero, we have, as in Art. 260, on multiplying (1), (2), (3) by $\frac{dx}{ds}$, $\frac{dy}{ds}$, $\frac{dz}{ds}$, adding and integrating,

$$T = C - \int m(X\,dx + Y\,dy + Z\,dz) \quad \text{........(5)}$$

$= C - V$, if the external forces are given by a potential function V.

Again, if from equations (1), (2), (3) we eliminate $\frac{dT}{ds}$ and R, we have

$$\left(T\frac{d^2x}{ds^2} + mX\right)\left(\frac{dy}{ds}n_1 - \frac{dz}{ds}m_1\right) + \text{two similar terms} = 0.$$

Substituting the above value of T in this we have a differential equation which, with (4), gives the curve in which the string rests.

262. If in the previous article the string rests under the action of no external forces, then $X = Y = Z = 0$, and hence, from (5), $T = \text{const.} = C$.

Hence (1), (2), and (3) give

$$\frac{\frac{d^2x}{ds^2}}{l_1} = \frac{\frac{d^2y}{ds^2}}{m_1} = \frac{\frac{d^2z}{ds^2}}{n_1}, \qquad \textit{i.e.} \quad \frac{\frac{d^2x}{ds^2}}{\frac{df}{dx}} = \frac{\frac{d^2y}{ds^2}}{\frac{df}{dy}} = \frac{\frac{d^2z}{ds^2}}{\frac{df}{dz}}.$$

The curve of the string is therefore such that its principal normal coincides with the normal to the surface, so that the osculating plane at every point of the string passes through the normal to the surface. Such a curve is called a Geodesic of the surface, and is such that any element PQ of it is the shortest distance on the surface between P and Q.

EXAMPLES

1. In the catenary of uniform strength, prove that

$$x=a\psi,\quad s=a\log(\sec\psi+\tan\psi),\quad \cos\psi\cosh\frac{s}{a}=1,\quad\text{and}\quad \rho=a\cosh\frac{s}{a},$$

where ρ is the radius of curvature and ψ the inclination of the tangent to the axis of x.

Hence shew that the mass per unit length at any point varies as the corresponding radius of curvature.

2. A catenary of uniform strength has a span of 50 feet and its total weight is 600 lbs.; the density of the material is 80 lbs. per cubic foot and the tension is 20 lbs. weight per square inch of its section; find the equation to its curve, and the areas of its cross sections at its lowest and highest points. $\left[\frac{y}{36}=\log_e\sec\frac{x}{36};\ 15\cot\frac{25}{36}\text{ and }15\operatorname{cosec}\frac{25}{36}\text{ sq. ins.}\right]$

3. If the density at any point of a cord vary as the radius of curvature of the curve in which it hangs, shew that this curve is the catenary of uniform strength.

4. Shew that the form of the curve of a suspension bridge when the weight of the rods is taken into account, but the weight of the rest of the bridge neglected, is the orthogonal projection of a catenary, the rods being supposed vertical and equidistant.

5. Find the form in which a chain hangs when the line-density is given by $\frac{T_0}{ga}\sec^2\frac{s}{a}$, T_0 being the tension at the lowest point and s being measured from this point. [A circle, of radius a.]

6. A string, of length πa, is fastened to two points of a straight line, distant $2a$ from one another, and is repelled by a force perpendicular to the line and rests in the form of a semi-circle. Shew that the force varies inversely as the square of the distance from the given line.

7. In a non-uniform string hanging under gravity the area of the cross section at any point is inversely proportional to the tension. Shew that the curve is an arc of a parabola with its axis vertical.

8. If a uniform string hangs in the form of a parabola, whose focus is S, under the action of normal forces only, shew that the force at any point P varies inversely as $(SP)^{\frac{3}{2}}$ and that the tension is constant.

263. ***Light inextensible string resting on a smooth plane curve.***

Let PQ be an element δs of the string, the arc OP being s and O a fixed point on the curve.

Let T and $T+\delta T$ be the tensions at P and Q, the tangents at which are inclined at angles ψ and $\psi+\delta\psi$ to any fixed line.

Let R be the reaction of the curve per unit of length of the element PQ, so that the reaction on the element may be taken to be $R\delta s$ acting along the normal at P drawn outwards.

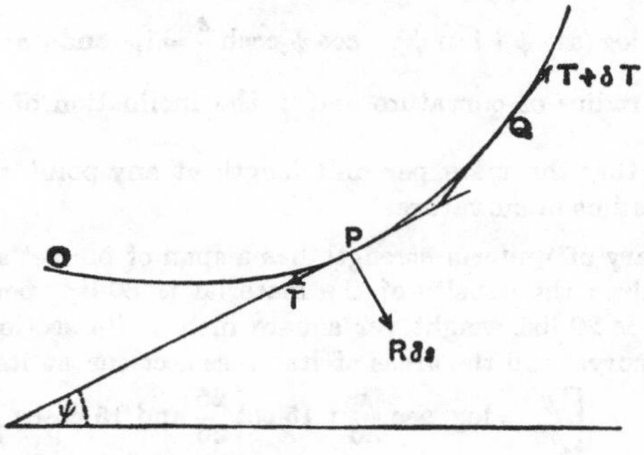

Resolving along the tangent and normal at P, we have

$$\left.\begin{aligned}(T+\delta T)\cos\delta\psi &= T\\ \text{and}\qquad (T+\delta T)\sin\delta\psi &= R\delta s\end{aligned}\right\}$$

To the first order of $\delta\psi$, we have $\cos\delta\psi = 1$, and $\sin\delta\psi = \delta\psi$. Hence, neglecting squares of small quantities, we have

$$\delta T = 0 \quad\text{.............................(1)},$$

and

$$T = R\frac{ds}{d\psi} = R\rho \quad\text{......................(2)},$$

where ρ is the radius of curvature at P. (1) gives T = constant.

Hence the tension of a light string passing round a smooth curve is constant throughout.

Also (2) gives $R \propto \frac{1}{\rho}$, *i.e.* the normal reaction varies as the curvature of the curve.

264. *Heavy string resting against a smooth curve.*

If the line from which ψ is measured be horizontal and taken as the axis of x we have, in addition to the forces of the previous article, a vertical force $w\delta s$ acting at P.

Hence, instead of the equations of the previous article, we obtain

$$\left.\begin{aligned}(T+\delta T)\cos\delta\psi &= T + w\,.\,\delta s\,.\sin\psi\\ \text{and}\qquad (T+\delta T)\sin\delta\psi &= R\delta s + w\,.\,\delta s\,.\cos\psi\end{aligned}\right\}.$$

These give, as before,

$$\delta T = w\delta s\,.\sin\psi = w\delta y \text{ ....(1)},$$

and

$$T\frac{d\psi}{ds} = R + w\cos\psi \text{ ....(2)}.$$

(1) gives $T = C + wy$, assuming w to be constant.

Hence, if T_1 and T_2 be the tensions at two points whose ordinates are y_1 and y_2,

$$T_1 - T_2 = w(y_1 - y_2),$$

i.e. if a heavy uniform string rests against a smooth curve the difference of the tensions at any two points is equal to the weight of a portion of the string whose length is equal to the vertical distance between the two points.

When T is known, then (2) gives

$$R = \frac{T}{\rho} - w\cos\psi,$$

where ρ is the radius of curvature at P.

265. Ex. *A heavy uniform string rests symmetrically on a smooth catenary whose axis is vertical and vertex upwards; find the tension and pressure on the curve at any point.*

The equations of the previous article become in this case

$$\frac{dT}{ds} + w\sin\psi = 0 \text{ ....(1)},$$

and

$$R = \frac{T}{\rho} + w\cos\psi \text{ ....(2)}.$$

But $s = c\tan\psi$, so that $\frac{dT}{d\psi} = -wc\frac{\sin\psi}{\cos^2\psi}$.

$$\therefore\; T = -\frac{wc}{\cos\psi} + A = wc[\sec\psi_0 - \sec\psi],$$

where ψ_0 is the inclination of the tangent at either of the free ends.

Hence (2) gives

$$R = \frac{T\cos^2\psi}{c} + w\cos\psi = w\sec\psi_0\,.\cos^2\psi = \frac{wc^2}{y^2}\,.\sec\psi_0.$$

Hence R varies inversely as the square of the distance below the directrix of the catenary.

266. *Light inextensible string resting in limiting equilibrium on a rough plane curve under the action of no external forces.*

Let PQ be an element of the string; let the arc OP be s, where O is a fixed point; let PQ be δs, and the angles that the tangents at P and Q make with some fixed line be ψ and $\psi + \delta\psi$.

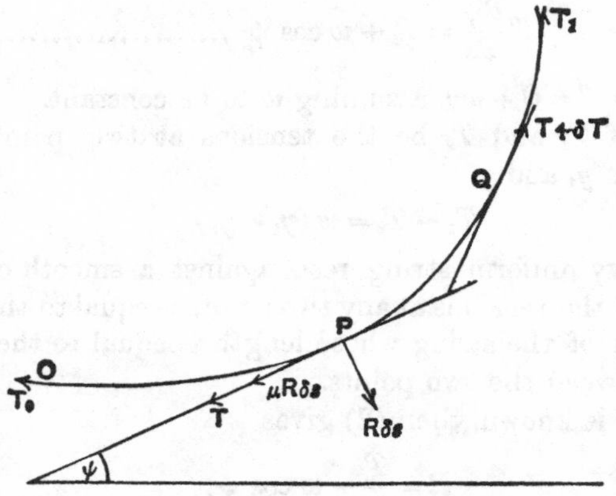

Let the tensions at the points where the string leaves the curve be T_0 and T_1, and suppose the tension T_1 to be on the point of overcoming T_0, so that the element PQ is on the point of motion in the direction PQ and hence the friction acts in the direction PT.

If R be the reaction on PQ per unit of length, the total normal reaction on PQ may be taken to be $R\delta s$ acting at P, and the tangential action is $\mu R\delta s$ acting in the direction PT.

Resolving the forces along the tangent and normal at P, we then have

$$(T + \delta T)\cos\delta\psi = T + \mu R\delta s,$$

and

$$(T + \delta T)\sin\delta\psi = R\delta s.$$

But $\cos\delta\psi = 1$, and $\sin\delta\psi = \delta\psi$, neglecting squares of $\delta\psi$.

These equations therefore give

$$\frac{dT}{ds} = \mu R, \quad \text{and} \quad T\frac{d\psi}{ds} = R.$$

Eliminating R, we have $\frac{dT}{T} = \mu d\psi$.

$$\therefore \quad \log T = \mu\psi + \text{const.},$$

i.e.
$$T = Ae^{\mu\psi}.$$

If ψ be measured from a line which is parallel to the direction at the point where the string leaves the curve, then $T = T_0$ when $\psi = 0$.

Hence $A = T_0$ and $T = T_0 . e^{\mu\psi}$, giving the tension at any point P in terms of the terminal tension and the angle through which the tangent at P has turned from the terminal tangent.

267. As a numerical example, take the case of a rope twisted through one complete revolution round a post. For an ordinary hemp rope round an oak post, μ is about $\frac{1}{2}$.

Hence $\frac{T}{T_0} = e^{\frac{1}{2}.2\pi} = e^{\pi} = (2\cdot718)^{3\cdot1416} =$ about 23, so that the tension of the rope is increased 23 times by its being twisted once round the post.

If it be twisted twice round, the ratio becomes about $e^{2\pi}$ or about 535.

268. Heavy String. If the string be heavy and w its weight per unit of length then, the angle ψ being measured from the horizontal, we have, in place of the equations of the previous article,

$$\left.\begin{aligned} (T + \delta T)\cos\delta\psi &= T + \mu R\,\delta s + w\delta s \sin\psi \\ \text{and} \qquad (T + \delta T)\sin\delta\psi &= R\,\delta s + w\delta s\cos\psi \end{aligned}\right\}.$$

Hence, in the limit,

$$\frac{dT}{ds} = \mu R + w\sin\psi \quad \text{..................(1)},$$

and
$$T\frac{d\psi}{ds} = R + w\cos\psi \quad \text{.....................(2)}.$$

$$\therefore \quad \frac{dT}{ds} - \mu T\frac{d\psi}{ds} = w(\sin\psi - \mu\cos\psi).$$

Since in our figure s and ψ increase together, $\frac{ds}{d\psi} = \rho$.

$$\therefore \quad \frac{dT}{d\psi} - \mu T = w\rho(\sin\psi - \mu\cos\psi).$$

To integrate this linear differential equation we, according to the usual rule, multiply by $e^{-\mu\psi}$ and have, on integration,

$$T \cdot e^{-\mu\psi} = C + \int w\rho\,(\sin\psi - \mu\cos\psi)\,e^{-\mu\psi}\,d\psi.$$

The curve on which the string rests being known, we can obtain ρ in terms of ψ and the integral on the right-hand side can be found. T is thus determined.

The reaction R is then found from either equation (1) or (2).

269. Ex. *A uniform inextensible string, of length l, hangs in limiting equilibrium over a fixed rough cylinder, of radius a, whose axis is horizontal; shew that the length of the greater of the two vertical portions is*

$$\frac{l-\pi a}{1+e^{-\mu\pi}} + \frac{2\mu a}{1+\mu^2}.$$

Let the shorter and longer portions hanging from points A and B, at the ends of the horizontal diameter AB, be y_1 and y_2, and let motion be about to ensue from A towards B. Then, T being the tension at a point P such that AP subtends an angle θ at the centre, we have, as in the last article,

$$(T+\delta T)\cos\delta\theta - T - \mu R\delta s - mg\cos\theta\delta s = 0,$$

and
$$(T+\delta T)\sin\delta\theta - R\delta s + mg\sin\theta\delta s = 0.$$

$$\therefore \quad \frac{1}{a}\frac{dT}{d\theta} = \mu R + mg\cos\theta,$$

and
$$\frac{T}{a} = R - mg\sin\theta.$$

Hence
$$\frac{dT}{d\theta} - \mu T = mga\,(\cos\theta + \mu\sin\theta).$$

$$\therefore \quad Te^{-\mu\theta} = mga\int(\cos\theta + \mu\sin\theta)e^{-\mu\theta}\,d\theta$$

$$= mgae^{-\mu\theta}\frac{(1-\mu^2)\sin\theta - 2\mu\cos\theta}{1+\mu^2} + C.$$

When $\theta = 0$, $T = mgy_1$, and when $\theta = \pi$, $T = mgy_2$.

$$\therefore \quad mgy_1 = -mga\frac{2\mu}{1+\mu^2} + C,$$

and
$$mgy_2 \cdot e^{-\mu\pi} = mga\frac{2\mu}{1+\mu^2}e^{-\mu\pi} + C.$$

Hence
$$y_2e^{-\mu\pi} - y_1 = \frac{2\mu a}{1+\mu^2}\left(1+e^{-\mu\pi}\right).$$

Also
$$y_1 + y_2 + \pi a = l.$$

Hence the result.

EXAMPLES

1. A single movable pulley, of weight W, is just supported by a power P which is applied to one end of a light cord which goes under the pulley and is then fastened to a fixed point; shew that, if ϕ be the angle subtended at the centre by the part of the string in contact with the pulley, then $P^2(1-2e^{\mu\phi}\cos\phi+e^{2\mu\phi})=W^2$.

2. A light string is passed over two rough pegs A and B in the same horizontal line at a distance $2a$ apart. The ends are fastened to a weight C, and in the position of limiting equilibrium AB subtends a right angle at C. Shew that the horizontal distance of C in this position from the middle point of AB is $a\tanh\left(\frac{3\mu\pi}{2}\right)$, where μ is the coefficient of friction.

3. A heavy particle is attached to an endless light inextensible string which passes over a rough pulley fixed in a vertical plane. If the straight parts of the string are inclined to each other at an angle a, prove that, for limiting equilibrium, their inclinations to the vertical are

$$\tan^{-1}\frac{\sin a}{\cos a+e^{\pm\mu(a+\pi)}}.$$

4. Four rough circular pegs are at the angular points of a square in a vertical plane with its sides horizontal and vertical. Over each peg passes a string supporting a weight W, and the other ends of these four strings are knotted together. Shew that the greatest weight that can be attached to this knot, so that it may remain in equilibrium at the centre of the square, is $2\sqrt{2}\,We^{\frac{\mu\pi}{4}}\sinh\left(\frac{\mu\pi}{2}\right)$.

5. Three equally rough pegs A, B, C of the same circular cross section are placed at the corners of an equilateral triangle, so that BC is horizontal and A above BC. Shew that the greatest weight which can be supported by a weight W tied to the end of a string, which is carried once round the pegs and does not completely surround each peg, is $We^{5\mu\pi}$, where μ is the coefficient of friction.

6. A circle rests in a vertical plane, being pressed against a perfectly rough vertical wall by a string fixed to a point in the wall above the circle. The string sustains a weight P, and the coefficient of friction between the string and the circle is μ. If W be the weight of the circle and θ the angle between the string and the wall, shew that, if the circle is on the point of sliding, then $P(1+\cos\theta)e^{\mu\theta}=W+2P$.

7. A weightless string lies stretched in one plane across a rough sphere of radius a; shew that the distance of the plane from the centre cannot exceed $a\sin\epsilon$, where ϵ is the angle of friction.

8. If a heavy uniform string passes round various smooth curves in the same vertical plane and its ends hang vertically, shew that they are in the same horizontal straight line.

9. A uniform heavy string rests on a smooth parabola, whose axis is vertical and vertex upwards, so that its ends are at the extremities of the latus rectum. Shew that the pressure on the curve, at the point where the tangent makes an angle ϕ with the horizontal, is

$$\frac{w}{2}(2\cos^3\phi+\cos\phi),$$

where w is the weight of the string per unit of length.

10. Upon a rough circle fixed vertically is placed a string which subtends an angle β at the centre. If the string is on the point of slipping off, prove that the angular distance a of its upper end from the highest point of the circle is determined by the equation

$$\frac{\cos(a+\beta-2\epsilon)}{\cos(a-2\epsilon)}=e^{\beta\tan\epsilon},$$

where ϵ is the angle of friction and a is measured in the direction towards which the string is slipping.

11. A uniform heavy string rests on the upper surface of a rough vertical circle of radius a, and partly hangs vertically. Prove that, if one end be at the highest point of the circle, the greatest length that can hang freely is $\dfrac{2\mu a+(\mu^2-1)\,a\,e^{\frac{\mu\pi}{2}}}{\mu^2+1}$.

12. A heavy chain, of length l, rests partly on a rough table and the remainder after passing over the smooth edge of the table, which is rounded off in the form of a cylinder of radius a, hangs freely down. If the coefficient of friction is μ, shew that the least length on the table is

$$\frac{1}{\mu+1}\left[l-\frac{\pi a}{2}+a\right].$$

13. A heavy uniform chain rests on a rough cycloid, whose axis is vertical and vertex upwards, one end of the chain being at the vertex and the other at a cusp; if the equilibrium be limiting, shew that

$$(1+\mu^2)\,e^{\frac{\mu\pi}{2}}=3.$$

14. A heavy uniform string is placed upon a rough catenary, whose axis is vertical and vertex upwards; the coefficient of friction being given by $e^{\mu\pi}=4$, shew that the string will be in limiting equilibrium with one end at the vertex if its length is equal to the parameter of the catenary.

15. A string rests on a rough semicircle, being acted on by a constant attractive force towards one of its extremities, and the friction is just sufficient to prevent motion. Shew that the coefficient of friction is given by $e^{\mu\pi}=\dfrac{3\mu}{1-2\mu^2}$.

16. An inextensible string, whose length is $2l$, passes over two equal smooth circular pulleys whose centres are in the same horizontal line and at a distance $2b$ apart. If a be the radius of the pulleys, and ψ the angle subtended at the centre of one of the pulleys by the portion of string in contact with it, prove that $b+a\cos\psi=\cot\frac{\psi}{2}(a\sin\psi+l-a\psi)\log\tan\frac{\psi}{2}$.

17. A string, whose length is l, is hung over two small rough pegs at a distance $2a$ apart in a horizontal line. If one free end of the string is as much as possible lower than the other, the inclination, θ, of the tangent to the vertical at either peg is given by the equation

$$\frac{l}{2a}\sin\theta\log\cot\frac{\theta}{2}=\cos\theta+\cosh\{\mu(\pi-\theta)\}.$$

Shew also that the lengths of the vertical portions are in the ratio $e^{2\mu\pi}:e^{2\mu\theta}$ and that the part of the string between the pegs is of length

$$2a\cot\theta\div\log\cot\frac{\theta}{2}.$$

270. Central Forces. *An inextensible string is in equilibrium in one plane under the action of forces, which vary according to some function of the distance from a given point O, and act either towards or from O; to find the curve of equilibrium.*

Let PQ be an element δs of the string, where s is the arc CP measured from some fixed point C of the curve. Let T and $T+\delta T$ be the tensions at P and Q, and ψ and $\psi+\delta\psi$ the angles which the tangents at P and Q make with some fixed line Ox.

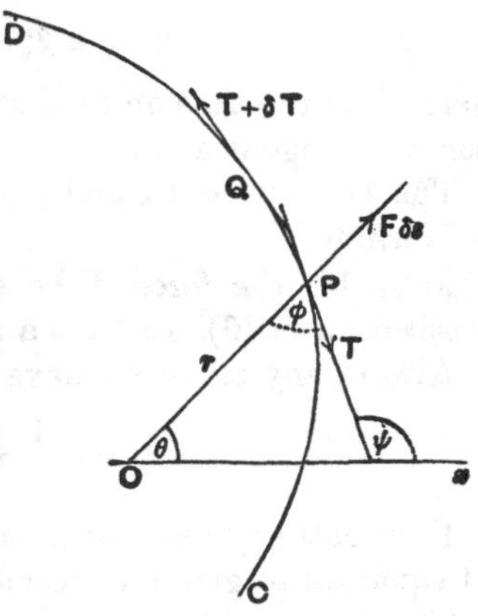

Let F be the force at P per unit of length in the direction OP, so that F = some function of r $=\phi(r)$. The resultant of the forces from O to the different points of the arc PQ may be taken to be $F\delta s$ along OP; for in the limit we shall take δs to be very small.

Resolving the forces on the element PQ along and perpendicular to the tangent at P, we have

$$(T+\delta T)\cos\delta\psi-T+Fm\delta s\cos\phi=0\qquad\text{.........(1)},$$

and

$$(T+\delta T)\sin\delta\psi-Fm\delta s\sin\phi=0\qquad\text{...........(2)},$$

where m is the mass of the string per unit of length.

Putting $\cos\delta\psi = 1$ and $\sin\delta\psi = \delta\psi$, and taking $\delta\psi$ to be indefinitely small, these give in the limit

$$\frac{dT}{ds} = -Fm\cos\phi = -Fm\frac{dr}{ds} \quad \text{..............(3)},$$

and $$T = Fm\rho\sin\phi = Fm \cdot r\frac{dr}{dp}\cdot\frac{p}{r} = Fm \cdot p\frac{dr}{dp} \quad \text{......(4)},$$

where p is the perpendicular from O upon the tangent at P.

(3) gives $$T = -\int mFdr + A \quad \text{..................(5)},$$

and, dividing (3) by (4), we have

$$\frac{dT}{T} = -\frac{dp}{p},$$

and hence $$T \cdot p = \text{const.} = B \quad \text{..................(6)}.$$

The equation (6) may be more easily obtained by considering the equilibrium of the finite portion, CP, of the string.

Take moments about O for all the forces acting on CP. All the central forces acting on it pass through O, and thus have no moments about O. Hence the moments about O of the tensions at P and C are equal. Hence

$$T \cdot p = T_0 \cdot p_0 = \text{constant} \quad \text{................(7)},$$

where T_0 is the tension at C and p_0 is the perpendicular from O upon the tangent at C.

The equations (5), and (6) or (7), give all the conditions of equilibrium.

First, let the force F be given; then (5) gives T, and, on substitution in (6), we have a relation between p and r.

Also in any curve we have

$$\frac{1}{p^2} = \frac{1}{r^2} + \frac{1}{r^4}\left(\frac{dr}{d\theta}\right)^2.$$

Eliminating p between these two relations, we have a differential equation to give r in terms of θ.

The result will contain three arbitrary constants. Two of them will be found since the two terminal points C and D of the string are given; the third will be determined from the fact that the length of the string between C and D is given.

Secondly, let the form of the string be given. We are thus given the relation between the p and r of the curve.

(6) then gives T in terms of r, and (3) gives $F=-\frac{1}{m}\frac{dT}{dr}$, and hence we have F in terms of r.

271. **Ex. 1.** *Shew that a string will rest in the form of a portion of a cardioid if it be acted upon by a force from its pole varying as*

$$\frac{1}{(distance)^{\frac{5}{2}}}.$$

The equation to a cardioid is $r=a(1+\cos\theta)$, so that

$$\frac{1}{p^2}=\frac{1}{r^2}+\frac{1}{r^4}\left(\frac{dr}{d\theta}\right)^2=\frac{1}{r^2}+\frac{1}{r^4}\cdot a^2\sin^2\theta=\frac{1}{r^4}\left[r^2+a^2-(r-a)^2\right]=\frac{2a}{r^3}.$$

Hence equation (6) gives $T=B.\sqrt{\frac{2a}{r^3}}=\mu r^{-\frac{3}{2}}$.

From equation (3) we then have

$$F=-\frac{1}{m}\frac{dT}{dr}=\frac{3\mu}{2m}.r^{-\frac{5}{2}}.$$

Ex. 2. *An infinite string passes through two small smooth rings A and B, and is acted upon by a force from a given fixed point O which varies inversely as the cube of the distance. Shew that the part of the string between the rings is in the form of the arc of a circle.*

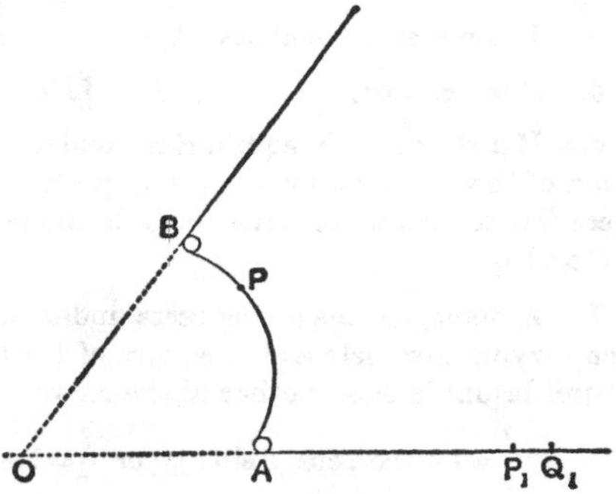

Let T_1 be the tension at any point P_1 of the straight portion of the string at a distance x from the centre O. Then

$$(T_1+\delta T_1)+\frac{\mu m}{x^3}\delta x-T_1=0,$$

giving $\frac{dT_1}{dx}=-\frac{\mu m}{x^3}$,

and $\therefore\ T_1=\frac{\mu m}{2x^2}+K$ ...(1).

Since the string must clearly have zero tension at infinity, we have $K=0$. Hence, if $OA=a$, the tension at A is $\frac{\mu m}{2a^2}$. Now the tension of the string is unaltered by passing through the smooth ring at A, so that the tension of the curved part at A is $\frac{\mu m}{2a^2}$ also.

For the curved part, equation (5) of the last article then gives

$$T=-\int\frac{m\mu}{r^3}dr=\frac{m\mu}{2r^2}+A.$$

Also, when $r=a$, we have seen that $T=\frac{m\mu}{2a^2}$, so that A is zero.

Equation (6) then gives

$$\frac{1}{p}=\frac{m\mu}{2r^2 . B}=\frac{\lambda}{r^2}, \text{ where } \lambda \text{ is some constant.}$$

$$\therefore \frac{1}{r^2}+\frac{1}{r^4}\left(\frac{dr}{d\theta}\right)^2=\frac{1}{p^2}=\frac{\lambda^2}{r^4}.$$

$$\therefore \theta=\int\frac{dr}{\sqrt{\lambda^2-r^2}}=\sin^{-1}\frac{r}{\lambda}+\gamma.$$

$$\therefore r=\lambda\sin(\theta-\gamma), \text{ which is the equation of a circle.}$$

If OA be the initial line, if $OB=b$ and $\angle AOB=a$, then the two points $(a, 0)$ and (b, a) lie on the curve, so that the equation becomes

$$r=a\cos\theta+\frac{b-a\cos a}{\sin a}\sin\theta.$$

EXAMPLES

Find the law of force in the case of strings resting in the form of the following curves under a central force F from the pole:

1. Parabola, focus the pole. $\left[F\propto r^{-\frac{3}{2}}.\right]$

2. Equiangular spiral, $r=ae^{\theta\cot a}$. $[F\propto r^{-2}.]$

3. Rectangular hyperbola, centre the pole. [F is constant and attractive.]

4. Lemniscate, $r^2=a^2\cos 2\theta$. $[F\propto r^{-4}.]$

5. $r^n\cos n\theta=a^n$. [$F\propto r^{n-2}$, and is attractive if $n>1$.]

6. If a string be in equilibrium under any central forces, the resultant action of these central forces on any portion PQ of the string is along OT, where O is the centre of force and T is the point of intersection of tangents at P and Q.

7. A homogeneous string rests under the action of a central repulsive force varying inversely as the square of the distance; verify that the form in equilibrium is one or other of the curves

$$\frac{l}{r}=1+\sec a\cos(\theta\sin a), \quad \text{or} \quad \frac{l}{r}=1+\operatorname{sech} a\cosh(\theta\sinh a).$$

8. A string, whose length is infinite, has one end attached to a fixed point O and after passing through a small smooth fixed ring goes to infinity. It is acted upon by a central repulsive force from O varying inversely as the nth power of the distance. Shew that the curved part of the string is given by the equation

$$r^{n-2}=a^{n-2}\cos(n-2)\theta.$$

If $n=2$, shew that the curved part is an equiangular spiral.

9. A string rests in the form of a plane curve under the action of a central repulsive force; if the force at any point be proportional to the curvature, prove that the curve is a parabola.

272. Extensible strings. The equations of equilibrium for extensible strings are formed as in the previous part of this chapter. The tension of any element of the string is connected, by means of Hooke's Law, with the stretched and unstretched lengths of the element. It must be carefully noted that a heavy elastic string when stretched is not of constant density, even if it were uniform when in the unstretched state.

273. *A uniform extensible string, of weight W and natural length l, is suspended from a fixed point and at the other end is hung a weight W'; if λ be the coefficient of elasticity, shew that the whole extension of the string is* $\frac{l}{\lambda}\left[\frac{W}{2}+W'\right]$.

Let T and $T+\delta T$ be the tensions of the string at depths x and $x+\delta x$, and let x_0 be the unstretched length of the part whose stretched length is x. Hence the weight of the part whose stretched length is $\delta x = \frac{W}{l}.\delta x_0$.

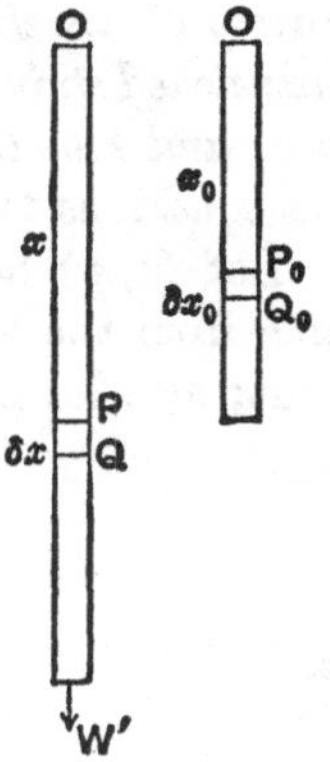

For the equilibrium of this element we have

$$T = T + \delta T + \frac{W}{l}.dx_0,$$

so that
$$\frac{dT}{dx_0} = -\frac{W}{l} \quad \text{..............(1).}$$

Also, by Hooke's Law,
$$T = \lambda\frac{\delta x - \delta x_0}{\delta x_0},$$

so that
$$\frac{dx}{dx_0} = 1 + \frac{T}{\lambda} \quad \text{......................(2).}$$

(1) and (2) give
$$\frac{d^2x}{dx_0^2} = -\frac{W}{\lambda l}.$$

$$\therefore \quad \frac{dx}{dx_0} = -\frac{W}{\lambda l}x_0 + A \quad \text{.............(3).}$$

Now, when $x_0 = l$, T must $= W'$, and hence then, by (2),

$$\frac{dx}{dx_0} = 1 + \frac{W'}{\lambda}.$$

Hence, from (3), $1 + \frac{W'}{\lambda} = -\frac{W}{\lambda} + A.$

$$\therefore \frac{dx}{dx_0} = -\frac{W}{\lambda l}x_0 + 1 + \frac{W + W'}{\lambda}.$$

$$\therefore x = -\frac{W}{\lambda l}\frac{x_0^2}{2} + \left(1 + \frac{W + W'}{\lambda}\right)x_0 \quad \text{........(4)},$$

the constant of integration being zero since x and x_0 vanish together.

(4) gives the stretched length corresponding to any unstretched length. When $x_0 = l$, we have the whole stretching

$$= -\frac{W}{\lambda l}\frac{l^2}{2} + \frac{W + W'}{\lambda}l = \frac{l}{\lambda}\left[\frac{W}{2} + W'\right].$$

274. *A heavy uniform elastic string is hung up under gravity as in the common catenary. If c be the length of the unstretched string whose weight is equal to the tension at the lowest point, and k be the ratio of this tension to the modulus of elasticity, find equations to give the form of the string.*

Let (x, y) be the coordinates of a point whose arcual distance from the lowest point is s, and let T be the tension at that point; also let s_0 be the unstretched length of this arc s, so that

$$\frac{T}{\lambda} = \frac{ds - ds_0}{ds_0},$$

i.e.
$$\frac{ds}{ds_0} = 1 + \frac{k}{wc}.T \quad \text{.....................(1)},$$

where w is the weight of unit length of unstretched string.

The equations of Art. 257 are then

$$-T\frac{dx}{ds} + \left\{T\frac{dx}{ds} + \frac{d}{ds}\left(T\frac{dx}{ds}\right)\delta s + \dots\right\} = 0,$$

and
$$-T\frac{dy}{ds} + \left\{T\frac{dy}{ds} + \frac{d}{ds}\left(T\frac{dy}{ds}\right)\delta s + \dots\right\} = w.\delta s_0,$$

i.e.
$$\frac{d}{ds}\left(T\frac{dx}{ds}\right) = 0 \quad \text{.......................(2)},$$

and
$$\frac{d}{ds}\left(T\frac{dy}{ds}\right) = w\frac{ds_0}{ds} \quad \text{...................(3)}.$$

These give $$T\frac{dx}{ds} = \text{const.} = wc$$

and $$T\frac{dy}{ds} = ws_0,$$

since s and s_0 vanish together.

Squaring and adding, we have $T = w\sqrt{c^2 + s_0^2}$.

Hence, by (1),

$$\frac{dx}{ds_0} = \frac{dx}{ds}\frac{ds}{ds_0} = wc\left[\frac{1}{T} + \frac{k}{wc}\right] = k + \frac{c}{\sqrt{c^2+s_0^2}} \quad \text{......(4)},$$

$$\frac{dy}{ds_0} = \frac{dy}{ds}\frac{ds}{ds_0} = ws_0\left[\frac{1}{T} + \frac{k}{wc}\right] = \frac{ks_0}{c} + \frac{s_0}{\sqrt{c^2+s_0^2}} \quad \text{......(5)},$$

and, by squaring and adding,

$$\frac{ds}{ds_0} = 1 + \frac{k}{c}\sqrt{c^2+s_0^2}.$$

Integrating these equations, we have

$$x = ks_0 + c\log\frac{s_0 + \sqrt{s_0^2+c^2}}{c} \quad \text{......................(6)},$$

$$y = k\frac{s_0^2}{2c} + \sqrt{s_0^2+c^2} - c \quad \text{..........................(7)},$$

and $$s = s_0 + \frac{k}{2c}\left[s_0\sqrt{c^2+s_0^2} + c^2\log\frac{s_0+\sqrt{s_0^2+c^2}}{c}\right] \quad \text{...(8)},$$

taking x and y to vanish with s.

(7) determines s_0 in terms of y and then (6) and (8) give x and s as functions of y.

The equations (4) and (5) give

$$\tan\psi = \frac{dy}{dx} = \frac{s_0}{c},$$

and hence, from (8),

$$s = c\tan\psi + \frac{kc}{2}\left[\sec\psi \,.\, \tan\psi + \log(\sec\psi + \tan\psi)\right].$$

If we put $s_0 = c\sinh u$, the equations (6), (7), and (8) can be written

$$\frac{x}{c} = u + k\sinh u,$$

$$\frac{y}{c} + 1 + \frac{k}{4} = \cosh u + \frac{k}{4}\cosh 2u,$$

and $$\frac{s}{c} = \sinh u + \frac{ku}{2} + \frac{k}{4}\sinh 2u.$$

275. **Ex.** *An extensible string is being wound very slowly on to the rim of a wheel, rough enough to prevent any sliding, and the other end is attached to a weight W which is on the ground at a depth l below the centre of the wheel, so that then the hanging part is initially vertical and unstretched. Shew that the work done in turning the wheel so as to just lift the weight off the ground is*

$$lW - l\lambda \log\left[1+\frac{W}{\lambda}\right],$$

where the weight of the string is neglected.

At any instant during the operation let x be the unstretched length of the string that is then vertical, and T its tension, so that, by Hooke's Law,

$$T=\lambda\frac{l-x}{x} \qquad \text{.............................(1).}$$

When l becomes $l+a\,.\,\delta\theta$, where a is the radius of the wheel, and $\delta\theta$ is the angle turned through by the wheel, then

$$T+\delta T=\lambda\frac{l-\dfrac{l}{l+a\,.\,\delta\theta}x}{\dfrac{l}{l+a\,.\,\delta\theta}x}=\lambda\frac{l+a\,.\,\delta\theta-x}{x}, \text{ so that } \delta T=\lambda\frac{a\,.\,\delta\theta}{x}.$$

Hence the work done during this infinitesimal stretching

$$=T\,.\,a\delta\theta=\frac{T\,.\,x\delta T}{\lambda}=\frac{lT\,.\,\delta T}{T+\lambda}.$$

Hence the whole work done until the weight rises, *i.e.* until T is equal to W

$$=\int_0^W\frac{lT\,.\,dT}{T+\lambda}=l\Big[T-\lambda\log(T+\lambda)\Big]_0^W=lW-\lambda l\log\frac{W+\lambda}{\lambda}.$$

EXAMPLES

1. When a uniform elastic string AB is hung up under gravity, prove that the upper half of the string lengthens three times as much as the lower half. If P is a point on it such that $AP:PB::\sqrt{2}-1:1$, shew also that the stretchings of the parts above and below P are equal.

2. A heavy elastic string of natural length $2l$, which would stretch to $4l$ if hung up by one end, rests on a smooth table of width $2a$ with its ends hanging over the sides of the table; find the whole extension of the string, and shew that the tension of the part in contact with the table is $\frac{1}{2}\sqrt{\frac{l-a}{l}}$ times the weight of the string.

3. An extensible string, uniform when unstretched and of length l, lies initially unstretched in a straight line on a horizontal plane. The string is then pulled at one end in the direction of its length produced, with a gradually increasing force, so that the acceleration is always infinitely small. Shew that when the force is F the extension of the string is $\frac{1}{2}F^2\frac{l}{\mu W\lambda}$, where W is the weight of the string, λ the coefficient of elasticity, μ the coefficient of friction and $F<\mu W$.

4. A heavy elastic string, of weight W and unstretched length c, is placed upon a smooth double inclined plane, the inclinations of whose faces to the horizontal are α and α'; shew that the total extension of the string is

$$\frac{Wc}{2\lambda}\frac{\sin\alpha\sin\alpha'}{\sin\alpha+\sin\alpha'}.$$

5. An elastic string rests on a rough inclined plane with the upper end fixed to the plane; shew that its extension will lie between the limits $\frac{Wl}{2\lambda}\frac{\sin(\alpha\pm\epsilon)}{\cos\epsilon}$, where α is the inclination of the plane, ϵ the angle of friction, W the weight of the string and λ its modulus of elasticity.

6. AB is an elastic cord whose natural length is 10 ft., whose mass is 5 lbs., and whose modulus of elasticity is 80 lbs. wt. It is suspended vertically from its end A, and a mass of 10 lbs. is attached to its end B; find the length to which it stretches if it is allowed to gradually reach its final position. Shew also that the density of the material at the middle point of AB is in the stretched cord diminished in the ratio 32 : 37.

7. If a uniform elastic string fixed at one end be acted upon at each point P by a force F in the direction of its length, so that the tension at P varies as its distance from the free end, prove that $\log F_0/F$ varies as the unstretched distance of P from the fixed end, F_0 being the value of F at the fixed end.

8. A uniform string, of weight w and modulus of elasticity λ, is lying in a stretched state on a rough horizontal table whose coefficient of friction is μ. If it be everywhere on the point of contracting, shew that its stretched length is $\left(1+\frac{\mu w}{4\lambda}\right)$ times its unstretched length.

9. A uniform heavy elastic string, of natural length $2a$, is stretched as much as possible and lies in limiting equilibrium on a rough inclined plane; shew that the direction of the friction changes at a point of the string whose natural distance from the upper end is $a[1+\tan\alpha\cot\epsilon]$, where ϵ is the angle of friction and α is the inclination of the plane to the horizon.

10. A uniform beam, of length l, rests along a line of greatest slope of a plane which is inclined to the horizon at an angle α. The beam is then subject to an extension from an increase of temperature and then to contraction from cooling to its original temperature. Find what points of the beam remain at rest during each of the two operations and shew that on the whole the beam descends through a distance along the plane equal to $\lambda l\tan\alpha\cot\epsilon$, where λ is the elongation of the beam per unit length for the extreme variation of temperature, and ϵ is the angle of friction.

11. An elastic string, of natural length a and weight mga, has one extremity fastened to a point in a smooth horizontal table and rotates on the table with uniform angular velocity ω; shew that the stretched length is $\frac{1}{\omega}\left(\frac{\lambda}{m}\right)^{\frac{1}{2}}\tan\left\{\omega a\left(\frac{m}{\lambda}\right)^{\frac{1}{2}}\right\}$, where λ is the modulus of elasticity.

12. A light elastic band, whose unstretched length is $2a$, is placed round four rough pegs A, B, C, D which are at the angular points of a square of side a. If it be taken hold of at a point P between A and B, and pulled in the direction AB, shew that it will begin to slip round both A and B at the same time if $\frac{AP}{PB}=e^{-\frac{\mu\pi}{2}}$.

13. A weight P just supports another weight Q by means of a fine elastic string which passes over a rough circular cylinder whose axis is horizontal; shew that the extension of the portion of the string in contact with the cylinder is $\frac{a}{\mu}\log\frac{Q+\lambda}{P+\lambda}$, where a is the radius of the cylinder, μ is the coefficient of friction and λ is the modulus of elasticity.

14. A spider hangs suspended by a light elastic thread from the ceiling, the modulus of the thread being equal to half the weight of the spider. Shew that, in climbing to the ceiling, the work done by the spider is one-third less than it would be if the thread were inelastic.

15. A heavy elastic string, whose unstretched length is l and whose mass is μl, lies loosely coiled on a horizontal table. If one end of the string be slowly lifted vertically until the whole string hangs just clear of the table, shew that the work done is

$$\tfrac{1}{2}\mu g l^2\left[1+\tfrac{2}{3}\frac{\mu g l}{\lambda}\right],$$

where λ is the modulus of elasticity of the string.

[When the end has been lifted a distance x, if x_0 be the unstretched length of x, then, by Art. 273, $x=x_0+\frac{\mu g}{2\lambda}x_0^2$. Also the tension T at the upper end $=\mu g x_0$, so that the work done, whilst x becomes $x+\delta x$, $=T\delta x=\mu g x_0\left[1+\frac{\mu g}{\lambda}x_0\right]\delta x_0$. Integrating this between limits 0 and l, we have the given result.]

16. If a uniform elastic string be at rest on a horizontal plane in its natural state with one end fixed to the edge, and if it then be allowed to hang freely from the point, prove that the loss of gravitational potential energy is $\frac{1}{6}\frac{w^2a}{\lambda}+\frac{1}{2}wa$, where w is the weight of the string, a is its natural length, and λ is Hooke's modulus.

MISCELLANEOUS EXAMPLES ON STRINGS AND CHAINS

1. A ring, of weight wb, is attached to the middle point C of a string, of length l and weight wl, which hangs symmetrically over two smooth pegs in the same horizontal line, the ends of the string hanging vertically. Shew that the parameter c of the catenary is given by the equation

$$b+l=e^{\frac{a}{c}}\left[b+\sqrt{4c^2+b^2}\right],$$

where $2a$ is the distance between the pegs, and that the least value of l for which equilibrium is possible occurs when $\dfrac{1}{(c-a)^2}-\dfrac{1}{c^2}=\dfrac{4}{b^2}$.

Shew also that the angle θ which the tangent at C makes with the vertical is given by the equation $b\tan\theta\log\left(\dfrac{l+b}{b}\,\dfrac{\cos\theta}{1+\cos\theta}\right)=2a$, and hence that its greatest value is $\cos^{-1}\dfrac{b}{l}$, since a cannot be negative.

2. A flexible string, of length $2l$ and line-density σ, has a heavy bead of mass $2\sigma k'$ knotted to it at its middle point; the ends of the string are fastened to two fixed points at a distance $2k$ apart slightly less than $2l$; shew that the parameter of the catenary of either half of the string is approximately

$$\left\{\frac{k\,(3kk'+k^2+3k'^2)}{6\,(l-k)}\right\}^{\frac{1}{2}}.$$

3. Two smooth circular cylinders, each of radius a, are placed with their axes parallel in a horizontal plane and at a distance $2b$ ($>2a$) apart. A uniform string is placed symmetrically across the cylinders with its ends hanging freely. Shew that the least possible length of the string is $2be+2a\,(2\tan^{-1}e-e)$. [Assume the result of Page 287, Ex. 16.]

4. A rod, of length $2b$, is suspended horizontally and symmetrically by two heavy strings attached to its ends and to two fixed points which are at a vertical distance a above it and at a distance $2(a+b)$ apart. If the length of each string be l, shew that the tension of the rod is equal to the weight of a piece of string whose length c is given by the equation

$$l^2=a^2+4c^2\sinh^2\frac{a}{2c}$$

5. A bar, of length $2a$, has its ends fastened to those of a heavy string, of length $2l$, by which it is hung symmetrically over a peg. The weight of the bar is n times and the horizontal tension $\frac{1}{2}m$ times the weight of the string. Prove that

$$m^2+n^2=\left\{(n+1)\operatorname{cosech}\frac{a}{ml}-n\coth\frac{a}{ml}\right\}^2$$

6. A uniform chain, of length s, hangs from two fixed points A, B on the same level at a distance $2a$ apart. Shew that, if s varies, the minimum depth of the directrix of the catenary is $\frac{a}{\sqrt{z^2-1}}$, where z is given by $z\tanh z=1$, and that the corresponding value of s is

$$\frac{2a}{z\sqrt{z^2-1}}.$$

Prove also that there are two values of s for which the directrix is at any given depth greater than this minimum depth, and that if s be slightly increased beyond the greater of these two values the directrix falls, whilst if it be slightly increased beyond the least of these two values the directrix rises.

7. A uniform chain has its ends attached to two pegs, one of which is distant $2a$ horizontally from the other and is at a depth $2b$ below it. Shew that, as the length of the chain alters, the parameter of the catenary with the highest directrix is determined by

$$\frac{a}{c}-\coth\frac{a}{c}=\frac{b^2a}{c^3}\operatorname{cosech}^4\frac{a}{c}.$$

8. A uniform chain, of length $2l$, hangs between two points A, B at the same level and the depth of its lowest point below AB is k. If the distance $AB\,(=a)$ be increased by the small quantity δa, prove that the vertex of the catenary will rise through the height $\delta a.\frac{k\cos\psi}{a-l\cos\psi}$, where ψ is the inclination to the horizontal of the tangent at A or B.

9. A given length l of uniform heavy chain is securely fastened to a fixed point at one end, and hangs over a smooth peg in the same horizontal at a distance $2a$ from it. Shew that there are two positions of equilibrium or none according as $\frac{l}{a}\gtrless\sqrt{\frac{3}{1-\xi^2}}$, where ξ is the positive root of the equation $3e^{2\xi}=\frac{1+\xi}{1-\xi}$.

Shew also that if there are two positions of equilibrium then the one for which the parameter of the catenary is the greater is stable.

10. A uniform chain of given length is fastened at its ends to two points in the same horizontal line, and passes over a smooth peg midway between these points. Shew that, if the symmetrical position of equilibrium be the only one, it is stable for displacements in the vertical plane; but if an unsymmetrical position also exist, the former is unstable.

11. If a chain of length a is held at its ends, and swung round, and the ends are then drawn apart till the chain is practically straight and its tension is equal to the weight of a length h of the chain, shew that the number of revolutions per second is $\sqrt{\frac{gh}{4a^2}}$.

12. A heavy uniform string hangs from one end in equilibrium in a wind blowing horizontally with uniform velocity. On the assumption that the wind exerts at any point of the string a normal force per unit length proportional to $\sin^2\psi$, shew that the equation giving the form of the string is

$$\rho(\cos\psi-\tan\alpha)^{2\cos^2\alpha}\times(\cos\psi+\cot\alpha)^{2\sin^2\alpha}=\text{constant},$$

where α is a constant such that the value of ψ at the free end of the string is $\cos^{-1}(\tan\alpha)$.

13. Find the velocity at which the power transmitted by a belt is a maximum. When this is the case, shew that the ratio of the tension on the tight side to that on the slack side is $2e^{\mu\alpha}+1:3$, where μ is the coefficient of friction and α is the angle of contact between the belt and the pulley.

[If a be the radius of the pulley, m the mass per unit length, and ω the angular velocity of the belt, then, by elementary Dynamics,

$$m\,\delta s\,\omega^2 a=(T+\delta T)\sin\delta\theta-R\,\delta s,\quad\text{and}\quad 0=(T+\delta T)\cos\delta\theta-T-\mu R\,\delta s.$$

Hence $\frac{dT}{d\theta}-\mu T=-\mu m\omega^2a^2$, so that $T=Ae^{\mu\theta}+m\omega^2a^2$. Hence, if T_1 and T_0 be the terminal tensions, we have easily $T=e^{\mu\theta}(T_0-m\omega^2a^2)+m\omega^2a^2$.

Hence the power transmitted

$$=(T_1-T_0)\,a\omega=(e^{\mu\alpha}-1)\,(T_0-m\omega^2a^2)\,a\omega,$$

and it is thus a maximum, for different values of ω, when $3m\omega^2a^2=T_0$, and then

$$\frac{T_1}{T_0}=\frac{2e^{\mu\alpha}+1}{3}.]$$

14. An endless string hangs symmetrically round a smooth right cylinder, of radius a, whose axis is horizontal. If the string have contact with the cylinder along three-quarters of the circumference, find its whole length and the position of its lowest point.

15. A heavy uniform string surrounds a vertical circle, being just so stretched that it is on the point of leaving the circle at the lowest point; shew that the tension at the highest point is three times that at the lowest.

16. A smooth elliptic disc (semi-axes a, b) is fixed in a vertical plane with its axes equally inclined to the vertical. A heavy string passes tightly round the disc and is gradually loosened. Shew that the eccentric angle ϕ of the point at which the string leaves the disc is given by

$$2a^2b\tan^3\phi-a(3a^2-b^2)\tan^2\phi+b(3b^2-a^2)\tan\phi-2ab^2=0.$$

[The point is determined by the fact that at it the pressure of the curve is zero and a minimum.]

17. A uniform string, the ends of which are fastened at a point A, surrounds a centre of force at O, which repels with a force varying inversely as the square of the distance. Shew that, if the length of the string is $4OA$, the internal angle between the two parts of the string at A is $120°$.

18. Shew that a uniform string will rest in the form of the arc of a circle if it be acted upon by a central force from a point on the circumference varying inversely as the cube of the distance.

19. $ABCD$ is a square of side b. A uniform string of line-density σ, fixed at B and D, is in equilibrium under a repulsive force μr^{-3} from A. If the tangents to the string at B and D are perpendicular to BD, and if the tension at each of these points is $\frac{\mu\sigma}{2b^2}$, shew that the shape of the string is the curve $r = b(\sin\theta + \cos\theta)$.

20. Shew that an arc of an equiangular spiral is a possible form for a catenary of uniform strength, whose ends are fixed, to assume under the action of a repulsive force from the pole varying inversely as the distance.

21. An inextensible endless string of given length is under the action of two central forces varying inversely as the cube of the distance from two fixed points; shew that a circle is a possible form of the string in equilibrium, and find the position of its centre.

22. Shew that a string can rest in equilibrium in the form of an ellipse under the action of two repulsive forces from its foci of magnitude $\mu r^{-\frac{3}{2}} r'^{-\frac{1}{2}}$ and $\mu r^{-\frac{1}{2}} r'^{-\frac{3}{2}}$, where r and r' are the focal distances of any point P, and prove that the tension at the point P is proportional to the perpendicular from the centre upon the tangent at P.

23. A smooth circular cylinder, of radius a, is fixed with its axis vertical and a smooth horizontal peg is fastened into it. An endless string, of length $2l$, is then thrown over the cylinder and catches on the peg. Prove that in the position of equilibrium the angle between the parts of the string at the peg is $2\cot^{-1}\frac{l}{c}$, where $c\sinh\frac{a\pi}{c} = l$.

[The origin being at the lowest point O of the string, the axis of z vertical, and x being the length of the circular arc which is the projection of any arc OP, we have, by resolving horizontally and vertically, $\frac{d}{ds}\left(T\frac{dz}{ds}\right) = w$, and $\frac{d}{ds}\left(T\frac{dx}{ds}\right) = 0$, so that $T\frac{dx}{ds} = \text{constant} = wc$. Hence $\frac{d}{ds}\left(\frac{dz}{dx}\right) = \frac{1}{c}$, *i.e.* $\frac{d^2z}{dx^2} = \frac{1}{c}\frac{ds}{dx}$. This is the differential equation of Art 251, and the same solution holds good. It follows that the string will not be disturbed if the cylinder be developed into a vertical plane.]

24. A uniform heavy string, of length $2l$, hangs in contact with a smooth vertical cylinder of radius r. It is fixed to two points, which are in a horizontal plane and also in a vertical plane through the axis of the cylinder, each being at distance a from the axis, where $a > r$. Shew that the depth y of the lowest point of the string below either support is given by

$$r\sin^{-1}\frac{r}{a} + \sqrt{a^2 - r^2} = \frac{l^2 - y^2}{2y}\log\frac{l+y}{l-y}.$$

25. A string rests on a smooth sphere cutting all the sections through a fixed diameter at a constant angle. Shew that it would so rest if acted on by a force varying inversely as the square of the distance perpendicular to the given diameter and that the tension varies inversely as that distance.

[Using polar coordinates (a, θ, ϕ) the curve of the string cuts the meridians at a constant angle β if $\frac{a\,d\theta}{ds} = \cos\beta$, and $\frac{a\sin\theta\,d\phi}{ds} = \sin\beta$. Hence easily

$$\frac{dx}{ds} = \cos\theta\cos\phi\cos\beta - \sin\phi\sin\beta,\quad \frac{dy}{ds} = \cos\theta\sin\phi\cos\beta + \cos\phi\sin\beta$$

and $\frac{dz}{ds} = -\sin\theta\cos\beta$. Taking moments about the fixed diameter, we have $T\sin\beta \,.\, a\sin\theta = \text{constant}$, so that $T = \frac{A}{\sin\theta}$.

The third equation of Art. 261 gives

$$R\cos\theta = \frac{d}{ds}\left(T\frac{dz}{ds}\right) = \frac{d}{ds}(-T\sin\theta\cos\beta) = 0, \text{ so that } R \text{ is zero.}$$

Hence, if F be the force perpendicular to the given diameter and outward, equation (1) of Art. 261 gives

$$F\cos\phi = -\frac{d}{ds}\left(T\frac{dx}{ds}\right) = -A\frac{d}{ds}\left(\cos\beta\cot\theta\cos\phi - \frac{\sin\phi\sin\beta}{\sin\theta}\right)$$

$$= \frac{A}{a}\frac{\cos\phi}{\sin^2\theta}, \text{ so that } F = \frac{A}{a\sin^2\theta}.]$$

26. The extremities of a string of length $r(e-1)\sqrt{1+\phi^2\sin^2\alpha}$ are attached to two points on the surface of a right circular cone at distances r and er respectively from its vertex, where 2α is the angle of the cone and ϕ is the angle between the planes through the axis and each of the points. If the string rest in equilibrium on the surface under the action of a repulsive centre of force at the vertex varying inversely as the square of the distance, then the curve of equilibrium will cut each generator at the same angle. [Here e is the base of the Napierian system of logarithms.]

27. Find the form of a smooth surface of revolution such that when its axis is vertical any uniform string resting upon it will cut all the meridian curves at the same angle. [The generating curve is a rectangular hyperbola.]

---

28. Two scale-pans, of weight λn, are connected by a weightless elastic string, of modulus λ, which hangs symmetrically over a fixed rough horizontal cylinder of radius a. Initially the string is uniformly stretched throughout. If one of the scale-pans is gradually loaded, prove that, before the other moves, the natural length of the additional vertical portion of the string supporting the first is $\frac{a}{\mu}\log\frac{1+ne^{\mu\pi}}{1+n} - \frac{\pi a n}{1+n}$.

29. A heavy elastic string, of length $2a$ and modulus of elasticity equal to its weight, rests in equilibrium on a smooth parabola of latus rectum $4a$. The axis of the parabola is vertical and the free end of the string is at the vertex which is the lowest point. Find the point on the parabola to which the upper end is attached, and shew that the tension there is $W(\sqrt{2}-1)$, where W is the weight of the string.

30. An elastic string, originally uniform and of length l, is fastened at two given points and is in equilibrium in the form of a portion of a circle under the action of a repulsive force tending from a given point in the circumference; find the law of force.

31. A heavy elastic string, which is uniform when unstretched, is placed round a smooth circular cylinder whose axis is horizontal and is just not in contact with the lowest point of the cylinder; if T be the tension at a point the radius to which is inclined at θ to the vertical, then $(T+\lambda)^2 = A\cos\theta + B$, where λ is the modulus of elasticity and A and B are constants depending on the weight of the string, the modulus of elasticity and the radius of the cylinder.

If w be equal to the weight of a length of unstretched string equal in length to the radius of the cylinder and if it be also equal to the modulus of elasticity, shew that the tension T_1 at the highest point is given by

$$T_1 + \frac{T_1^2}{2w} = \frac{9+\sqrt{5}}{4}w.$$

32. A heavy elastic string, uniform when unstretched, rests on the convex side of a smooth vertical circle, one end being fastened to the highest point of the circle. If in the position of equilibrium the whole length is equal to a quadrant of the circle, prove that the unstretched length equals $a\sqrt{2}\log_e(\sqrt{2}+1)$, where a is the radius of the circle, $2wa$ is the modulus of elasticity, and w is the weight of a unit length of unstretched string.

33. An elastic string, uniform when unstretched, lies at rest in a smooth circular tube under the action of an attracting force, equal to μ times the distance, tending to a point on the circumference of the tube just opposite to the middle point of the string. If the string when in equilibrium just occupies a semi-circle, shew that the greatest tension is $\sqrt{\lambda(\lambda+2\mu\rho a^2)}-\lambda$, where λ is the modulus of elasticity, a is the radius of the tube, and ρ the mass of unit length of unstretched string.

34. A heavy elastic string, of natural length l, hangs from a fixed point in a state of equilibrium, and its total extension is μl. The string is now enclosed in a smooth fixed helical tube, the tangent to which at any point makes a constant angle α with the horizontal. The highest point of the string is attached to the tube, and the string takes up its position of equilibrium. Shew that the total extension is now $\mu l\sin\alpha$.

35. A uniform string in one plane is in equilibrium under the action of a central force; shew that the latter varies as $\frac{d}{dr}\left(\frac{1}{p}\right)$.

What is the kinetic analogue?

If the string be elastic, shew that, in order that the string may assume a given form, the central force must vary as $\frac{d}{dr}\left(\frac{1}{p}+\frac{a}{p^2}\right)$, where a is a constant for every point of the same string.

36. Over a fixed sphere, of radius a, there rests horizontally a heavy elastic ring of natural radius c. Shew that, in its equilibrium position, the angle 2α which a diameter of the ring subtends at the centre of the sphere is given by $\tan\alpha = 2\pi k\frac{\sin\alpha - \sin\beta}{\sin\beta}$, where $c = a\sin\beta$, and k times the weight of the ring is equal to the modulus of elasticity.

If the sphere be slightly rough (coefficient of friction μ), shew that, before equilibrium is broken, the ring can be lowered till the angle 2α is increased, approximately, by $2\mu\frac{\sin\alpha - \sin\beta}{\sin\beta - \sin^3\alpha}$.

37. A rough surface of revolution is fixed with its axis vertical; an endless elastic string rests on it lying in a horizontal plane and being uniformly stretched. If the string be on the point of slipping up wherever it is placed, find the equation of the generating curve of the surface and shew that it reduces to $x - \mu y = a(1+\mu^2)\log\frac{x}{c}$, where λ is the modulus of elasticity and μ is the coefficient of friction, $2\pi\lambda\mu$ is the weight and $2\pi a$ the natural length of the string, and the y-axis is the axis of revolution.

38. An endless slightly extensible strap is stretched over two equal pulleys; shew that the maximum couple which the strap can exert on either pulley is

$$\frac{2a(c+\pi a)}{c\coth\frac{\mu\pi}{2}+\frac{2a}{\mu}}T,$$

where a is the radius of either pulley, c the distance of their centres, μ the coefficient of friction, and T the tension with which the strap is put on.

39. If a bicycle tyre be regarded as a thin elastic band, stretched within a smooth groove of depth d in a rim whose outer radius is a, shew that the work required to remove the tyre from the rim is

$$\frac{8\lambda a^2}{l}(\sin\phi - \phi\cos\phi)\left\{(\pi-\phi)\cos\phi + \sin\phi - \frac{l}{2a}\right\},$$

where l is the unstretched length of the band, λ is its coefficient of elasticity, $\cos\phi = \frac{a-d}{a}$ and the breadth of the wheel is neglected.

CHAPTER XIV

ATTRACTIONS AND POTENTIAL

276. THE law of attraction between particles of matter, known as the Newtonian Law of Gravitation, is as follows: *Every particle of matter attracts every other particle of matter with a force which varies directly as the product of the masses of the particles and inversely as the square of the distance between them.*

Hence the attraction between two particles, of m_1 and m_2 grammes, placed at a distance of x centimetres is $\gamma \dfrac{m_1 m_2}{x^2}$ dynes, where γ is a constant whose value will be found in a later article. This quantity γ is called the Constant of Attraction.

At the present stage of the Student's reading the above law must be looked upon as an hypothesis, and no proof can be given here. But he may assume that this law is found to hold good throughout the Universe, and that it is sufficient to account for the motion of the heavenly bodies. The verification of its truth is obtained from Dynamical considerations.

We shall have a few examples in which some other law of attraction, besides that of the inverse square, is assumed, but these assumptions do not correspond to any such attractions as we meet with in the physical universe.

277. *To find the attraction of a thin uniform rod AB upon an external point P.*

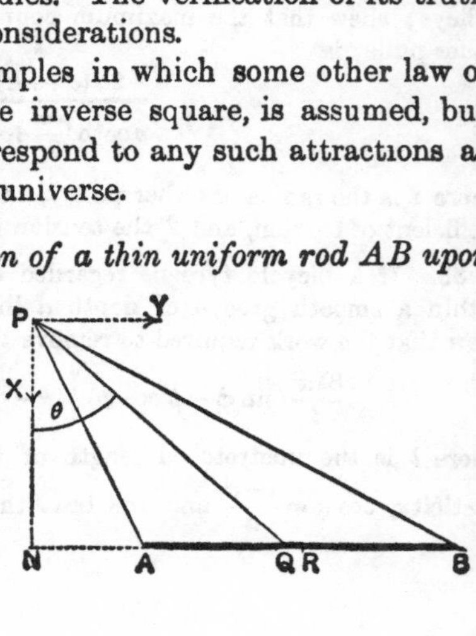

Draw $PN, (=p)$, perpendicular to the rod. Let Q be any point of the rod such that $NQ = x$ and $\angle NPQ = \theta$, and let QR be an element δx.

If k be the cross section

and ρ the density of the rod, the attraction of the element QR on a unit of mass at P

$$= \frac{\gamma k\rho \, . \, \delta x}{PQ^2} = \frac{\gamma k\rho \, . \, \delta x}{p^2 \sec^2\theta} = \frac{\gamma k\rho \, . \, \delta\theta}{p},$$

since $x = p \tan\theta$ and hence $\delta x = p \sec^2\theta \, . \, \delta\theta$.

The direction of this attraction ultimately, when QR is very small, is in the direction PQ.

Its components along and perpendicular to PN are $\frac{\gamma k\rho \cos\theta \, . \, \delta\theta}{p}$ and $\frac{\gamma k\rho \sin\theta \, . \, \delta\theta}{p}$.

Hence, if X and Y are the component attractions of the whole rod along and perpendicular to PN, and if $\angle NPA = \alpha$ and $\angle NPB = \beta$, we have

$$X = \int_\alpha^\beta \frac{\gamma k\rho \cos\theta \, d\theta}{p} = \frac{\gamma k\rho}{p}\Big[\sin\theta\Big]_\alpha^\beta = \frac{\gamma k\rho}{p}[\sin\beta - \sin\alpha] \text{...(1)},$$

and

$$Y = \int_\alpha^\beta \frac{\gamma k\rho \sin\theta \, d\theta}{p} = \frac{\gamma k\rho}{p}\Big[-\cos\theta\Big]_\alpha^\beta = \frac{\gamma k\rho}{p}[\cos\alpha - \cos\beta] \text{...(2)}.$$

If R be the resultant attraction inclined at ϕ to PN, then

$$R = \sqrt{X^2 + Y^2} = \frac{\gamma k\rho}{p}\sqrt{(\sin\beta - \sin\alpha)^2 + (\cos\alpha - \cos\beta)^2}$$

$$= \frac{\gamma k\rho}{p}\sqrt{2 - 2\cos(\beta - \alpha)} = \frac{2\gamma k\rho}{p}\sin\frac{\beta - \alpha}{2}$$

$$= \frac{2\gamma k\rho}{p}\sin(\tfrac{1}{2}APB) \text{..............(3)},$$

and $$\tan\phi = \frac{Y}{X} = \frac{\cos\alpha - \cos\beta}{\sin\beta - \sin\alpha} = \tan\frac{\alpha + \beta}{2},$$

so that $$\phi = \frac{\alpha + \beta}{2} \text{ ....(4)},$$

and hence the direction of the resultant attraction bisects the angle APB.

Cor. If the rod AB be infinite in length, then $\alpha = -90°$ and $\beta = 90°$.

Hence $R = \frac{2\gamma k\rho}{p}$, so that the attraction of an infinite rod on an external point varies inversely as the distance of the point from the rod.

278. If in the result (3) of the previous article we put $p = 0$, *i.e.* take P on the surface of the rod, we have infinity as the result. This is clearly impossible. The reason for the apparent failure of our analysis is that in our working we assumed each element of the cross section at Q to be at the same distance from P. Now if P be on the bar, the distances from P of points of the cross section through P vary from zero to the diameter of the bar and hence cannot be taken to be equal.

This case is further considered in Ex. 2 of Art. 285.

279. *To shew that the attraction of a rod AB at an external point P is the same as that of the arc of a circle, of like material, with centre P and of radius equal to the perpendicular from P on AB, which is intercepted by the lines PA and PB.*

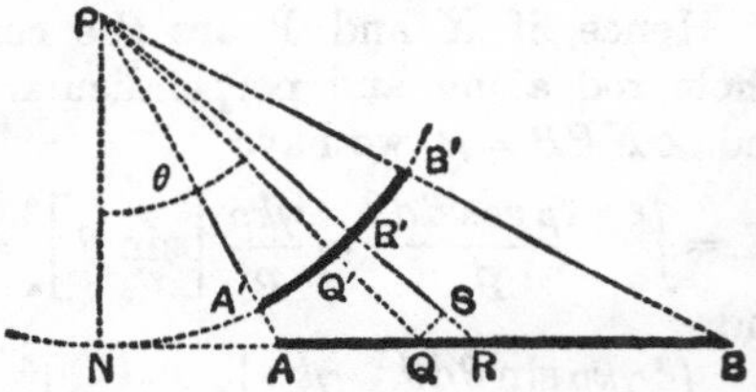

QR being an element of the rod, let $Q'R'$ be the corresponding element of the circular arc.

Then
$$\frac{\text{attraction of } Q'R' \text{ at } P}{\text{attraction of } QR \text{ at } P} = \frac{Q'R'}{PQ'^2} \div \frac{QR}{PQ^2}$$
$$= \frac{Q'R'}{QR} \cdot \frac{PQ^2}{PN^2} = \frac{Q'R'}{QS} \cdot \frac{QS}{QR} \cdot \sec^2\theta,$$

where QS is perpendicular to PR,
$$= \frac{PQ'}{PQ} \cdot \cos SQR \cdot \sec^2\theta = \frac{PN}{PQ} \cdot \cos\theta \cdot \sec^2\theta = 1.$$

Hence the attractions of corresponding elements of the arc and rod are the same and in the same direction.

The resultant attractions must thus be the same in the two cases.

280. ***Attraction of a uniform thin rectangular plate upon a unit mass situated on a perpendicular to the plate through its centre.***

Let $ABCD$ be the plate, of thickness k and density ρ; let $AB = 2a$, $AD = 2b$, and $OP = h$, where P is the attracted unit mass.

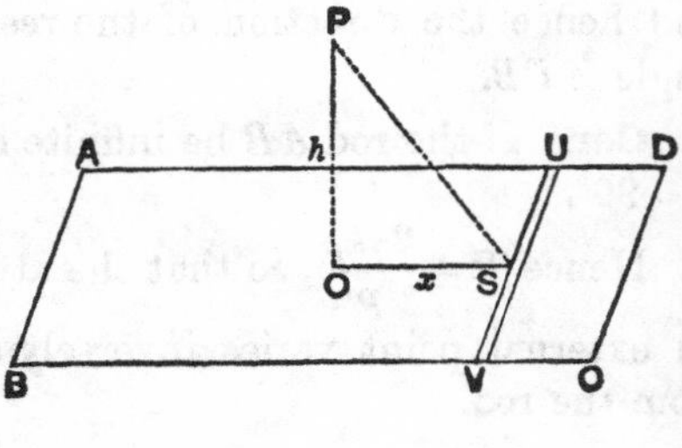

If UV be a section of the plate parallel to AB at a distance x from O, whose breadth is δx, its attraction on P by Art. 277, if S be the middle point of UV,

$$=\frac{2\gamma k\rho\,.\,\delta x}{PS}\sin\frac{VPU}{2}=\frac{2\gamma\rho k\,.\,\delta x}{\sqrt{h^2+x^2}}\cdot\frac{a}{\sqrt{h^2+x^2+a^2}},$$

and acts in the direction PS.

Hence the total attraction X along PO

$$=\int_{-b}^{+b}\frac{2\gamma\rho k\,.\,dx}{\sqrt{h^2+x^2}}\cdot\frac{a}{\sqrt{h^2+x^2+a^2}}\times\frac{h}{\sqrt{h^2+x^2}}.$$

$$\therefore\ \frac{X}{4\gamma\rho kah}=\int_0^b\frac{dx}{(h^2+x^2)\sqrt{h^2+x^2+a^2}}.$$

Put $x=h\tan\theta$, so that the limits for θ are zero and

$$\tan^{-1}\frac{b}{h},\quad i.e.\quad \sin^{-1}\frac{b}{\sqrt{b^2+h^2}}.$$

$$\therefore\ \frac{X}{4\gamma\rho kah}=\frac{1}{h}\int_0^{\sin^{-1}\frac{b}{\sqrt{b^2+h^2}}}\frac{\cos\theta\,d\theta}{\sqrt{h^2+a^2-a^2\sin^2\theta}}$$

$$=\frac{1}{ah}\left[\sin^{-1}\frac{a\sin\theta}{\sqrt{h^2+a^2}}\right]_0^{\sin^{-1}\frac{b}{\sqrt{b^2+h^2}}}=\frac{1}{ah}\sin^{-1}\frac{ab}{\sqrt{(h^2+a^2)(h^2+b^2)}}.$$

$$\therefore\ X=\frac{\gamma M}{ab}\sin^{-1}\frac{ab}{\sqrt{(h^2+a^2)(h^2+b^2)}},$$ where M is the mass of the plate.

EXAMPLES

1. A triangular framework of three rods, of uniform mass per unit of length, attracts according to the law of Nature; shew that a particle will be in equilibrium under their attraction if placed at the in-centre of the triangle.

[Use the theorem of Art. 279.]

2. Shew that the attraction of a uniform rod AB on a unit mass at P, in the direction parallel to AB, varies as $\frac{1}{PA}-\frac{1}{PB}$.

3. Two straight wires, of length l and l' and of masses m and m', are placed so that they are in the same straight line and the distance between the ends next one another is c; shew that the force of attraction between the two wires is $\gamma\frac{mm'}{ll'}\log\frac{(l+c)(l'+c)}{c(l+l'+c)}$.

4. Shew that the attraction of a thin uniform straight bar, of line density σ and length l, on another uniform thin straight bar, of line density σ' and length l', placed parallel to it and symmetrically with respect to it at a distance h, is $\gamma\frac{\sigma\sigma'}{h}\{\sqrt{(l+l')^2+4h^2}-\sqrt{(l-l')^2+4h^2}\}$.

Explain the meaning of this expression when h is zero.

5. Two thin straight uniform rods, AB and CD, are pivoted together at their middle points. Shew that the attraction between them reduces to a couple of moment $2\gamma mm'(AC \sim BC)\operatorname{cosec}\alpha$, where m and m' are their line densities and α is the angle between them.

6. Two uniform non-intersecting straight bars, of infinite length and of line densities σ and σ', attract according to the Newtonian law. Shew that the resultant force between the bars is $2\gamma\pi\sigma\sigma'\operatorname{cosec}\alpha$, where α is the angle between the bars.

[Let OO', $=c$, be the shortest distance between the bars; then, by symmetry, the resultant force is along OO'. Let P be any point on the first bar such that $OP=\xi$, and p the perpendicular from P on the second bar, so that $p^2=c^2+\xi^2\sin^2\alpha$. The attraction at P of the second bar, by Art. 277, Cor., $=\dfrac{2\gamma\sigma'}{p}$, and the resultant of this along $OO'=\dfrac{2\gamma\sigma'.c}{p^2}$.

Hence the whole attraction $=\displaystyle\int_{-\infty}^{+\infty}\frac{2\gamma\sigma' c}{c^2+\xi^2\sin^2\alpha}\,.\,\sigma d\xi=$ etc.]

281. *To find the attraction of a uniform circular plate, of radius a and small thickness k, upon a point P which is on the axis of the plate at a distance p from its centre.*

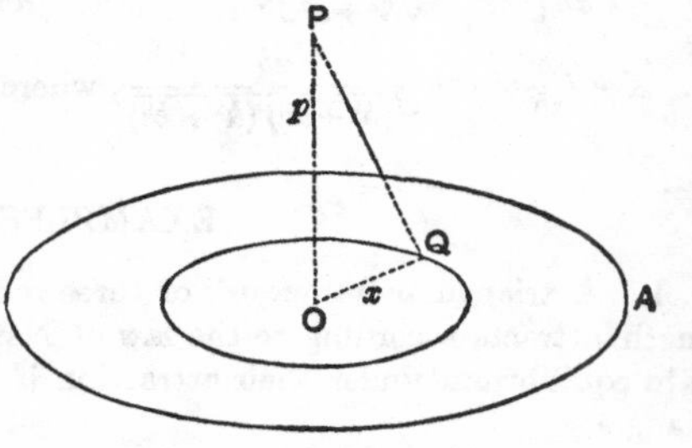

Consider the portion of the plate included between two concentric circles of radii x and $x+\delta x$.

The attraction of any point Q of it upon a unit mass at P is along the line PQ, and its component in the direction PO

$$=\gamma\,.\,\frac{\text{mass of the element at } Q}{PQ^2}\cos OPQ$$

$$=\gamma\,.\,\frac{\text{mass of the element at } Q}{PQ^2}\,.\,\frac{p}{PQ}.$$

The same is true for each point of the elementary area.

Hence the resultant attraction of this elementary area

$$=\gamma\,.\,\frac{2\pi x\,.\,\delta x\,.\,k\rho}{PQ^3}\,.\,p=2\pi\gamma k\rho p\,\frac{x\,\delta x}{(x^2+p^2)^{\frac{3}{2}}},$$

where ρ is the density of the plate per unit of area.

The attraction of the whole plate therefore

$$= 2\pi\gamma k\rho p \int_0^a \frac{x\,dx}{(x^2+p^2)^{\frac{3}{2}}} = 2\pi\gamma k\rho p \left[-\frac{1}{(x^2+p^2)^{\frac{1}{2}}}\right]_0^a$$

$$= 2\pi\gamma k\rho p \left[\frac{1}{p} - \frac{1}{(a^2+p^2)^{\frac{1}{2}}}\right] = 2\pi\gamma k\rho \left[1 - \frac{p}{\sqrt{a^2+p^2}}\right].$$

If α be the angle that any radius OA of the plate subtends at P, then

$$\cos\alpha = \frac{OP}{PA} = \frac{p}{\sqrt{a^2+p^2}},$$

and the resultant attraction of the plate, which is clearly in the direction PO,

$$= 2\pi\gamma k\rho\,[1 - \cos\alpha].$$

Cor. Let the radius a of the plate become infinite, and hence the angle α equal to $90°$; the resultant attraction then is $2\pi\gamma k\rho$, which is independent of p, the distance of the attracted point from the plate.

Hence the attraction of an infinite thin plate upon a point P, situated at a finite distance from the plate, is independent of the distance of P and is equal to $2\pi\gamma \times$ mass of the plate per unit of area.

282. *Change in the attraction of a thin attracting surface on a unit mass as the latter crosses the surface normally from one side to the other.*

Let P and P' be two points, on opposite sides of, and indefinitely close to, the surface, so that PP' is normal to the surface.

Round PP' as axis describe on the surface a small circle of area A, and let the rest of the plate be called B.

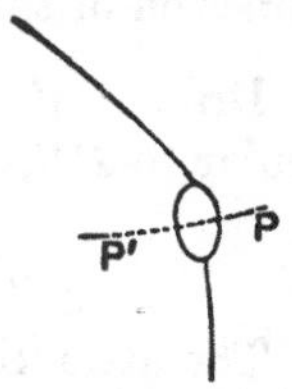

Then

Attraction at P = attraction of A at P together with the attraction of B at P ...(1),

and

Attraction at P' = attraction of A at P' together with the attraction of B at P'........................(2).

Now P and P' being indefinitely close together, the attraction of B at P = the attraction of B at P' in the limit.

Also A is to P as far as its attraction is concerned as an infinite plate is to a point at a finite distance.

Hence, by the corollary to the last article, the attraction of A at $P = 2\pi\gamma\rho k$, and the attraction of A at $P' = -2\pi\gamma\rho k$.

Hence, from (1) and (2) by subtraction,

$$\text{Attraction at } P - \text{attraction at } P' = 4\pi\gamma\rho k.$$

Thus the change in the attraction on the unit mass as it passes normally from a position indefinitely close to the attracting surface on one side to a position indefinitely close on the other side is $4\pi\gamma\rho k$, and therefore depends only on the thickness and density of the plate at the point of crossing.

283. *To shew that the component attraction in a direction normal to its plane of a uniform plane lamina, of any shape, at any point P is $\gamma m\omega$, where m is the mass per unit area of the lamina, and ω is the solid angle subtended at P by the lamina.*

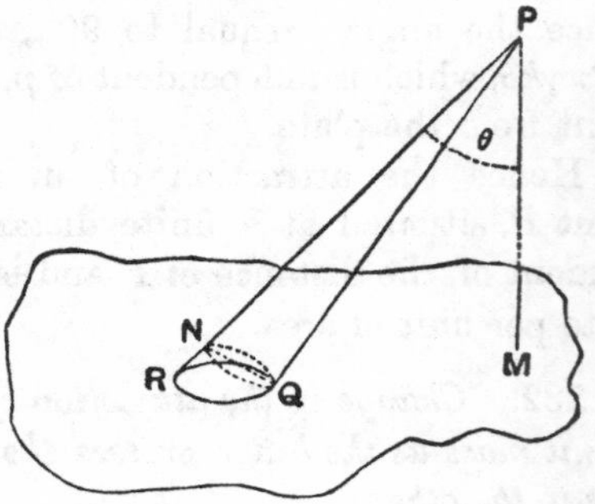

Let QR be any very small element of the lamina which subtends an angle $\delta\omega$ at P, and let PQ or PR be r.

Then, in the limit, the attraction of this element on $P = \gamma m \,.\, \dfrac{\text{area } RQ}{PQ^2}$.

Draw PM perpendicular to the lamina, and QN perpendicular to PR, so that

$$\angle RQN = 90° - \angle QRN = \angle RPM = \theta.$$

The attraction of QR on P resolved along PM

$$= \gamma m \,.\, \frac{\text{area } RQ \,.\, \cos\theta}{PQ^2} = \gamma m \,.\, \frac{\text{area } QN}{PQ^2} = \gamma m \,.\, \delta\omega.$$

Hence the resultant attraction on P in the direction normal to the lamina $= \Sigma\gamma m \,.\, \delta\omega = \gamma m\,\omega$, where ω is the whole solid angle subtended by the lamina at P.

284. *All frustra of a uniform cone, of the same thickness and with their plane faces parallel to the base of the cone, exert equal attractions at the vertex of the cone.*

Let AB and CD be two sections of the cone, of the same small thickness t, which are parallel to the base of the cone.

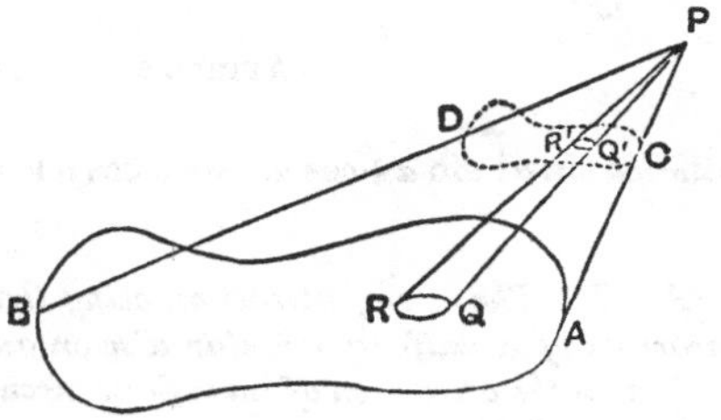

Let any cone, of vertex P, and very small vertical angle at P cut these sections in the very small curves QR and $Q'R'$.

Since QR and $Q'R'$ are similar curves, their areas, and hence also their masses since they are of equal thickness, are proportional to the squares of their distances from the vertex.

$$\text{Hence }\frac{\text{attraction of } QR \text{ at } P}{\text{attraction of } Q'R' \text{ at } P}=\frac{\text{area } QR}{PQ^2}\div\frac{\text{area } Q'R'}{PQ'^2}$$

$$=\frac{\text{area } QR}{\text{area } Q'R'}\cdot\frac{PQ'^2}{PQ^2}=\frac{PQ^2}{PQ'^2}\cdot\frac{PQ'^2}{PQ^2}=1.$$

Since the attractions of corresponding elements QR and $Q'R$, are equal, the attractions of the whole areas AB and CD are the same both in magnitude and direction.

Hence, by summation, the attractions of any two frustra of the same finite thickness are the same both in magnitude and direction.

285. Ex. 1. *Find the attraction of a uniform solid right circular cone, of height h and vertical angle 2a, at the centre O of its plane base.*

The plane section at a height x above the base subtends an angle 2β at O, where

$$\cos\beta=\frac{x}{\sqrt{x^2+(h-x)^2\tan^2 a}}=\frac{x\cos a}{\sqrt{x^2-2hx\sin^2 a+h^2\sin^2 a}}.$$

The attraction of this section of thickness δx

$$=2\pi\gamma\rho\,\delta x\left[1-\frac{x\cos a}{\sqrt{x^2-2hx\sin^2 a+h^2\sin^2 a}}\right],\text{ by Art. 281.}$$

Hence the attraction of the whole cone

$$=2\pi\gamma\rho\int_0^h\left[1-\frac{x\cos a}{\sqrt{(x-h\sin^2 a)^2+h^2\sin^2 a\cos^2 a}}\right]dx.$$

On putting $y = x - h\sin^2 a$, this

$$=2\pi\gamma\rho\int_{-h\sin^2 a}^{h\cos^2 a}\left[1-\frac{(y+h\sin^2 a)\cos a}{\sqrt{y^2+h^2\sin^2 a\cos^2 a}}\right]dy$$

$$=2\pi\gamma\rho\Big[y-\cos a\sqrt{y^2+h^2\sin^2 a\cos^2 a}$$

$$-h\sin^2 a\cos a\log\left(y+\sqrt{y^2+h^2\sin^2 a\cos^2 a}\right)\Big]_{-h\sin^2 a}^{h\cos^2 a}$$

$$=2\pi\gamma\rho h\sin a\left[\sin a+\cos a-\sin a\cos a\log\frac{\cos a+\cos^2 a}{\sin a-\sin^2 a}\right].$$

Ex. 2. *The law of attraction being that of the inverse distance, find the attraction of a uniform circular disc on an external point in its own plane.*

Deduce the attraction of an infinite circular cylinder, attracting according to the law of Nature.

Let a be the radius, ρ the density and k the thickness of the disc, and c the distance of the given point P from the centre O. Let θ be the angle any radius vector through P makes with OP.

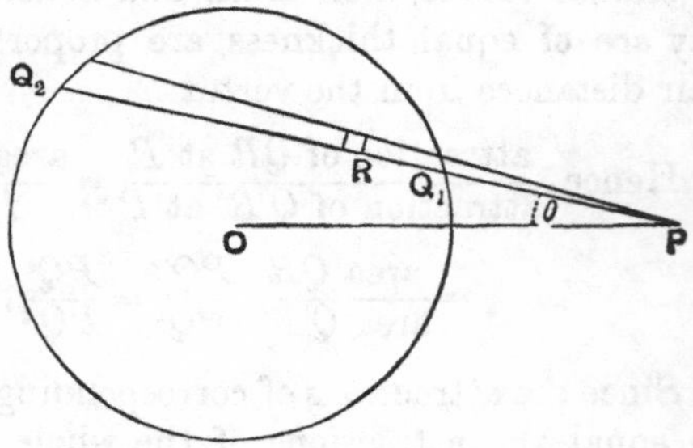

Then the total attraction of the disc

$$=2\iint\gamma\cdot\frac{r\,d\theta\,dr\,k\rho}{r}\cos\theta,$$

the limits of r being PQ_1 and PQ_2,

i.e. $$c\cos\theta-\sqrt{a^2-c^2\sin^2\theta}\quad\text{and}\quad c\cos\theta+\sqrt{a^2-c^2\sin^2\theta},$$

and those of θ being zero and $\sin^{-1}\dfrac{a}{c}$.

Hence the attraction $=4\gamma k\rho\int\sqrt{a^2-c^2\sin^2\theta}\,.\cos\theta\,d\theta$

$$=\frac{4\gamma k\rho a^2}{c}\int_0^{\frac{\pi}{2}}\cos^2\phi\,d\phi,\quad\text{if } c\sin\theta=a\sin\phi.$$

The attraction therefore $=\dfrac{\pi\gamma k\rho a^2}{c}=\gamma\dfrac{\text{Mass of the disc}}{\text{Distance from the centre}}$, so that P is attracted as it would be if the whole mass of the disc were collected at its centre.

Next, let the circle be the normal cross section of an infinite cylinder. The attraction of the filament of infinite length through R perpendicular to the plane of the paper is, by Art. 277, Cor., equal to $\dfrac{2\gamma\,.\,r\delta\theta\,.\,\delta r\,.\,\rho}{PR}$

in the direction PR. The attraction of the infinite cylinder is thus the same as that of the above disc, if for k we put 2, and so $= \frac{2\pi\gamma\rho a^2}{c}$.

If P be actually on the surface of the cylinder, so that $c=a$, this attraction becomes $2\pi\gamma\rho a$. We hence have the attraction of a thin rod on a point upon its surface [cf. Art. 278].

If P be inside the cylinder, the attraction on it $=2\times\iint 2\gamma\rho\, dr\, d\theta \,.\cos\theta$, the limits for r being zero to $c\cos\theta+\sqrt{a^2-c^2\sin^2\theta}$ and for θ from 0 to π.

This easily gives $2\pi\gamma\rho c$ as the attraction at P.

EXAMPLES

1. Shew that the attraction of a uniform cylinder of height h, radius a, and density ρ, at a point on its axis at a distance c from its end and outside it, is $2\pi\gamma\rho\,[h-\sqrt{a^2+(c+h)^2}+\sqrt{a^2+c^2}]$.

2. Shew that the attraction of a spherical segment on a unit particle at its vertex is $2\pi\gamma\rho h\left[1-\frac{1}{3}\sqrt{\frac{2h}{a}}\right]$, and that on a unit particle at the centre of its base is $\frac{2\pi\gamma\rho h}{3(a-h)^2}[3a^2-3ah+h^2-h^{\frac{1}{2}}(2a-h)^{\frac{3}{2}}]$, where a is the radius and ρ the density of the sphere and h is the height of the segment.

3. Prove that a solid uniform hemisphere, of radius a, exerts no resultant attraction at a point on its axis at a distance c from the centre given by the equation $12c^4-8a^3c+3a^4=0$.

Shew that $c=\frac{3a}{7}$ approximately.

4. A right circular cone, of uniform density ρ, has its axis vertical and vertex upwards. Shew that its attraction at a point P on the axis, at a distance c above the vertex, is

$$2\pi\gamma\rho c\sin\alpha\cos\alpha\left\{\frac{\cos\beta-\cos\alpha}{\sin\beta}-\sin\alpha\log\left(\tan\frac{\alpha}{2}\cot\frac{\beta}{2}\right)\right\},$$

where 2α is the vertical angle of the cone and $\alpha-\beta$ the angle which the radius of its base subtends at P.

5. A frustum of a uniform thin hollow cone attracts a particle placed at the vertex; shew that the attraction is $2\pi\gamma\sigma\sin\alpha\cos\alpha\log\frac{R}{r}$, where R and r are the radii of the circular ends, α is the semivertical angle of the cone, and σ is the superficial density of the cone.

6. A homogeneous right circular cylinder is of infinite length in one direction, and at the other end the section is perpendicular to the generators; prove that the attraction of the cylinder on a particle at the centre of this end is $\frac{2\gamma M}{a}$, where M is the mass of the cylinder per unit of length.

7. A vertical solid cylinder of height a, radius r, and density ρ, bounded by plane ends perpendicular to its axis, is divided by a plane through the axis into two parts. Shew that the horizontal attraction of one part on a particle at the centre of the base is $2\gamma a\rho \log \frac{r+\sqrt{r^2+a^2}}{a}$, and find the angle the resultant attraction makes with the axis.

8. A homogeneous prism, infinite in length, whose cross section is an equilateral triangle ABC attracts a particle at A; shew that the resultant attraction is $\frac{4\gamma\pi M}{3a}$, where M is the mass of a unit length of the prism and a is the length of a side of the triangle.

9. Shew that the attraction of a thin elliptic disc, of uniform thickness k and uniform density ρ, at the focus is $2\pi\gamma k\rho\sqrt{\frac{a-b}{a+b}}$, where $2a$ and $2b$ are its semi-axes.

[The attraction required $=\iint \frac{\gamma k\rho . r d\theta dr}{r^2}\cos\theta$, the limits being zero to $\frac{l}{1-e\cos\theta}$ for r, and from zero to 2π for θ. We thus introduce the infinite quantity $\log r$, when r is zero. To avoid this, take any circle, of radius λ, $[<a(1-e)]$ surrounding the focus. The resultant attraction of this circle is zero by symmetry. Evaluate the above integral between limits λ and $\frac{l}{1-e\cos\theta}$ for r, and zero and 2π for θ.]

10. An elliptic disc, of mass M and semi-axes a and b, attracts according to the law $\frac{\mu}{\text{distance}}$; shew that its component attractions at an internal point (x, y) in directions parallel to the axes are

$$\frac{2\mu M}{a+b}\cdot\frac{x}{a} \text{ and } \frac{2\mu M}{a+b}\cdot\frac{y}{b}.$$

Deduce that in the case of an infinite homogeneous elliptic cylinder of density ρ, which attracts according to the law of Nature, the components are $\frac{4\gamma\pi\rho ab}{a+b}\cdot\frac{x}{a}$ and $\frac{4\pi\gamma\rho ab}{a+b}\cdot\frac{y}{b}$, where a and b are the semi-axes of its cross section.

11. Shew that the attraction of a circular disc of radius a, whose law of attraction is

$$\frac{\mu}{(\text{distance})^3}, \text{ is } \frac{M\mu}{c(c^2-a^2)} \text{ or } \frac{M\mu c}{a^2(a^2-c^2)}, \text{ according as } c \gtrless a,$$

where M is the mass of the disc and c the distance from the centre of the attracted point which is in the plane of the disc.

12. Find the attraction of a lamina in the shape of a uniform circular annulus on an external point in its plane, the law of attraction being that of the inverse seventh power of the distance.

286. ***Attraction of a thin uniform spherical shell on an external or internal point P.***

Let a be the radius of the spherical shell, k and ρ its thickness and density, and c the distance of P from its centre O.

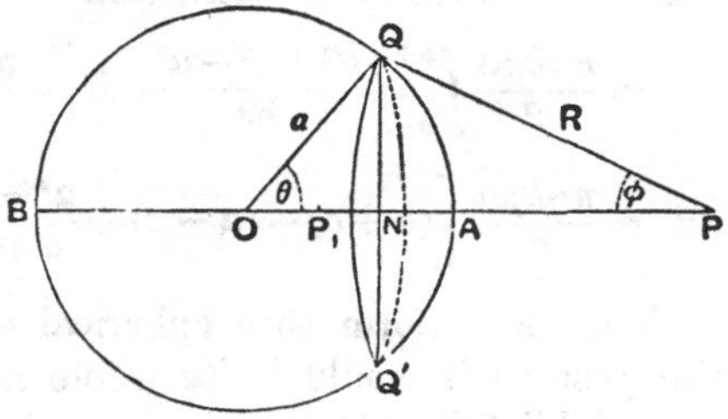

If Q be any point of the shell, QN the perpendicular on OP, and $\theta = \angle POQ$, then all the points on a circle with NQ as radius are at equal distances from P, and the attraction of each on P resolved along $PO \propto \dfrac{\cos\phi}{PQ^2}$.

Hence the resultant attraction of the portion of the shell generated by the arc $a.\delta\theta$

$$= \gamma k\rho \frac{a\,\delta\theta \,.\, 2\pi a \sin\theta}{PQ^2} \cos\phi.$$

Now $$R^2 = a^2 + c^2 - 2ac\cos\theta,$$

so that $$R.\delta R = ac\sin\theta.\delta\theta.$$

Hence the attraction of this elementary portion

$$= 2\pi\gamma k\rho \,.\, \frac{a}{c} \,.\, \frac{\cos\phi}{R}\,\delta R = \pi\gamma k\rho \,.\, \frac{a}{c^2}\left(\frac{R^2 + c^2 - a^2}{R^2}\right)\delta R.$$

First, let P be outside the shell so that $c > a$. If we integrate the quantity thus found for values of R between PA and PB, *i.e.* between $(c-a)$ and $(c+a)$, we have the resultant attraction of the whole shell.

$$\therefore \text{ resultant attraction} = \int_{c-a}^{c+a} \pi\gamma k\rho \frac{a}{c^2} \,.\, \frac{R^2 + c^2 - a^2}{R^2} \,.\, dR$$

$$= \pi\gamma k\rho \frac{a}{c^2}\left[R - \frac{c^2 - a^2}{R}\right]_{c-a}^{c+a}$$

$$= \pi\gamma k\rho \frac{a}{c^2}\left[(c+a) - (c-a) - \frac{c^2 - a^2}{c+a} + \frac{c^2 - a^2}{c-a}\right]$$

$$= \frac{4\pi\gamma k\rho a^2}{c^2} = \gamma \,.\, \frac{\text{Mass of the shell}}{OP^2},$$

and is thus the same as it would be if the whole mass of the shell were concentrated at O.

Secondly, let P be inside the shell as at P_1, so that $c < a$. The limits of the integration are now P_1A and P_1B, *i.e.* $a-c$ and $a+c$.

Hence the resultant attraction

$$=\frac{\pi\gamma k\rho a}{c^2}\int_{a-c}^{a+c}\frac{R^2+c^2-a^2}{R^2}dR=\frac{\pi\gamma k\rho a}{c^2}\left[R+\frac{a^2-c^2}{R}\right]_{a-c}^{a+c}$$

$$=\frac{\pi\gamma k\rho a}{c^2}\left[(a+c)-(a-c)+\frac{a^2-c^2}{a+c}-\frac{a^2-c^2}{a-c}\right]=0.$$

Hence a uniform thin spherical shell attracts an external point just as it would if its whole mass were collected at its centre, whilst its attraction on an internal point is zero.

287. *Attraction of a uniform solid sphere at an external or internal point.*

Conceive the sphere as made up of an infinite number of concentric shells, each of indefinitely small thickness.

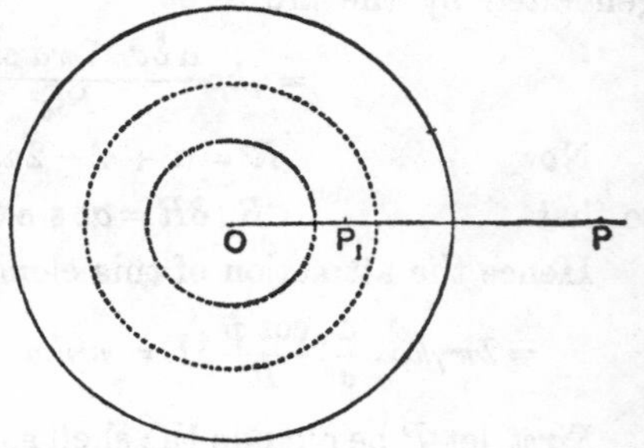

If P be outside the sphere, it is outside each of these shells, and hence the attraction of each shell on a unit mass at P

$$=\gamma.\frac{\text{Mass of the shell}}{OP^2}.$$

Hence the total attraction at P

$$=\gamma.\frac{\text{Sum of the masses of the shells}}{OP^2}$$

$$=\gamma.\frac{\text{Mass of the whole sphere}}{OP^2},$$

and is therefore the same as it would be if the whole mass of the sphere were concentrated at its centre O.

If the point be within the substance of the sphere as at P_1, then for all the shells of a radius greater than OP_1, P_1 is an internal point and hence, by Art. 286, the attraction of all such shells is zero. We need only therefore consider shells of a radius y which is less than OP_1, *i.e.* c.

For any such shell P_1 is an external point and its attraction at $P_1 = \gamma . \frac{\text{Mass of the shell}}{OP_1^2} = \gamma . \frac{4\pi y^2 \delta y . \rho}{c^2}$, and we must integrate this for values of y from zero to c.

Hence the resultant attraction at P_1

$$= \frac{4\pi\gamma\rho}{c^2} \int_0^c y^2 dy = \frac{4}{3} \pi\gamma\rho c.$$

Hence for a point inside the sphere the attraction varies directly as the distance of the attracted point from the centre.

288. *Attraction of a spherical shell. Geometrical Proof.*

Let P be an external point and Q its inverse, so that $CQ . CP = CA^2 = a^2$.

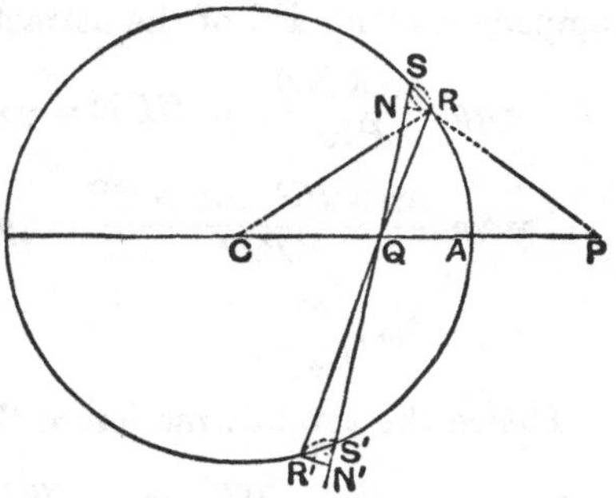

Through Q as vertex draw a very slender cone to cut the sphere in very small areas RS and $R'S'$.

Since

$$CP . CQ = a^2 = CR^2,$$

$$\therefore \frac{CQ}{CR} = \frac{CR}{CP},$$

so that the triangles CQR, CRP are similar.

So also the triangles CQR', $CR'P$ are similar.

Hence $$\frac{QR}{RP} = \frac{CR}{CP} = \frac{CR'}{CP} = \frac{QR'}{R'P} \quad \text{.................(1)},$$

and $$\angle CPR = \angle CRQ = \angle CR'Q = \angle CPR' \quad \text{.........(2)}.$$

Thus

$$\frac{\text{Attraction of } RS \text{ at } P}{\text{Attraction of } R'S' \text{ at } P} = \frac{\text{area } RS}{RP^2} \div \frac{\text{area } R'S'}{R'P^2}$$

$$= \frac{\text{area } RS}{\text{area } R'S'} . \frac{R'P^2}{RP^2}.$$

Draw RN, $R'N'$ perpendicular to SQS'.

Then area RS = cross area $RN \times \sec NRS$

= cross area $RN \times \sec CRQ$,

since CR, RQ are respectively perpendicular to RS and RN.

So area $R'S'$ = cross area $R'N' \times \sec CR'Q$.

$$\therefore \frac{\text{area } RS}{\text{area } R'S'} = \frac{\text{cross area } RN}{\text{cross area } R'N'} = \frac{QR^2}{QR'^2}.$$

Hence

$$\frac{\text{Attraction of } RS \text{ at } P}{\text{Attraction of } R'S' \text{ at } P} = \frac{QR^2}{QR'^2}\frac{R'P^2}{RP^2} = 1, \text{ by equation (1).}$$

Thus the attraction of these elementary areas at P are equal and, by (2), they are equally inclined to CP; hence their resultant is along CP.

Also, if $\delta\omega$ be the solid angle of the slender cone at Q and ρ be the mass of the shell per unit of area, then,

component along PC of the attraction of RS

$$= \gamma\rho \,.\, \frac{\text{area } RS}{PR^2} \,.\, \cos CPR = \gamma\rho \,.\, \frac{\text{area } RS \times \cos CRQ}{PR^2}$$

$$= \gamma\rho \,.\, \frac{\text{area } RS \,.\, \cos NRS}{PR^2} = \gamma\rho \,.\, \frac{\text{area } RN}{PR^2} = \gamma\rho \,.\, \frac{QR^2 \,.\, \delta\omega}{PR^2}$$

$$= \gamma\rho \,.\, \delta\omega \,.\, \frac{a^2}{CP^2}.$$

Hence the total attraction of the shell

$$= \Sigma\gamma\rho \quad \delta\omega \,.\, \frac{a^2}{CP^2} = \frac{\gamma\rho a^2}{CP^2}\Sigma\delta\omega = \frac{\gamma\rho a^2}{CP^2} \,.\, 4\pi = \gamma \,.\, \frac{\text{Mass of the shell}}{CP^2}.$$

Since the attraction of RS and $R'S'$ are equal, it follows that the portion of the shell to the right of the inverse point Q, and the part to the left of it, attract P equally.

Also the plane through Q perpendicular to CA contains all the points of contact of tangents to the shell drawn from P, *i.e.* it is the polar plane of P. Hence the polar plane of P divides the shell into two parts whose attractions at P are equal.

Secondly, consider the attraction at Q. Then

$$\frac{\text{Attraction of } RS \text{ at } Q}{\text{Attraction of } R'S' \text{ at } Q} = \frac{\text{area } RS}{QR^2} \div \frac{\text{area } R'S'}{QR'^2} = \frac{QR^2}{QR'^2} \,.\, \frac{QR'^2}{QR^2} = 1.$$

Hence the resultant attraction of RS and $R'S'$ at Q is zero. So for all such slender cones. Therefore the attraction of the whole shell on Q is zero.

Also the part of the shell to the right of a plane through Q perpendicular to CA and the part to the left attract Q equally.

289. *Value of gravity on the top of a tableland of height x above the Earth's surface.*

If a be the radius of the Earth and g the attraction due to gravity at its surface, then the value of gravity at a height x above the surface $=\dfrac{\mu}{(a+x)^2}$, where $g=\dfrac{\mu}{a^2}$, and hence this value

$$=g\frac{a^2}{(a+x)^2}=g\left[1-\frac{2x}{a}\right], \text{ if } \frac{x}{a} \text{ is small.}$$

If σ be the density of the material of the tableland (assumed to be homogeneous), its attraction at a point close to its surface is $2\pi\gamma x\sigma$, by Art. 281.

Now, if ρ be the mean density of the Earth,

$$g=\gamma\,.\,\frac{\frac{4}{3}\pi a^3\,.\,\rho}{a^2}=\frac{4\pi\gamma\rho a}{3}.$$

Hence the attraction of the tableland $=\dfrac{3}{2}\dfrac{x\sigma}{a\rho}g$.

The total attraction, g', at the top of the tableland thus

$$=g\left[1-\frac{2x}{a}\right]+\frac{3}{2}\frac{x\sigma}{a\rho}g=g\left[1-\frac{2x}{a}+\frac{3}{2}\frac{x\sigma}{a\rho}\right].$$

If we assume, as an approximation, that the mean density σ of the rocks near the Earth's surface is about one-half that of the mean density ρ of the whole Earth, this gives

$$g'=g\left[1-\frac{5x}{4a}\right].$$

290. *Value of the constant of gravitation.*

By the use of the proposition of Art. 287, and the known value of the acceleration due to gravity at the Earth's surface, we can obtain an approximate value for the constant of gravitation.

For, if E be the mass of the Earth and R its radius in any system of units, its attraction on a unit mass at the surface

$$=\gamma\,.\,\frac{E\,.\,1}{R^2}.$$

Hence $$g=\gamma\,.\,\frac{E}{R^2} \qquad\qquad (1).$$

Centimetre-Gramme-Second Units.

In this system $E = \frac{4}{3}\pi R^3 \times$ mean density,

and $$R = 6{\cdot}37 \times 10^8 \text{ cms.}$$

Now the mean density of the Earth, according to the most recent determination by C. V. Boys, is 5·527.

Hence (1) gives

$$981 = \gamma \,.\, \tfrac{4}{3}\pi R \times 5{\cdot}527 = \gamma \,.\, \frac{4\pi}{3} \times 6{\cdot}37 \times 10^8 \times 5{\cdot}527.$$

Hence $\gamma = 6{\cdot}66 \times 10^{-8}$, *i.e.* the force of attraction between two concentrated masses, each equal to one gramme, placed at a distance of one centimetre is $6{\cdot}66 \times 10^{-8}$ dynes.

It is easily seen that the force of attraction between two concentrated masses each equal to about 3877 grammes, placed at a distance of one centimetre, is one dyne.

Ft.-Lb.-Sec. Units.

As a rough approximation, taking the Earth to be a sphere of 4000 miles radius, (1) gives

$$32{\cdot}2 = \gamma \,.\, \frac{4\pi}{3} \times 4000 \times 5280 \times 5{\cdot}527 \times 62\tfrac{1}{2},$$

since the mean density of the Earth is 5·527 times that of water, *i.e.* is $5{\cdot}527 \times 62\frac{1}{2}$ lbs. per cubic foot.

Hence $\gamma = 1{\cdot}05 \times 10^{-9}$ nearly.

The force of attraction between two uniform spheres, each of mass one pound, whose centres are one foot apart, is thus $1{\cdot}05 \times 10^{-9}$ poundals approximately.

Dimensions of γ. If $[\Gamma]$ denote the dimensions of γ and $[M]$, $[L]$, $[T]$ the units of mass, length and time, then equation (1) gives

$$[L]\,[T]^{-2} = [\Gamma]\,\frac{[M]}{[L]^2}.$$

$$\therefore\ [\Gamma] = [M]^{-1}\,[L]^3\,[T]^{-2}.$$

EXAMPLES

1. The centre of a sphere of silver, of radius 5·5 cms. and sp. gr. $10\frac{1}{2}$, is distant 17 cms. from the centre of a sphere of gold, of radius 3 cms. and sp. gr. $19\frac{1}{4}$; shew that the attraction due to the two spheres is zero at a point between them distant 11 cms. from the centre of the silver sphere.

2. Draw a graph shewing the weight of a particle in its different positions as it is brought up from the centre of the Earth and taken to infinity.

3. Shew that to bring any mass from the centre of the Earth, treated as homogeneous, to the surface requires half as much work as that required to remove it from the surface to infinity against the Earth's attraction.

4. If the Earth (supposed spherical) were covered by an ocean of uniform depth h, prove that the value of gravity at the bottom of the ocean would exceed that at the top by $4\pi\gamma h(\frac{2}{3}\rho - \sigma)$ approximately, where σ is the density of the ocean and ρ is the mean density of the Earth.

5. If half the mass of the Earth were concentrated in an extremely thin uniform external crust, shew that at the centre of a circular gap in the crust the intensity of gravity would be less than its normal value by one-fourth.

6. The density of a sphere varies as the depth below the surface; shew that the resultant attraction is greatest at a depth equal to $\frac{1}{3}$ of the radius, and that its value there is $\frac{4}{3}$ of the value at the surface.

7. If a sphere consist of concentric layers, of uniform density, shew that its attraction is the same at any point of its volume if the density at each point varies inversely as the distance from the centre

8. Determine at external and internal points the attraction of a solid sphere of radius a, given that its density at a distance r from the centre is $k\left(\frac{r}{a}\right)^n$.

9. If the density of a solid sphere is a function of the distance from the centre, shew that its attraction begins to increase as we penetrate into the sphere, if the density at the surface is less than two-thirds of the mean density of the sphere.

10. Shew that the attraction of a uniform thin hemispherical shell, of mass M and radius a, at a point on the diameter perpendicular to the plane of the rim of the shell and at a distance x from its centre, is

$$\frac{\gamma M}{x^2}\left[1 - \frac{a}{\sqrt{a^2+x^2}}\right].$$

11. *Find the attraction of a solid homogeneous hemisphere at a point O on the edge of its plane base.*

Take O as the origin of coordinates; Ox as the line through the centre C of the base, Oz perpendicular to the base. Then using polar coordinates r, θ, ϕ, the attraction Z along Oz

$$= \iiint \gamma\rho \,.\, \frac{dr\,.\,r\,d\theta\,.\,r\sin\theta\,d\phi}{r^2}\cos\theta.$$

The limits for r are 0 to $2a\cos\phi\sin\theta$, since the equation to the surface of the sphere is $(x-a)^2+y^2+z^2=a^2$, *i.e.* $r^2=2ax=2ar\cos\phi\sin\theta$.

The limits for θ are zero to $\frac{\pi}{2}$, and those for ϕ are $-\frac{\pi}{2}$ to $\frac{\pi}{2}$.

Hence $$Z=\gamma\rho\iint 2a\cos\phi\sin^2\theta\cos\theta\,d\theta\,d\phi=\frac{4a\gamma\rho}{3}.$$

The component attraction X towards the centre

$$=\iiint\gamma\rho\,.\,\frac{dr\,.\,rd\theta\,.\,r\sin\theta\,d\phi}{r^2}\cos\phi\sin\theta=\frac{2\pi\gamma\rho a}{3}.$$

[This result follows also because clearly the resultant attraction of the hemisphere along OC must be half that of the complete sphere.]

By symmetry, the attraction along Oy vanishes.

Hence the resultant attraction is $\frac{2\gamma\rho a}{3}\sqrt{4+\pi^2}$, inclined at

$$\tan^{-1}\frac{2}{\pi}\text{ to }OC.$$

12. Shew that at the southern base of a hemispherical hill, of radius a and density ρ, the apparent latitude is diminished by $\frac{1}{2}\frac{\rho a}{\sigma r}$, where σ is the mean density and r the radius of the Earth.

13. If the northern and southern hemispheres of the Earth had been of uniform densities ρ and σ respectively, the mean density being as at present, prove that gravity at the equator would be greater than it is now in the ratio

$$\sqrt{1+\frac{4}{\pi^2}\left(\frac{\rho-\sigma}{\rho+\sigma}\right)^2}:1,$$

and that the deviation of the plumb-line from the zenith at any point of the equator would be

$$\tan^{-1}\left\{\frac{2}{\pi}\frac{\rho\sim\sigma}{\rho+\sigma}\right\}.$$

14. If a mountain, in the form of an enveloping cone of semivertical angle a, were added to a sphere of uniform density, then gravity at its summit would be

$$g\frac{1+\sin^3 a-\cos^3 a}{2\sin a},$$

where g is the value of gravity at all points of the surface of the sphere and the mountain is supposed to be of the same density as the sphere.

15. *Find if there is any law of attraction, in addition to that of the inverse square of the distance, by which the attraction of a thin uniform spherical shell would be zero at an internal point.*

Let the law of attraction be $\frac{f(r)}{r^2}$, so that, by Art. 286, we have given

$$\int_{a-c}^{a+c}\frac{R^2+c^2-a^2}{R^2}f(R)\,dR=0 \quad \text{.....................(1)},$$

for all values of c less than a.

Differentiating this equation with respect to c, we obtain

$$-\frac{f(a+c)}{a+c}+\frac{f(a-c)}{a-c}=\int_{a-c}^{a+c}\frac{f(R)}{R^2}\,dR.$$

Again, differentiating (1) with respect to a, we have

$$\frac{c}{a+c}f(a+c)+\frac{c}{a-c}f(a-c)=a\int_{a-c}^{a+c}\frac{f(R)}{R^2}\,dR.$$

Eliminating the integral from these two equations, we obtain

$$f(a+c)=f(a-c).$$

This relation is true for all values of a and for all values of c less than a. It follows that $f(r)$ is the same for all values of r, *i.e.* that $f(r)$ is constant $=A$ (say).

Hence the only law of force admissible

$$=\frac{f(r)}{r^2}=\frac{A}{r^2}.$$

16. *Find if there is any other law of attraction, in addition to that of the inverse square of the distance, by which the attraction of a thin uniform shell of radius a would be the same at all external points as if its mass were collected at its centre.*

Let the law of attraction be $\frac{f(r)}{r^2}$ so that, by Art. 286, we have given

$$\pi\gamma k\rho\,\frac{a}{c^2}\int_{c-a}^{c+a}\frac{R^2+c^2-a^2}{R^2}f(R)\,dR=4\pi\gamma k\rho a^2.\frac{f(c)}{c^2},$$

and hence $$\frac{1}{a}\int_{c-a}^{c+a}\frac{R^2+c^2-a^2}{R^2}f(R)\,dR=4f(c) \quad\ldots\ldots\ldots\ldots(1).$$

This is to be true for all values of a and all values of c greater than a. Differentiate with respect to a, and we have

$$-\frac{1}{a^2}\int_{c-a}^{c+a}\frac{R^2+c^2-a^2}{R^2}f(R)\,dR$$
$$+\frac{1}{a}\left[\frac{2c}{c+a}f(c+a)+\frac{2c}{c-a}f(c-a)-2a\int\frac{f(R)}{R^2}\,dR\right]=0.$$

$$\therefore\ 2ac\left[\frac{f(c+a)}{c+a}+\frac{f(c-a)}{c-a}\right]=\int_{c-a}^{c+a}\frac{R^2+c^2+a^2}{R^2}f(R)\,dR.$$

Differentiate this with respect to c and a respectively; thus we obtain

$$2c\left[-\frac{f(c+a)}{c+a}+\frac{f(c-a)}{c-a}\right]+2ac\left[\frac{f'(c+a)}{c+a}+\frac{f'(c-a)}{c-a}\right]=2c\int_{c-a}^{c+a}\frac{f(R)}{R^2}\,dR,$$

and

$$2a\left[-\frac{f(c+a)}{c+a}+\frac{f(c-a)}{c-a}\right]+2ac\left[\frac{f'(c+a)}{c+a}-\frac{f'(c-a)}{c-a}\right]=2a\int_{c-a}^{c+a}\frac{f(R)}{R^2}\,dR.$$

Hence, on elimination of the integral, $\frac{f'(c+a)}{(c+a)^2}=\frac{f'(c-a)}{(c-a)^2}$, for all values of a and for all values of c greater than a.

Hence $\frac{f(r)}{r^3}$ must be constant. $\therefore f(r) = Ar^3 + B$, where A and B are arbitrary constants; hence the law of force required $= Ar + \frac{B}{r^2}$.

Thus the only possible laws are those of the direct distance, and inverse square, or a combination of them.

THE POTENTIAL.

291. The potential at a point P of an attracting mass M is the work done on a unit mass as it moves from an infinite distance, by any path, curved or straight, to the point P.

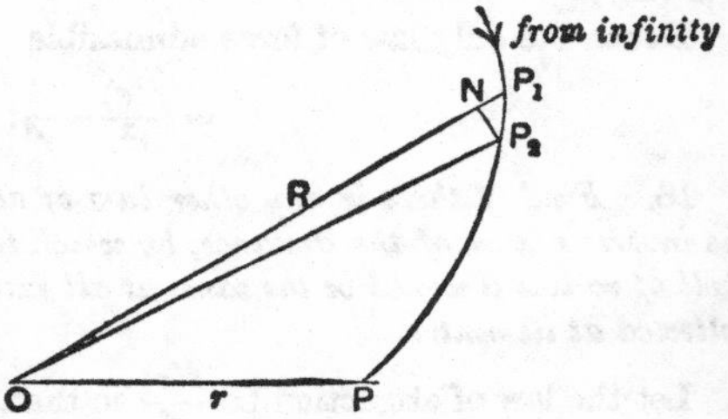

Consider any elementary portion, m, of the attracting mass at O and let P_1P_2 be any elementary arc of the path of the particle, where

$$OP_1 = R \text{ and } OP_2 = R + \delta R.$$

Draw P_2N perpendicular to OP_1. Then, in the limit,

$$ON = OP_2 = R + \delta R.$$

$$\therefore P_1N = OP_1 - ON = R - (R + \delta R) = -\delta R.$$

Hence the work done by the attraction of m as the unit mass moves from P_1 to P_2

$$= \gamma \cdot \frac{m}{OP_1^2} \cdot P_1N = \frac{\gamma m}{R^2}(-\delta R).$$

Hence the whole work done by this attraction as the unit mass moves from infinity to P, where OP is r,

$$= \int_\infty^r \left(-\frac{\gamma m}{R^2}\right) dR = \left[\frac{\gamma m}{R}\right]_\infty^r$$

$$= \gamma m \left[\frac{1}{r} - \frac{1}{\infty}\right] = \frac{\gamma m}{r}.$$

A similar result is true for elements m_1, m_2, ... of the attracting mass at O_1, O_2, ...

Hence the total work done by the whole attracting mass

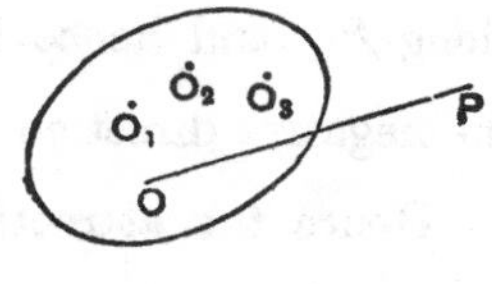

$$=\frac{\gamma m}{PO}+\frac{\gamma m_1}{PO_1}+\frac{\gamma m_2}{PO_2}+\dots$$

$$=\int\frac{\gamma}{r}\,dm.$$

Thus the potential of a mass M at any point P is obtained as follows; *Let dm be any element of M whose distance from P is r, then the potential at $P=\gamma\int\frac{dm}{r}$, where the integral is taken throughout the attracting mass.*

This quantity is usually denoted by V.

292. If V be the potential of an attracting mass at any point P, whose coordinates are x, y, z, we can shew that $\frac{dV}{dx}$ is the component attraction at P parallel to the axis of x in the direction of x increasing, and similarly for $\frac{dV}{dy}$ and $\frac{dV}{dz}$.

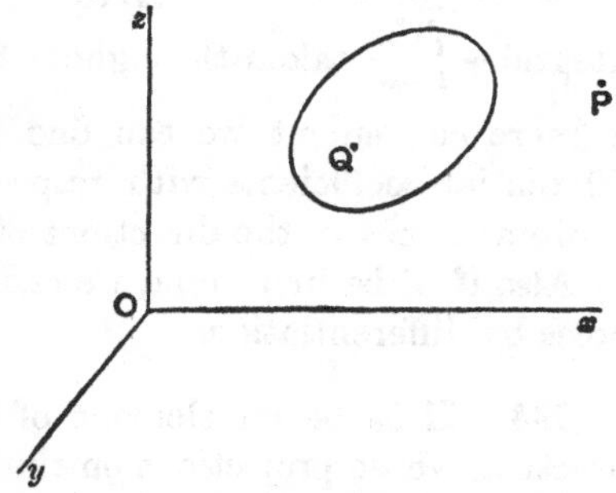

For let dm be an element of the attracting mass at the point Q, whose coordinates are x', y', z'. Then, by the definition of the preceding article,

$$V=\gamma\int\frac{dm}{PQ}=\gamma\int\frac{dm}{\sqrt{(x-x')^2+(y-y')^2+(z-z')^2}}.$$

$$\therefore\ \frac{dV}{dx}=-\gamma\int\frac{dm}{\{(x-x')^2+(y-y')^2+(z-z')^2\}^{\frac{3}{2}}}(x-x')$$

$$=-\gamma\int\frac{dm}{PQ^2}\cdot\frac{x-x'}{PQ}=-\gamma\int\frac{dm}{PQ^2}\times\cos\theta\quad\text{............(1)},$$

where θ is the inclination of QP to the axis of x.

Now the attraction of the element dm on P is $\gamma \cdot \frac{dm}{PQ^2}$ along PQ, and hence the resolved part of this attraction along the negative direction of the axis of $x = \gamma \cdot \frac{dm}{PQ^2} \cos\theta$.

Hence the attraction of the whole mass resolved parallel to the positive direction of the axis of $x = -\gamma \int \frac{dm}{PQ^2} \cos\theta$.

Therefore, from (1), it follows that

$\frac{dV}{dx}$ = the resultant attraction of the whole mass parallel to the axis of x in the direction of x increasing.

Similarly $\frac{dV}{dy}$ and $\frac{dV}{dz}$ are the resultant attractions in the directions of the axes of y and z.

293. From the preceding articles we thus see that we may obtain the potential at any point by finding the value of the integral $\gamma \int \frac{dm}{r}$ taken throughout the mass of the body, or, if it be more convenient, we can find it from the property that its differential coefficients with respect to x, y, z are equal to the resultant forces in the directions of these coordinates.

Also if V be first found we can easily obtain the component forces by differentiation.

294. If δs be an element of a line drawn from P in any direction, whose projections on the axis of x, y, z are $\delta x, \delta y$, and δz, and whose direction cosines are therefore

$$\frac{\delta x}{\delta s}, \; \frac{\delta y}{\delta s} \text{ and } \frac{\delta z}{\delta s},$$

then the resultant force along the element δs

$$= \frac{dV}{dx} \cdot \frac{dx}{ds} + \frac{dV}{dy} \frac{dy}{ds} + \frac{dV}{dz} \frac{dz}{ds} = \frac{dV}{ds}.$$

If this element δs be PP', this expresses the fact that the force in the direction PP'

$$= \text{Lt.} \frac{\text{Potential at } P' - \text{Potential at } P}{PP'}.$$

295. From the preceding article it follows that if the position of P be given by the usual polar coordinates r, θ, ϕ, the resultant attractions are

$$\frac{dV}{dr} \text{ in the direction of } r,$$

$$\frac{1}{r}\frac{dV}{d\theta} \text{ perpendicular to } r \text{ in the plane of } \theta,$$

and $$\frac{1}{r\sin\theta}\frac{dV}{d\phi} \text{ perpendicular to the plane of } \theta.$$

296. When the attracted point P is inside the attracted mass, in which case some of the values of r are zero, it seems at first sight as if the value of the potential $\int \gamma . \frac{dm}{r}$ might be infinite.

But it can easily be shewn that this is not so. For take P as the origin of polar coordinates (r, θ, and ϕ) in three dimensions, in which case $\delta m = \delta r . r\delta\theta . r\sin\theta\,\delta\phi . \rho$.

Hence the potential $V = \gamma \iiint \rho . r\sin\theta\, dr\, d\theta\, d\phi$, and hence is finite even if r be zero for some of the elements.

Similarly, by (1) of Art. 292, $\frac{dV}{dx} = -\gamma \int \frac{dm}{PQ^2} . \cos QPx$

$$= -\gamma \iiint \rho . \frac{r^2 \sin\theta\, dr\, d\theta\, d\phi}{r^2} \cos\phi \sin\theta = -\gamma \iiint \rho \sin^2\theta \cos\phi\, dr\, d\theta\, d\phi,$$

and hence no element of it becomes infinite even if r be zero.

The potential and components of attraction therefore are continuous functions, if every element of the attracting mass is of finite volume density.

But the same will not be true of the second differential coefficients of the potential. For differentiating the expression of Art. 292 with respect to x, we have

$$\frac{d^2V}{dx^2} = -\gamma \int dm \left[\frac{1}{PQ^3} - 3\frac{(x-x')^2}{PQ^5}\right].$$

Taking the attracted point (x, y, z) as the origin as above, and substituting in polar coordinates, this gives

$$\frac{d^2V}{dx^2} = \gamma \iiint \rho . \frac{3\cos^2\phi \sin^2\theta - 1}{r} . \sin\theta\, dr\, d\theta\, d\phi.$$

Here the quantity under the integral sign is infinite, when

r vanishes, *i.e.* for some values of r when the attracted point is inside the attracting mass.

Hence the second differential coefficient is not continuous as we pass from outside to inside the attracting mass.

The above results will not necessarily hold for other laws of attraction than that of Nature. For instance, if the law were that of the inverse cube of the distance, the above integral for $\frac{dV}{dx}$ would have an r in its denominator, and so some of its elements would become infinite if the attracted particle were within the attracting mass.

297. *Potential for laws of attraction other than that of the inverse square.*

If the law of attraction be $\gamma \cdot \frac{m_1 m_2}{(\text{distance})^n}$, the potential of a mass m at a distance r, as in Art. 291,

$$= \int_\infty^r \left(\frac{-\gamma m}{R^n}\right) dR = \frac{\gamma m}{n-1}\left[\frac{1}{R^{n-1}}\right]_\infty^r = \frac{\gamma m}{n-1}\frac{1}{r^{n-1}}.$$

Hence the potential of the whole mass $M = \frac{\gamma}{n-1}\int \frac{dm}{r^{n-1}}$.

This holds so long as n is positive and greater than unity.

If the law be that of the inverse distance, *i.e.* if $n = 1$, the potential $= \int_\infty^r \left(\frac{-\gamma m}{R}\right) dR = -\gamma m \left[\log R\right]_\infty^r = C - \gamma m \log r$, where C is an infinite constant.

Also the potential of the whole attracting mass

$$= C_1 - \gamma \int \log r dm,$$

where C_1 is also an infinite constant.

298. *Potential of a thin uniform rod at any external point.*

With the figure and notation of Art. 277, the potential of the rod AB at P

$$= \int \frac{\gamma k\rho \, . \, dx}{PQ} = \int_\alpha^\beta \frac{\gamma k\rho p \sec^2\theta \, d\theta}{p \sec\theta}$$

$$= \gamma k\rho \int_\alpha^\beta \sec\theta \, d\theta = \gamma k\rho \left[\log\tan\left(\frac{\pi}{4} + \frac{\theta}{2}\right)\right]_\alpha^\beta$$

$$= \gamma k\rho \log \frac{\tan\left(\frac{\pi}{4} + \frac{\beta}{2}\right)}{\tan\left(\frac{\pi}{4} + \frac{\alpha}{2}\right)}.$$

Now $\angle PAB = \frac{\pi}{2} + \alpha$, so that $\tan\left(\frac{\pi}{4} + \frac{\alpha}{2}\right) = \tan\frac{PAB}{2}$.

Also $\angle PBA = \frac{\pi}{2} - \beta$, so that

$$\tan\left(\frac{\pi}{4} + \frac{\beta}{2}\right) = \cot\left(\frac{\pi}{4} - \frac{\beta}{2}\right) = \cot\frac{PBA}{2}.$$

Hence the potential at P

$$= \gamma k\rho \log\left[\cot\frac{PAB}{2} \cdot \cot\frac{PBA}{2}\right].$$

If $AB = a$, $AP = r_1$, and $BP = r_2$, then, by the ordinary formulae in Trigonometry,

$$\cot\frac{PAB}{2} \cdot \cot\frac{PBA}{2} = \sqrt{\frac{s(s-r_2)}{(s-r_1)(s-a)}}\sqrt{\frac{s(s-r_1)}{(s-r_2)(s-a)}}$$

$$= \frac{s}{s-a} = \frac{r_1 + r_2 + a}{r_1 + r_2 - a}.$$

Hence the potential at P

$$= \gamma k\rho \log\frac{r_1 + r_2 + a}{r_1 + r_2 - a}.$$

Cor. 1. It follows that the potential is constant for all points for which $r_1 + r_2$ is constant, *i.e.* for all points which lie on an ellipse whose foci are A and B.

Hence, in the case of a thin rod AB, the equi-potential curves are ellipses whose foci are A and B.

Cor. 2. If the rod be of infinite length in both directions, the potential V

$$= 2\int_0^\infty \frac{\gamma k\rho \, . \, dx}{PQ} = 2\gamma k\rho \int_0^\infty \frac{dx}{\sqrt{p^2 + x^2}}$$

$$= 2\gamma k\rho \left[\log\{x + \sqrt{x^2 + p^2}\}\right]_0^\infty$$

$$= C - 2\gamma k\rho \log p,$$

where C is an infinite constant.

This result may also be obtained from the result of Art. 277, Cor.; for $\frac{dV}{dp} = -$ force of attraction $= -\frac{2\gamma k\rho}{p}$.

$$\therefore \quad V = C - 2\gamma k\rho \log p.$$

Cor. 3. If the rod be of infinite length in the direction AP produced, but ends at A, the potential at P

$$=\gamma k\rho\left[\log\tan\left(\frac{\pi}{4}+\frac{\theta}{2}\right)\right]_{\alpha}^{\frac{\pi}{2}}=\gamma k\rho\left[C'-\log\tan\left(\frac{\pi}{4}+\frac{\alpha}{2}\right)\right],$$

where C' is an infinite constant,

$$=\gamma k\rho\left[C'-\log\frac{1+\sin\alpha}{\cos\alpha}\right].$$

299. *Potential of a uniform thin circular plate at a point on its axis.*

With the figure and notation of Art. 281, the potential at P

$$=\int\frac{\gamma\,.\,2\pi x dx\,.\,k\rho}{PQ}=2\pi\gamma k\rho\int_0^a\frac{x}{\sqrt{p^2+x^2}}\,dx$$

$$=2\pi\gamma k\rho\left[\sqrt{p^2+x^2}\right]_0^a$$

$$=2\pi\gamma k\rho\,[\sqrt{p^2+a^2}-p].$$

Or the potential V may be obtained from the result of Art. 281. For

$$\frac{dV}{dp}=\text{attraction in the direction } OP$$

$$=-\,2\pi\gamma k\rho\left[1-\frac{p}{\sqrt{a^2+p^2}}\right].$$

$$\therefore\quad V=2\pi\gamma k\rho\,[\sqrt{a^2+p^2}-p]+C.$$

The constant C is zero; for when p is zero,

$$V=\text{the potential of the plate at its centre}$$

$$=\int_0^a\frac{\gamma\,.\,2\pi x dx\,.\,k\rho}{x}=2\pi\gamma k\rho a.$$

EXAMPLES

1. If a uniform rod be of infinite length, shew that the work done in removing a unit particle against its attraction, from a perpendicular distance y_1 to a perpendicular distance y_2, is

$$2\gamma k\rho\log\frac{y_2}{y_1}.$$

2. Two equal uniform bars, of masses m_1 and m_2 and length l, are symmetrically placed so as to be parallel and at a distance y_1. Shew that the work done against their mutual attractions in pulling one away symmetrically, till it is at a distance y_2 from the other, is

$$2\gamma\frac{m_1m_2}{l^2}\left[y-\sqrt{y^2+l^2}-l\log\frac{\sqrt{y^2+l^2}-l}{y}\right]_{y_1}^{y_2}.$$

[Use the result of Ex. 4, page 307.]

3. The potential of a distribution of matter is given by

$$V=\mu\log\frac{x-a+\sqrt{y^2+z^2+(a-x)^2}}{a+x+\sqrt{y^2+z^2+(a+x)^2}};$$

find the most compact distribution of matter that will produce it.

4. A number n of equal, infinitely long, homogeneous straight filaments lie on a cylinder of radius a, and are at equal distances from one another. Shew that the potential at any point P can be put in the form

$$C-\gamma m\log(r^{2n}-2a^nr^n\cos n\theta+a^{2n}),$$

where r and θ are the polar coordinates of P referred to an origin which is the intersection of the axis of the cylinder with a plane through P perpendicular to it.

5. Shew that the potential of the surfaces of a cube at the centre of the cube is

$$\frac{\gamma M}{a}[6\log(2+\sqrt{3})-\pi],$$

where $2a$ is the length of a side and M is the mass of a face supposed indefinitely thin.

Deduce that the potential of a cube at its centre is

$$\frac{\gamma M}{4a}[6\log(2+\sqrt{3})-\pi],$$

where M is its mass and $2a$ the length of one of the edges of the cube.

Deduce also that the value of the potential of the cube at one of its corners is one-half the value at its centre.

[Start with the result of Art. 280.]

6. Shew that the potential of a thin homogeneous ring, of mass m and radius a, at a point in its plane distant c from its centre is $C-\gamma m\log c$ or $C-\gamma m\log a$, according as c is $\gtrless a$, the law of force being inversely as the distance.

7. Shew that the potential of an infinite uniform thin cylindrical shell at a point P is

$$C-4\pi\gamma aM\log a \quad\text{or}\quad C-4\pi\gamma aM\log r,$$

according as P is inside or outside the cylinder, the mass per unit area being M, the radius of the shell a, and r being the distance of P from its axis.

8. In the plane of a thin uniform circular ring, of radius a cms. and mass M grammes, a point O is taken at a distance c from the centre ($c > a$); shew that the potential at O due to the gravitation of the ring is

$$\gamma \frac{M}{c+a}\left\{1+\left(\frac{1}{2}\right)^2 k^2+\left(\frac{1.3}{2.4}\right)^2 k^4+\left(\frac{1.3.5}{2.4.6}\right)^2 k^6+\dots\right\},$$

where
$$k=\frac{2\sqrt{ac}}{a+c}.$$

9. Shew that the potential of a uniform circular disc, of mass M and radius a, at a point in its plane distant c from its centre, is

$$\frac{4\gamma M}{\pi a^2}\int_0^{\frac{\pi}{2}}\sqrt{a^2-c^2\sin^2\theta}\,d\theta \quad\text{or}\quad \frac{4\gamma M}{\pi a^2}\int_0^{\sin^{-1}\frac{a}{c}}\sqrt{a^2-c^2\sin^2\theta}\,d\theta,$$

according as c is less or greater than a.

10. Find the potential of a uniform lamina, bounded by two concentric circles whose radii are a and b, at a point in its plane distant r from the common centre, the law of attraction being $\frac{\mu}{(\text{distance})^5}$.

11. The density of an elliptic lamina varies as the distance from the major axis, the mass of a unit element of area at unit distance being μ. Shew that the potential due to the lamina at a focus is $2\gamma\mu b^2$.

12. O is the centre of a homogeneous hemisphere and A is the other end of the radius perpendicular to its base. If the radius be a cms., the density ρ grammes per cub. cm., and γ be the constant of gravitation, shew that the work done against the attraction of the hemisphere in carrying one gramme from O to A is

$$2\pi\gamma\rho a^2\left[1-\frac{2}{3}\sqrt{2}\right]\text{ ergs.}$$

13. Shew that the quantity of work necessary to move one condensed unit of mass along any path from the middle point of the base of a homogeneous solid cone to the vertex is

$$\pi\gamma\rho h^2\sin^2\alpha\cos\alpha\left[\log\frac{\cos\alpha(1+\cos\alpha)}{\sin\alpha(1-\sin\alpha)}+\frac{\cos^3\alpha+\sin^3\alpha-1}{\sin^2\alpha\cos^2\alpha}\right],$$

where h is the height and 2α the vertical angle of the cone, and ρ is its density.

300. *Potential of a thin uniform spherical shell at an external or internal point.*

With the figure and notation of Art. 286 the potential at an external point P

$$=\int\gamma k\rho\,.\,\frac{a\,d\theta\,.\,2\pi a\sin\theta}{PQ}.$$

Also $R.\delta R = ac\sin\theta.\delta\theta$, as in that article.

Hence the potential

$$= 2\pi\gamma k\rho\frac{a}{c}\int_{c-a}^{c+a}\frac{RdR}{R} = 2\pi\gamma k\rho\frac{a}{c}\Big[R\Big]_{c-a}^{c+a} = \frac{4\pi\gamma k\rho a^2}{c}$$

$$= \gamma.\frac{\text{Mass of the shell}}{OP},$$

so that, for an external point, the potential of the shell is the same as it would be if the whole mass of the shell were collected at its centre.

For an internal point P_1 the limits for the integration are from $R = P_1A$ to P_1B, *i.e.* from $a - c$ to $a + c$. Hence the potential at P_1

$$= 2\pi\gamma k\rho\frac{a}{c}\Big[R\Big]_{a-c}^{a+c} = 4\pi\gamma k\rho a = \gamma.\frac{\text{Mass of the shell}}{\text{Its radius}}.$$

Hence, for an internal point P_1, the potential is constant and equal therefore to its value for a point at the centre.

301. *Potential of a uniform solid sphere at an external or internal point.*

Take the figure and notation of Art. 287, and conceive the sphere as made up of an infinite number of thin concentric spherical shells as in that article.

If P be outside the sphere, it is outside each of these shells, for any one of which the potential at $P = \gamma.\dfrac{\text{Mass of the shell}}{OP}$, by the last article.

Hence the total potential at an external point P

$$= \gamma.\frac{\text{Sum of the masses of the shells}}{OP}$$

$$= \gamma.\frac{\text{Mass of the sphere}}{OP},$$

and is therefore the same as it would be if the whole mass of the sphere were concentrated at its centre.

If the point be inside the sphere, then for all the shells of radius y less than OP_1, P_1 is an external point, and hence, by Art. 300, the potential for such a shell $= \gamma.\dfrac{4\pi\rho y^2\delta y}{OP_1}$, and this must be integrated between limits 0 and a.

For all the shells of radius y greater than OP_1, P_1 is an internal point, and hence, by Art. 300, the potential for such a shell $= \gamma . \frac{4\pi\rho y^2 \delta y}{y}$, and this must be integrated between limits c and a.

Hence, finally, the potential at an internal point P_1

$$= \int_0^c \gamma . \frac{4\pi\rho y^2 dy}{c} + \int_c^a \gamma . \frac{4\pi\rho y^2 dy}{y}$$

$$= \frac{4\pi\rho\gamma}{3c} . c^3 + 2\pi\rho\gamma (a^2 - c^2)$$

$$= 2\pi\gamma\rho \left(a^2 - \frac{c^2}{3}\right).$$

302. ***Potential and attraction of a spherical shell, of finite thickness, bounded by spheres of radii a and b.***

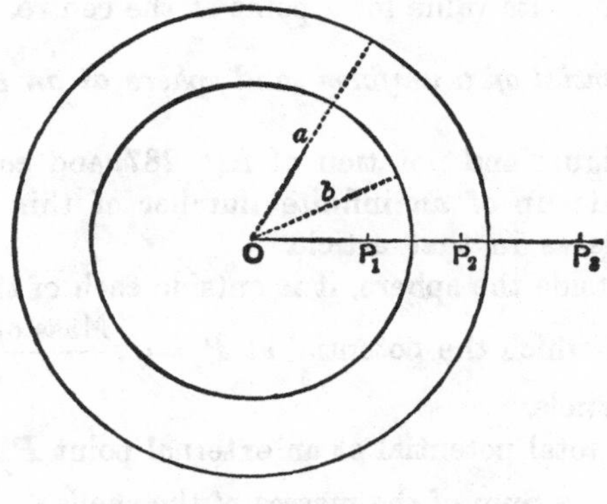

Let O be the centre and ρ the density of the shell. Conceive this shell of finite thickness to be composed of an infinite number of thin shells, and apply the results of Art. 300.

First; for a point P_1 $(OP_1 = x)$ within the inner surface of the shell, the second case applies, and the potential at it

$$= \int_b^a \gamma . \frac{4\pi\rho y^2 dy}{y} = 2\pi\gamma\rho (a^2 - b^2).$$

Secondly; for a point P_2 $(OP_2 = x)$ between the two bounding surfaces; for the shells of a less radius than x, P_2 is an external point and the first case of Art. 300 holds; whilst, for shells of a greater radius, P_2 is an internal point and the second case applies. Hence the potential at P_2

$$= \int_b^x \gamma . \frac{4\pi\rho y^2 dy}{x} + \int_x^a \gamma . \frac{4\pi\rho y^2 dy}{y}$$

$$= \frac{4}{3}\pi\rho\gamma\left(\frac{x^3 - b^3}{x}\right) + 2\pi\rho\gamma(a^2 - x^2)$$

$$= 2\pi\rho\gamma\left[a^2 - \frac{x^2}{3} - \frac{2}{3}\frac{b^3}{x}\right].$$

Thirdly; for a point P_3 $(OP_3 = x)$, external to the whole given shell and therefore to all the thin shells, the potential V

$$= \int_b^a \gamma . \frac{4\pi\rho y^2 dy}{x} = \frac{4\pi\rho\gamma}{3}\frac{a^3 - b^3}{x}.$$

We thus have the following results for V and its differential coefficients:

	$x<b$	$b<x<a$	$x>a$
V	$2\pi\gamma\rho(a^2 - b^2)$	$2\pi\gamma\rho\left[a^2 - \frac{x^2}{3} - \frac{2}{3}\frac{b^3}{x}\right]$	$\frac{4\pi\rho\gamma}{3}\frac{a^3 - b^3}{x}$
$\frac{dV}{dx}$	0	$-\frac{4\pi\gamma\rho}{3}\left[x - \frac{b^3}{x^2}\right]$	$-\frac{4\pi\rho\gamma}{3}\frac{a^3 - b^3}{x^2}$
$\frac{d^2V}{dx^2}$	0	$-\frac{4\pi\gamma\rho}{3}\left[1 + \frac{2b^3}{x^3}\right]$	$\frac{8\pi\rho\gamma}{3}\frac{a^3 - b^3}{x^3}$

It will be noted that, if we conceive the point P as travelling from the centre O outwards through the positions $P_1, P_2, P_3, \ldots$, the values of V and $\frac{dV}{dx}$ are always continuous, and in particular are continuous at the values $x = b$ and $x = a$, *i.e.* when the point P passes into, and out of, the attracting matter.

But the value of $\frac{d^2V}{dx^2}$ is discontinuous at these values.

At the value $x = b$, $\frac{d^2V}{dx^2}$ suddenly changes from 0 to $-4\pi\gamma\rho$, and at the value $x = a$, it suddenly changes from

$$-\frac{4\pi\gamma\rho}{3}\left(1 + \frac{2b^3}{a^3}\right) \text{ to } \frac{8\pi\rho\gamma}{3}\frac{a^3 - b^3}{a^3}.$$

The accompanying figure, taken from Thomson and Tait's *Natural Philosophy*, illustrates the variations of V, $\frac{dV}{dx}$, and $\frac{d^2V}{dx^2}$. OE, ON are the radii b and a of the bounding surfaces of the shell.

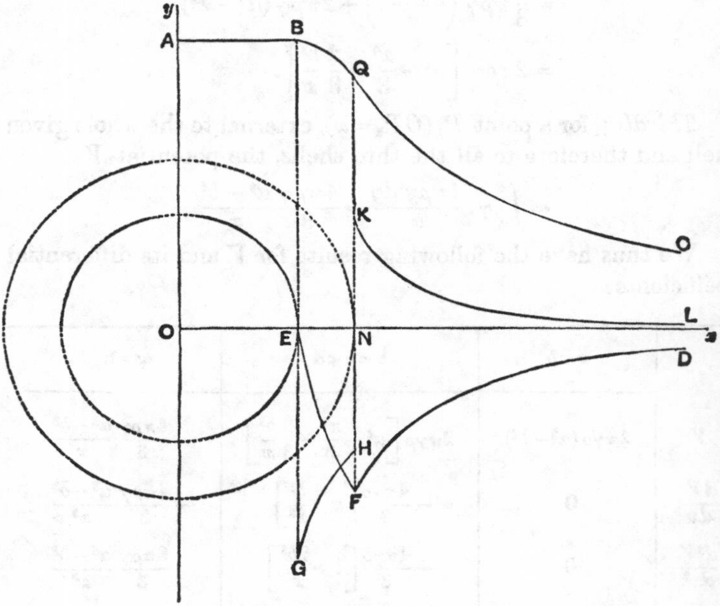

V is represented by the continuous curve $ABQC$, which has no abrupt change of direction; $\frac{dV}{dx}$ by the continuous curve $OEFD$ which changes its direction abruptly at E and F; $\frac{d^2V}{dx^2}$ by the discontinuous curve consisting of the three portions OE, GH and KL.

EXAMPLES

1. Shew that the potential of a zone of a thin homogeneous spherical shell at any point P on the axis of the zone is $\frac{2\gamma M}{r_1+r_2}$, where M is the mass of the zone and r_1, r_2 are the distances of the point P from the bounding edges of the zone.

2. Shew that half of the potential of a uniform spherical shell at an external point O is due to that portion of the sphere which is nearer to O than the centre is.

3. If the radius of a sphere be a, its density ρ, and the distance of an internal point from the centre O be b, shew that the difference of the potentials at P due to the two portions into which the sphere is divided by a plane through P perpendicular to OP is

$$\frac{4\pi\rho\gamma}{3b}\left[a^3-(a^2-b^2)^{\frac{3}{2}}\right].$$

4. Find the potential of a solid homogeneous attracting sphere, of radius a and density $\dfrac{3\mu}{2\pi}$, together with a uniform distribution on the surface of the sphere of repelling matter of surface density $\dfrac{\mu a}{2\pi}$, at any point inside the sphere, and also at any point outside the sphere.

[$\mu\gamma(a^2-x^2)$ inside; zero outside.]

5. Shew that the potential of a solid hemisphere, of radius a and density ρ, at an external point P situated on the axis at a distance ξ from the centre is

$$\frac{2\pi\gamma\rho}{3\xi}\left[a^3 \pm \left\{(a^2+\xi^2)^{\frac{3}{2}}-\xi^3-\frac{3}{2}a^2\xi\right\}\right],$$

the upper or lower sign being taken according as P is on the convex or plane side of the body.

6. If the density of a sphere at a point distant x from its centre is $\dfrac{k}{x}\sin\dfrac{x}{c}$, where k and c are given constants, prove that the attraction at an internal point distant x from the centre is

$$4\pi\gamma kc\left(\frac{c}{x^2}\sin\frac{x}{c}-\frac{1}{x}\cos\frac{x}{c}\right),$$

and find the potential.

7. Shew that the potential of a uniform spherical shell, of small thickness k, and of density ρ and radius a, at an external point, distant c from its centre, is

$$\frac{-2\pi\gamma\rho ka}{(n+1)(n+3)c}\left[(c+a)^{n+3}-(c-a)^{n+3}\right],$$

if the law of force be that of the nth power of the distance.

8. Shew that the mean value, taken for all points on a spherical surface, of the potential of attracting matter outside the surface is equal to the potential of the attracting matter at the centre of the spherical surface.

If the attracting matter lie within the spherical surface, shew that the corresponding mean value of the potential is equal to γ multiplied by the quotient of the mass by the radius of the sphere.

9. Shew that the mean value, taken for all points on the surface of a circular cylinder of infinite length, of the potential of an attracting mass external to the cylinder is equal to the mean value of the potential of the same mass for points taken on the axis of the cylinder.

MISCELLANEOUS EXAMPLES ON ATTRACTIONS AND POTENTIAL

1. *NS is a magnet and P a magnetic particle, so that P is acted on by forces, towards N and from S, varying inversely as the square of the distance. If O be the middle point of NS, if $NOP=\theta$, and if OP be great compared with the dimensions of the magnet, prove that the resultant attraction at P makes an angle $\tan^{-1}(\frac{1}{2}\tan\theta)$ with PO.*

Let $OP=r$, $ON=OS=a$, and $\angle NOP=\theta$. The attraction of the magnet is equivalent to that of equal quantities of positive and negative magnetism at its poles N and S.

Hence the potential V at $P=\mu\left[\frac{1}{NP}-\frac{1}{SP}\right]$.

Now $NP^2=r^2+a^2-2ar\cos\theta$.

Hence, neglecting the square of a,

$$\frac{1}{NP}=\frac{1}{r}\left[1-\frac{2a}{r}\cos\theta\right]^{-\frac{1}{2}}=\frac{1}{r}\left[1+\frac{a}{r}\cos\theta\right].$$

So
$$\frac{1}{SP}=\frac{1}{r}\left[1-\frac{a}{r}\cos\theta\right].$$

$$\therefore\quad V=\frac{2\mu a}{r^2}\cos\theta.$$

Hence, if X and Y are the forces along PO and perpendicular to PO in the direction of θ decreasing,

$$X=-\frac{dV}{dr}=\frac{4\mu a}{r^3}\cos\theta,$$

and
$$Y=-\frac{1}{r}\frac{dV}{d\theta}=\frac{2\mu a}{r^3}\sin\theta.$$

Hence the required angle $=\tan^{-1}\frac{Y}{X}=\tan^{-1}\left[\frac{1}{2}\tan\theta\right]$.

2. Iron filings are spread on a piece of paper on which is placed a magnet whose poles are S and N; shew that the curves in which the filings arrange themselves are given by the equations $\cos\theta-\cos\theta'=$const., where θ and θ' are the angles PSX and PNX and X is a point in SN produced.

Shew also that all the filings which point towards a given point O on SN lie on a circle.

[Each filing is a little magnet, and thus must set itself in the direction of the resultant force on either pole; otherwise it would be acted upon by a couple.]

3. *A very thin uniform circular ring is composed of attracting matter; shew that a particle constrained to remain in its plane will be in unstable equilibrium at its centre.*

Let O be the centre of the ring, P the attracted particle at a small distance c from O. Then, OP being the initial line, the attraction of the ring in the direction OP

$$=2\int_0^\pi \frac{\gamma k\rho a\,d\theta}{a^2+c^2-2ac\cos\theta}\cdot\frac{a\cos\theta-c}{\sqrt{a^2+c^2-2ac\cos\theta}}$$

$$=\frac{2\gamma k\rho}{a^3}\int_0^\pi\left[1+\frac{3c}{a}\cos\theta\right][a\cos\theta-c]\,d\theta$$

$$=\frac{2\gamma k\rho}{a^3}\int_0^\pi(a\cos\theta-c+3c\cos^2\theta)\,d\theta,$$

squares of c being neglected,

$$=\frac{2\gamma k\rho}{a^3}\left[a\sin\theta+\frac{c}{2}\theta+\frac{3c}{4}\sin 2\theta\right]_0^\pi=\frac{\gamma\pi k\rho}{a^3}\cdot c.$$

The attraction therefore tends to increase c, *i.e.* to increase the distance of the attracted particle from the centre. The equilibrium is thus unstable.

4. n equal centres of forces are ranged symmetrically round the circumference of a circle; each force is repulsive and varies inversely as the mth power of the distance. Shew that a particle placed at the centre of the circle is in stable equilibrium, except when m is unity.

5. Eight central forces, the centres of which are at the corners of a cube, attract, according to the same law and with the same absolute intensity, a particle placed very near the centre of the cube; shew that their resultant action passes through the centre of the cube, unless the law of force be that of the inverse square.

6. A particle is attracted according to the inverse cube of the distance by an infinite number of equal masses arranged at distances $\frac{2\pi}{m}$ along a straight line distant y from the particle. Shew that the smallest angle which the direction of the resultant action can make with the line is

$$\cos^{-1}\left[\frac{my}{\sinh my}\right].$$

[If V be the potential at any point (x, y), the origin being at any one of the attracting masses, we obtain $V=\frac{m}{2y}\frac{\sinh my}{\cosh my-\cos mx}$.]

7. If every particle of matter attracted every other particle with a force proportional to the nth power of the distance, shew that, at any point within the matter, the attraction would be infinite if $n<-2$.

8. An infinite series of parallel infinite long rods, of uniform line-density μ, are placed at equal intervals c in a plane. Shew that the resultant attraction at a point in the plane, whose distance from the nearest rod is a, is $\frac{2\gamma\mu\pi}{c}\cot\frac{\pi a}{c}$.

[Use the expression for $\sin\theta$ in factors, and, by logarithmic differentiation, obtain a series for $\cot\theta$.]

9. A uniform wire of infinite length attracts according to the inverse nth power of the distance; shew that the resulting attraction is

$$\gamma\sqrt{\pi}\frac{m}{c^{n-1}}\frac{\Gamma\left(\frac{n}{2}\right)}{\Gamma\left(\frac{n}{2}+\frac{1}{2}\right)},$$

where m is the mass per unit length of the wire and c is its least distance from the attracted point.

10. Shew that the attraction of a uniform cube, of density ρ, at a point distant r from its centre is $\frac{4}{3}\gamma\pi\rho r$ towards the centre, if r be small.

11. A uniform cube attracts according to the law of nature. Shew that the attraction at a point situated at a corner consists of three components along the edges equal to

$$\gamma\rho a\left[\log_e(3+2\sqrt{2})(2-\sqrt{3})+\frac{\pi}{6}\right],$$

where ρ is the density and a the length of an edge.

12. An infinitely long homogeneous prism, of density ρ, has a rectangular cross section, of length a and breadth b. Shew that, at any point on one of the edges, the components of the attraction along the sides a and b of the cross section through the point are

$$\gamma\rho\left[2a\tan^{-1}\frac{b}{a}+b\log\frac{a^2+b^2}{b^2}\right] \text{ and } \gamma\rho\left[2b\tan^{-1}\frac{a}{b}+a\log\frac{a^2+b^2}{a^2}\right].$$

13. From the preceding, shew that the apparent latitude of a point on one edge of a long deep narrow crevasse of breadth a, running east and west, is altered by the angle $\frac{3\rho_0 a}{4\rho r}$ nearly by the presence of the crevasse, where ρ_0 and ρ are respectively the surface density and the mean density of the Earth and r is its radius.

Prove also that, if the depth h of the crevasse be small compared with its breadth a, then the alteration is $\frac{3}{2}\cdot\frac{\rho_0 h}{\pi\rho r}\left[1+\log\frac{a}{h}\right]$ nearly.

14. Shew that gravity is diminished by $\frac{3}{4}\frac{\pi+2\log_e 2}{\pi}\cdot\frac{(1-n)a}{r}$ of itself at the middle point of the surface of a canal of rectangular section whose length is great compared with its depth a; the breadth being $2a$, r being the radius of the Earth, and n the ratio of the density of water to the mean density of the Earth.

15. A lamina bounded internally and externally by concentric circles, of radii b and a respectively, is formed of material attracting according to the law $(\text{distance})^{-5}$. Shew that the resultant attraction vanishes at points distant

$$a^{\frac{1}{3}}b^{\frac{1}{3}}\sqrt{\frac{a^{\frac{4}{3}}+b^{\frac{4}{3}}}{a^{\frac{2}{3}}+b^{\frac{2}{3}}}}$$

from its centre.

16. Shew that the attraction at any internal point of a homogeneous sphere of radius a, every element of which attracts with a force proportional to its mass and inversely proportional to the cube of the distance, is

$$\frac{\pi\mu a}{x}-\pi\mu\frac{a^2+x^2}{2x^2}\log\frac{a+x}{a-x},$$

where x is the distance of the point from the centre of the sphere and μ is the attraction of unit mass at unit distance.

17. The matter of a spherical shell attracts with a force varying as the inverse fifth power of the distance. Shew that the attraction on an external point P is $\gamma M\cdot\frac{CP}{PT^6}$, where M is the mass of the shell, C its centre, and PT is the tangent from P.

What does this become when P is inside the shell?

18. If the law of force be the inverse fifth power of the distance, shew that the attraction of a uniform solid sphere, of density ρ and radius a, at an external point distant c from the centre is

$$\gamma\frac{\pi\rho}{4c^2}\left\{\frac{2ca(c^2+a^2)}{(c^2-a^2)^2}+\log\frac{c-a}{c+a}\right\}.$$

19. Shew that the attraction of a solid oblate spheroid of small eccentricity and whose semi-axes are a, a, b is $\frac{4}{3}\gamma\pi\rho\left(1-\frac{2\epsilon}{5}\right)a$ on a unit particle at the end of the axis a, and $\frac{4}{3}\gamma\pi\rho\left(1+\frac{4\epsilon}{5}\right)b$ at the end of the axis b, where $b=(1-\epsilon)a$.

20. Shew that the attraction of a uniform hemispherical shell, of radius a, at a point in the plane of its rim distant $r\,(>a)$ from the centre, is made up of a force $\frac{M}{r^2}$ towards the centre, and a force

$$\frac{2Ma}{\pi r^2}\int_0^{\frac{\pi}{2}}\frac{\sin^2\theta\,d\theta}{\sqrt{r^2-a^2\sin^2\theta}}$$

perpendicular to the plane.

Hence find the attraction of a solid hemisphere at a point on its rim.

21. A solid, of density ρ, is formed by the revolution about the axis of x of the part cut off from the parabola $y^2 = 4ax$ by its evolute

$$27ay^2 = 4(x-2a)^3.$$

Shew that the attraction at the cusp of the evolute is

$$\pi\gamma\rho a\,[4\sinh^{-1}4 + 5\sqrt{17} - 27].$$

22. Shew that the attraction at the focus S of a segment of a paraboloid of revolution, bounded by a plane perpendicular to the axis at a distance b from the vertex, is

$$4\pi\gamma\rho a \log_e \frac{a+b}{ae},$$

where $4a$ is the latus rectum of the generating parabola.

23. Shew that the resultant attraction of one-half of a solid homogeneous oblate spheroid cut off by an equatoreal plane, at a point on the rim of the base, is inclined to the plane of the base at an angle whose tangent is

$$\frac{4c\,(\tanh^{-1}e - e)}{\pi\,(a\sin^{-1}e - ce)},$$

where a and c are the semi-axes and e is the eccentricity of the meridian section.

24. If a lamina contains the origin and is bounded by the hyperbola $\frac{x^2}{a^2} - \frac{y^2}{b^2} = 1$, shew that the z-component of its attraction at any point on the ellipse

$$\frac{x^2}{a^2+b^2} + \frac{z^2}{b^2} = 1,\ y = 0 \text{ is } \frac{2\pi\gamma mah}{\sqrt{a^2h^2+b^4}},$$

where h is the z-coordinate of the point, and m is the mass of the lamina per unit of area.

25. Shew that the attraction at a pole of a solid prolate spheroid due to the matter on the far side of the equatoreal plane is

$$2\pi\gamma\rho a\left[1 - \frac{1}{e^2}(2 - \sqrt{2-e^2}) + \frac{1-e^2}{e^3}\log\frac{(1+e)^2}{1+e\sqrt{2-e^2}}\right],$$

where a is the semi-axis, e the eccentricity of the meridian, and ρ is the density.

26. The arc of a curve attracts a particle placed at its pole with a law of attraction equal to $\mu \div (\text{distance})^n$; if the resultant attraction of the arc always bisects the angle between the radii vectores to its ends, shew that the equation to the curve is $r^{n-1}\sin\{(n-1)\theta\} = \text{const.}$

27. If a homogeneous solid of revolution, whose mass M and density ρ are given, be such that its attraction at a point O on the axis of revolution is a maximum, shew that the solid is generated by the revolution of a curve whose equation is

$$r^2 = a^2\cos\theta.$$

[Let Ox be the axis of revolution. It is clear that O must lie on the solid. The surface, which is such that the attraction resolved along Ox of a particle, placed anywhere on it, is always the same, is clearly

$$\frac{\cos\theta}{r^2} = \text{const.}, \quad i.e. \quad r^2 = a^2 \cos\theta \text{.....................(1).}$$

Choose a so that this surface just includes all the mass M of the given matter, *i.e.* so that M = mass of the surface of revolution given by (1) $= \frac{4\pi\rho}{15} a^3$, as is easily found by integration. The surface thus obtained is the required one. For suppose we remove an element m_1 of the mass from a point P_1 inside this surface to a point P_2 outside it, and let OP_1 and OP_2 meet the surface in Q_1 and Q_2. Then clearly attraction of m_1 at P_1 on a particle at $O >$ its attraction when at Q_1, and its attraction when at $P_2 <$ its attraction when at Q_2, whilst its attractions when at Q_1 and Q_2 resolved along Ox are the same. Hence, by removing the mass m_1 from any point inside the above surface to any point outside, we have lessened the attraction along Ox. Hence the proposition.]

28. *A uniform solid sphere, of mass M, is cut in two by a plane through its centre; shew that the reaction between the halves due to their mutual attraction is $\frac{3}{16}\gamma\frac{M^2}{a^2}$, where a is the radius of the sphere.*

Clearly the attraction of one half on itself is zero; for it is the resultant of pairs of equal and opposite forces. Hence the attraction of one half on the other half is equal to the attraction of the whole sphere on that half. Let P be any point of this half, $OP = r$, $\angle POZ = \theta$, where O is the centre and OZ perpendicular to the cutting plane.

The attraction of the whole sphere at $P = \gamma \cdot \frac{4}{3}\pi\rho r$ [Art. 287], and the element of volume having this attraction

$$= r\,\delta\theta\,\delta r \times 2\pi r \sin\theta.$$

Hence the attraction of the whole sphere on the hemisphere resolved perpendicular to the plane base

$$= \int_0^a \int_0^{\frac{\pi}{2}} \rho \,.\, r\,d\theta\,dr \,.\, 2\pi r \sin\theta \left[\gamma \,.\, \frac{4}{3}\pi\rho r\right] \cos\theta$$

$$= \frac{1}{3}\gamma\pi^2\rho^2 a^4 = \frac{3\gamma}{16} \cdot \frac{M^2}{a^2}.$$

The resultant reaction between the two halves is clearly equal to the resultant attraction between them.

Aliter. Consider a sphere of fluid at rest under its own attraction. The attraction at a point distant r from the centre $= \gamma \,.\, \frac{4}{3}\pi\rho r$.

Hence the fundamental equation of Hydrostatics gives

$$\frac{\delta p}{\rho} = -\frac{4\pi\rho\gamma}{3} r\,\delta r.$$

$$\therefore\ p = C - \frac{2}{3}\pi\rho^2\gamma r^2 = \frac{2}{3}\pi\rho^2\gamma(a^2 - r^2),$$

since the pressure clearly vanishes at the surface of the sphere.

The resultant pressure across a plane through the centre then

$$= \int_0^a 2\pi r\,dr \,.\, p = \frac{4}{3}\pi^2\rho^2\gamma \int_0^a (a^2 r - r^3)\,dr = \frac{\pi^2\rho^2\gamma a^4}{3}.$$

If one of the hemispheres be now made rigid, it will be in equilibrium under the same forces as before, and hence the required reaction

$$= \frac{\pi^2\rho^2\gamma a^4}{3} = \frac{3\gamma}{16}\cdot\frac{M^2}{a^2}.$$

29. If a self-attracting shell of mass M, bounded by concentric spheres of radii a and b, be cut by a plane through the centre, prove that the pressure between the halves is

$$\frac{3\gamma M^2}{16}\,\frac{a^2 + 2ab + 3b^2}{(a^2 + ab + b^2)^2}.$$

30. Shew that the force required to separate the two parts of a solid uniform sphere, of radius a, divided in any manner is $\gamma\,\frac{MM'}{a^3}\,.\,PP'$, where M, M' are the masses and P, P' the centres of gravity of the two parts.

31. The mass of a unit length of an infinite homogeneous cylinder, of radius a, is M. It is divided into two parts by a plane through its axis. Shew that the pressure between the two parts due to their mutual attractions is $\frac{4\gamma M^2}{3\pi a}$ per unit length of the cylinder.

32. A lune is divided off from a thin spherical shell by two great circles whose planes cut at an angle 2α. Shew that the attraction of the rest of the shell on the lune is $\gamma\,\frac{M^2}{4\,a^2}\sin\alpha$, where M is the mass of the shell and a is its radius.

33. A solid homogeneous sphere is laid on a thin uniform circular plate so as to touch it at its centre, and the sphere and plate have their radii and masses equal. Shew that the reaction between them due to their mutual gravitation is $4\lambda\mu\sin^2\frac{\pi}{8}$ of the weight of either, where λ is the ratio of the radius of either to the radius of the Earth, and μ is the ratio of the density of the sphere to the mean density of the Earth.

34. A homogeneous sphere, of radius a and mass M_1, is in contact with the centre of the plane face of a homogeneous hemisphere, of radius a and mass M_2. Shew that the pressure between them due to their mutual attraction is $\frac{\gamma M_1 M_2}{a^2}(\sqrt{2} - 1)$.

35. A uniform circular plate, of mass M, rests in contact with a fixed rough gravitating sphere of the same radius. A small mass m is then fastened to the rim of the plate. Shew that the plate will turn through an angle whose circular measure is $\frac{m}{M}$, if the squares of this ratio be neglected.

36. Shew that the potential of a uniform thin equilateral triangle ABC at a point P, situated on a perpendicular to its plane drawn through its centre O, is

$$\frac{2\gamma M \cot a}{a}\left[\tan a \,.\, \log \frac{2+\sqrt{3}\sin a}{2-\sqrt{3}\sin a} - \frac{4\pi}{3} + 4\tan^{-1}(\sqrt{3}\cos a)\right],$$

where M is the mass of the triangle, a is its side, and $\angle OPA = a$.

[Consider the triangle as made up of straight lines parallel to its base.]

37. Shew that the potential of a uniform regular tetrahedron, formed of gravitational matter, at its centre is

$$\gamma \frac{M}{a}\left\{6 \log(\sqrt{3}+\sqrt{2}) - \frac{\pi}{\sqrt{2}}\right\},$$

where M is the mass of the tetrahedron and a the length of an edge.

[Use the result of the previous question.]

38. A particle is let fall from an angular point of a regular tetrahedron whose opposite face consists of matter of surface density σ attracting as the inverse square. Shew that, when it strikes this face, the square of its velocity is

$$2\gamma\sigma a\sqrt{3}\left[\log\left(1+\frac{2}{\sqrt{3}}\right) + 2\sqrt{2}\cot^{-1}\sqrt{2} - \frac{\pi\sqrt{2}}{3}\right],$$

where a is the length of a side and the remaining faces exert no attraction.

39. If M and M' be any two masses, and if V' be the potential of M' at any element dM of M, and V be the potential of M at any element dM' of M', shew that

$$\int V dM' = \int V' dM.$$

40. If V_n be the potential of any point due to a distribution of matter attracting according to the nth power of the distance, and V_{n-2} the potential due to the same distribution attracting as the $(n-2)$th power of the distance, shew that

$$\nabla^2 V_n = (n-1)(n+2) V_{n-2}.$$

41. If $\phi(x, y, z)$ be the potential at an internal point P (x, y, z) of a thin heterogeneous spherical shell, then the potential at an external point P' (x', y', z') is

$$\frac{a}{r'}\phi\left(\frac{a^2x'}{r'^2}, \frac{a^2y'}{r'^2}, \frac{a^2z'}{r'^2}\right),$$

where a is the radius of the shell and r' the distance of P' from its centre.

CHAPTER XV

ATTRACTIONS AND POTENTIAL (*continued*). GENERAL THEOREMS

303. Surface Integral of Normal Attraction over any closed surface (Gauss' Theorem).

If N be the normal attraction at any point of the element δS of any closed surface, measured positively along the normal outwards, due to any attracting mass, then $\int N \, . \, dS = -4\gamma\pi M$, where M is the amount of the attracting mass within the surface, the integral being taken over the whole surface.

Let O be the position of any element m of the attracting mass within the closed surface.

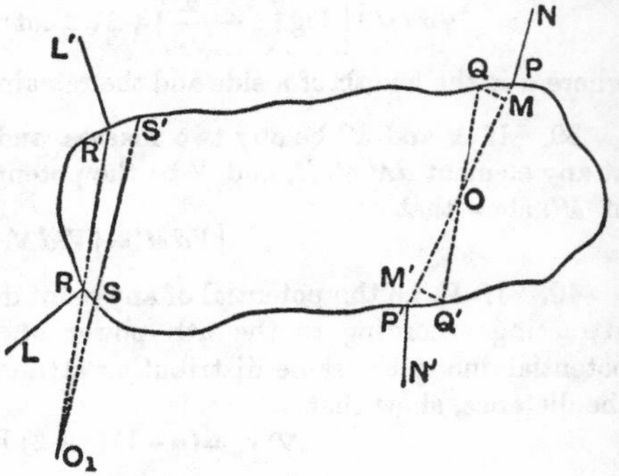

Through O draw a cone of very small vertical angle and let it cut the surface in the elements PQ and $P'Q'$, whose areas are δS and $\delta S'$.

The attractions of the mass m at these elements are

$$\gamma m \, . \, \frac{\delta S}{OP^2} \cos OPN \text{ along } PN,$$

and $$\gamma m \, . \, \frac{\delta S'}{OP'^2} \cos OP'N' \text{ along } P'N',$$

where PN and $P'N'$ are the outward drawn normals at P and P'.

Through Q and Q' draw normal sections QM and $Q'M'$ of this slender cone, let $\delta\omega$ be the solid angle of the cone, and let θ be the angle between the elements QM and QP, so that $\theta =$ the supplement of the angle between the normals OM and $PN = \pi - OPN$.

Then $$\delta\omega = \frac{\text{area } QM}{OQ^2} = \frac{\delta S \cdot \cos\theta}{OQ^2} = -\frac{dS \cdot \cos OPN}{OP^2},$$

in the limit when PQ is very small, and similarly

$$\delta\omega = -\frac{\delta S' \cdot \cos OP'N'}{OP'^2}.$$

Hence the normal attractions for the elements δS and $\delta S'$ at P and P' are each $-\gamma m \cdot \delta\omega$.

Hence the total normal attraction for the whole surface

$$= -\gamma m \cdot \Sigma\delta\omega = -\gamma m \cdot 4\pi,$$

i.e. for the single particle m at O

$$\int N \cdot dS = -4\gamma\pi m.$$

Similarly for all other particles of the attracting mass inside the surface. Hence finally, for the whole mass,

$$\int N \cdot dS = -4\gamma\pi M.$$

Next, let O_1 be the position of a particle m_1 of the attracting mass outside the closed surface, and draw similarly a slender cone cutting the surface in RS and $R'S'$. Then, just as before, the normal attractions at RS and $R'S'$ are $\gamma m_1 \cdot \delta\omega_1$ in magnitude. But their sign is opposite. For at R the attraction is positive measured along the outward drawn normal RL, and at R' it is negative. Hence the elements of the normal attraction for the surfaces RS and $R'S'$ are $\gamma m_1 \cdot \delta\omega_1$ and $-\gamma m_1 \cdot \delta\omega_1$, so that their sum is zero.

The same result is true for all such slender cones drawn through O_1. Hence the surface integral of normal attraction is zero for any such elementary mass as m_1 outside the closed surface, and so it is zero for the total mass M_1 which lies outside the closed surface.

[In the above figure it will be noted that, when the attracting mass is inside the surface as at O, both the angles OPN and $OP'N'$ are obtuse; when it is outside, as at O_1, one of the angles $O_1R'L'$ is obtuse and the other O_1RL is acute.]

304. When the closed curve is cut by the slender cone in more than two sections as in the following figure, the same result is easily seen to be true.

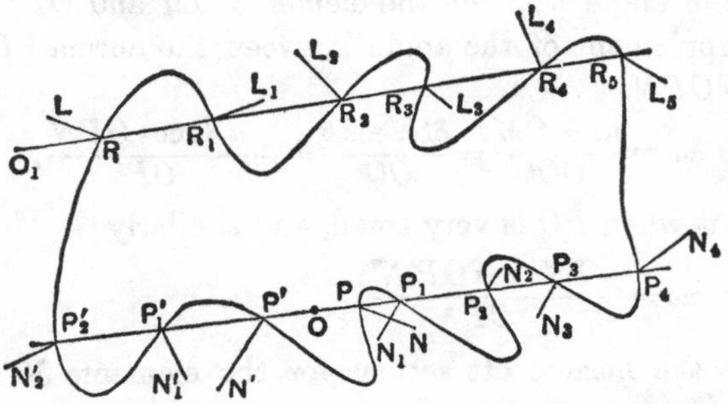

For the angles OPN, OP_2N_2, OP_4N_4, $OP'N'$ and $OP_2'N_2'$ are all obtuse, and hence the element of the integral for each is $-\gamma m\,\delta\omega$; also the angles OP_1N_1, OP_3N_3, and $OP_1'N_1'$ are all acute, and hence the corresponding elements are $+\gamma m\,\delta\omega$; so that for all these points the sum of the elements

$$= -5\gamma m\,.\,\delta\omega + 3\gamma m\,.\,\delta\omega = -2\gamma m\,.\,\delta\omega,$$

as in the first figure. Hence, as before, for the whole surface

$$\int N\,.\,dS = -\gamma M\,.\,4\pi.$$

Similarly, starting with the point O_1, the angles $O_1R_1L_1$, $O_1R_3L_3$, and $O_1R_5L_5$ are obtuse and the corresponding elements of the integral each $-\gamma m_1\,.\,\delta\omega_1$; also the three angles O_1RL, $O_1R_2L_2$, $O_1R_4L_4$ are acute and the corresponding elements $+\gamma m_1\,.\,\delta\omega_1$; so that for this slender cone the total surface integral of normal attraction is zero.

Hence, as in the case of the first figure, $\int N\,.\,dS = 0$.

305. When the point O is on the surface, *i.e.* when the element m is on the surface, the slender cone through O either meets the surface in one point, or in an odd number of points. In either case, the element of the surface integral due to it is $-\gamma m\,.\,d\omega$. Hence the whole surface integral due to m is $-\int \gamma m\,d\omega$, where $d\omega$ refers to the solid angle on one side only of the tangent plane at O, so that $-\int \gamma m\,d\omega = -\gamma m\,.\,2\pi$.

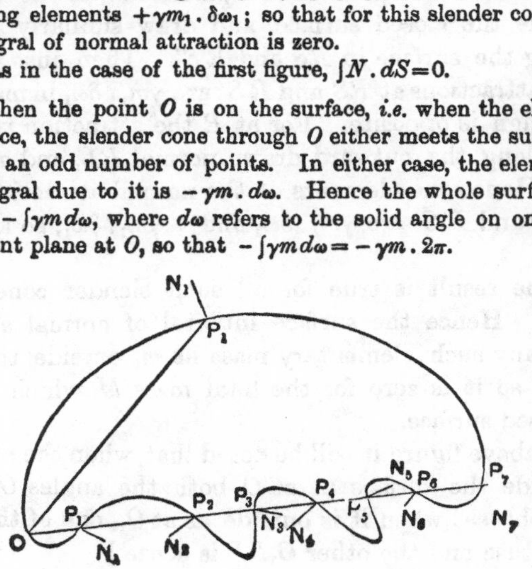

Similarly for any other element of mass on the surface.

Hence, if M_1 be the attracting mass on the closed surface, the surface integral due to it is $-2\pi\gamma M_1$.

306. Laplace's and Poisson's Equations.

If N be the normal attraction at any point of a closed surface measured outwards, then, by Art. 303, we have

$$\int N \,.\, dS = -4\pi\gamma M \text{..................}(1),$$

where M is the amount of attracting matter contained by the closed surface.

Take as the closed surface the small rectangular parallelopiped one of whose angular points P is the point (x, y, z), and whose edges PQ, PR, PS are parallel to the axes and of lengths δx, δy, δz respectively.

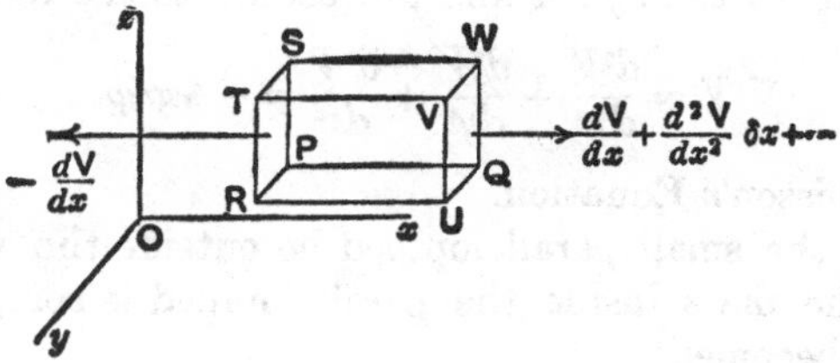

The face $PRTS$ being very small, the force at each point of it is ultimately the same and equal to $-\frac{dV}{dx}$ towards the negative direction of Ox.

Hence the component of $\int N \,.\, dS$ due to this face is

$$-\frac{dV}{dx} \,.\, \delta y \,.\, \delta z.$$

Also, if $\frac{dV}{dx} = f(x)$, the component force parallel to the positive direction of Ox at each point of $QUVW$

$= f(x + \delta x) = f(x) + \delta x \,.\, f'(x) + \dots$

$= \frac{dV}{dx} + \frac{d^2V}{dx^2}\delta x +$ terms containing higher powers of δx.

Hence the component of $\int N \,.\, dS$ due to this face

$$= \left(\frac{dV}{dx} + \frac{d^2V}{dx^2}\delta x + \dots\right)\delta y \,.\, \delta z.$$

Hence the component of $\int N.dS$ due to these two faces

$$= \delta x . \delta y . \delta z \left[\frac{d^2V}{dx^2} + \ldots\right].$$

So the components for the faces perpendicular to y and z are

$$\delta x . \delta y . \delta z \left[\frac{d^2V}{dy^2} + \ldots\right] \text{ and } \delta x . \delta y . \delta z \left[\frac{d^2V}{dz^2} + \ldots\right]$$

Also, if the small parallelopiped be inside the attracting mass, then $M =$ mass of the parallelopiped $= \delta x . \delta y . \delta z . \rho$.

Hence equation (1) gives

$$\delta x . \delta y . \delta z \left[\frac{d^2V}{dx^2} + \frac{d^2V}{dy^2} + \frac{d^2V}{dz^2} + \text{small quantities}\right]$$
$$= -4\pi\gamma \times \delta x . \delta y . \delta z . \rho,$$

i.e. on dividing by $\delta x . \delta y . \delta z$ and proceeding to the limit,

$$\nabla^2 V = \frac{d^2V}{dx^2} + \frac{d^2V}{dy^2} + \frac{d^2V}{dz^2} = -4\pi\gamma\rho.$$

This is Poisson's Equation.

If P and the small parallelopiped be outside the attracting mass, then the mass inside the parallelopiped is zero, and the equation (1) becomes

$$\nabla^2 V = \frac{d^2V}{dx^2} + \frac{d^2V}{dy^2} + \frac{d^2V}{dz^2} = 0.$$

This is Laplace's Equation.

This may also be proved by simple differentiation. For as in Art. 296, differentiating the expressions of Art. 292, we have

$$\frac{d^2V}{dx^2} + \frac{d^2V}{dy^2} + \frac{d^2V}{dz^2}$$
$$= -\gamma \int dm \left[\frac{3}{PQ^3} - 3\frac{(x-x')^2 + (y-y')^2 + (z-z')^2}{PQ^5}\right].$$

If P be outside the attracting mass, so that PQ never vanishes, this gives

$$\frac{d^2V}{dx^2} + \frac{d^2V}{dy^2} + \frac{d^2V}{dz^2} = 0.$$

307. By the ordinary methods of the Differential Calculus for changing the coordinates x, y, z into polar coordinates r, θ, ϕ (*i.e.* where $x = r\cos\phi\sin\theta$, $y = r\sin\phi\sin\theta$, $z = r\cos\theta$), or, by a method similar to the previous article, Poisson's equation may be put into the form

$$\frac{d}{dr}\left[r\delta\theta \,.\, r\sin\theta\delta\phi \,.\, \frac{dV}{dr}\right]\delta r + \frac{d}{rd\theta}\left[r\sin\theta\delta\phi \,.\, \delta r \,.\, \frac{dV}{rd\theta}\right] r\delta\theta$$
$$+ \frac{d}{r\sin\theta d\phi}\left[r\delta\theta \,.\, \delta r \,.\, \frac{dV}{r\sin\theta d\phi}\right] r\sin\theta\delta\phi$$
$$= -4\pi\gamma\rho \,.\, \delta r \,.\, r\delta\theta \,.\, r\sin\theta\delta\phi,$$

i.e. $$\frac{1}{r^2}\left[\frac{d}{dr}\left(r^2\frac{dV}{dr}\right) + \frac{1}{\sin\theta}\frac{d}{d\theta}\left(\sin\theta\frac{dV}{d\theta}\right) + \frac{1}{\sin^2\theta}\frac{d}{d\phi}\left(\frac{dV}{d\phi}\right)\right]$$
$$= -4\pi\gamma\rho \quad \text{.........(1)},$$

i.e. $$\frac{d^2V}{dr^2} + \frac{2}{r}\frac{dV}{dr} + \frac{1}{r^2}\frac{d^2V}{d\theta^2} + \frac{\cot\theta}{r^2}\frac{dV}{d\theta} + \frac{1}{r^2\sin^2\theta}\frac{d^2V}{d\phi^2}$$
$$= -4\pi\gamma\rho \quad \text{.........(2)}.$$

Again, if we use cylindrical coordinates ϖ, θ, and z (so that $x = \varpi\cos\theta$ and $y = \varpi\sin\theta$), the equation becomes

$$\frac{d}{d\varpi}\left(\varpi\frac{dV}{d\varpi}\right) + \frac{d}{d\theta}\left(\frac{1}{\varpi}\frac{dV}{d\theta}\right) + \frac{d}{dz}\left(\varpi\frac{dV}{dz}\right) = -4\pi\gamma\rho\varpi,$$

i.e. $$\frac{d^2V}{d\varpi^2} + \frac{1}{\varpi}\frac{dV}{d\varpi} + \frac{1}{\varpi^2}\frac{d^2V}{d\theta^2} + \frac{d^2V}{dz^2} = -4\pi\gamma\rho \quad \text{.........(3)}.$$

If the point considered be not within the attracting mass, the right-hand members of (1), (2), and (3) are zero, and we have the corresponding forms of Laplace's equation.

308. By the use of the equations of the previous article we may at once obtain the values of V in some simple cases.

Spherical Shell. Clearly V depends only on the distance r of the point P considered from the centre of the shell, and is independent of θ and ϕ.

Hence equation (1) of Art. 307 gives $\frac{d}{dr}\left(r^2\frac{dV}{dr}\right) = 0.$

$$\therefore\ r^2\frac{dV}{dr} = \text{const.},\quad \textit{i.e.}\quad V = \frac{A}{r} + B.$$

(i) If P be inside the shell, then clearly the resultant attraction $\frac{dV}{dr}$ vanishes at the centre, so that $A = 0$.

Hence V is constant all through the inside of the shell and $=$ its value at the centre $= \gamma \,.\, \frac{\text{Mass}}{\text{Radius}}$.

(ii) If P be outside the shell, then, since V must vanish at infinity, $B=0$.

Also, V being continuous, its value must at the surface of the shell agree with that of the internal potential, so that

$$\frac{A}{a}=\gamma\,.\frac{\text{Mass}}{\text{Radius}},\quad i.e.\ A=\gamma\,.M.$$

Hence outside the shell the potential $=\gamma\,.\dfrac{M}{r}$

Solid Sphere. Internal point. Here, since V is independent of θ and ϕ, Poisson's equation gives

$$\frac{1}{r^2}\frac{d}{dr}\left[r^2\frac{dV}{dr}\right]=-4\pi\gamma\rho.$$

$$\therefore\ r^2\frac{dV}{dr}=-\frac{4\pi}{3}\gamma\rho r^3+C.$$

Now $\left(\dfrac{dV}{dr}\right)_0=$ the resultant attraction at the centre $=0$, so that $C=0$.

$$\therefore\ \frac{dV}{dr}=-\frac{4\pi}{3}\gamma\rho r,\quad i.e.\ V=-\tfrac{2}{3}\pi\gamma\rho r^2+B.$$

But clearly V at the centre $=\displaystyle\int_0^a\frac{4\pi\gamma\rho y^2dy}{y}=2\pi\gamma\rho a^2=B.$

$$\therefore\ V=2\pi\gamma\rho a^2-\tfrac{2}{3}\pi\gamma\rho r^2.$$

309. Ex. 1. ***Matter is distributed between the infinite cylinders $r=\frac{1}{2}a$ and $r=a$ in such a way that the density is proportional to $\dfrac{a}{r}-\dfrac{r}{a}$, and the mass per unit length parallel to the axis is M. Find the law of potential within the matter, and prove that the difference between the potentials of the distribution at the outer and inner surfaces is***

$$\tfrac{2}{15}\gamma M(29-33\log_e 2).$$

If the density be $\lambda\left(\dfrac{a}{r}-\dfrac{r}{a}\right)$, then

$$M=\int_0^{2\pi}\int_{\frac{a}{2}}^{a}r\,d\theta\,dr\,\lambda\left(\frac{a}{r}-\frac{r}{a}\right)=\frac{5}{12}\pi\lambda a^2\ \ \ldots\ldots\ldots\ldots(1).$$

The potential V is clearly a function of r only, so that Poisson's equation [(3) of Art. 307] becomes

$$\frac{d^2V}{dr^2}+\frac{1}{r}\frac{dV}{dr}+4\pi\gamma\lambda\frac{a^2-r^2}{ar}=0.$$

$$\therefore\ r\frac{dV}{dr}+\frac{4\pi\gamma\lambda}{a}\left(a^2r-\frac{r^3}{3}\right)=A\ \ldots\ldots\ldots\ldots\ldots\ldots\ldots(2).$$

Now $\int \frac{dV}{dr} dS$ taken over the inner surface is zero, by Gauss' Theorem, and $\frac{dV}{dr}$ is by symmetry constant over this inner surface.

$$\therefore \frac{dV}{dr} = 0 \text{ when } r = \frac{a}{2}.$$

Hence $$A = \frac{11}{6} \pi\gamma\lambda a^2 \quad\text{.................................(3).}$$

$\therefore$ (2) gives $$V = A \log r - \frac{4\pi\gamma\lambda}{a}\left(a^2 r - \frac{r^3}{9}\right) + B.$$

[Also Laplace's equation gives, for a point outside the larger cylinder, $V = -C \log r + D$, where D is an infinite constant, since V must clearly be zero at infinity. Since the potentials within and without the mass have the same value at the outer surface of the cylinder, it follows that B is also an infinite constant.]

$$\therefore V_{\frac{a}{2}} - V_a = A \log\frac{1}{2} - \frac{4\pi\gamma\lambda}{a}\left[\frac{a^3}{2} - \frac{a^3}{72} - a^3 + \frac{a^3}{9}\right]$$

$$= -A \log_e 2 + \frac{29}{18}\pi\gamma\lambda a^2 = \frac{2\gamma M}{15}(29 - 33 \log_e 2).$$

Ex. 2. If one value of V satisfying the equation

$$\frac{d^2 V}{dx^2} + \frac{d^2 V}{dy^2} + \frac{d^2 V}{dz^2} = 0 \text{ be } r^n f(\theta, \phi)$$

expressed in polar coordinates, then $r^{-(n+1)} f(\theta, \phi)$ is another value of V satisfying the equation.

310. *Equipotential Surfaces.* For any attracting mass M the potential V at any point P, (x, y, z), will be some function $\phi(x, y, z)$ of the coordinates of the point, so that $V = \phi(x, y, z)$

V will have a constant value C for all points P whose coordinates satisfy the equation

$$\phi(x, y, z) = C \quad\text{.........................(1).}$$

The surface given by (1) will be such that the potential at any point of it for the given attracting mass is constant. It is hence called an Equipotential Surface. By giving a series of different values to C, we get a series of equipotential surfaces.

The same result will clearly be true whatever be the coordinates in terms of which V is expressed.

Thus in the case of the rod AB, (Art. 298), the potential at P is

$$\gamma k\rho \log\frac{r_1 + r_2 + a}{r_1 + r_2 - a},$$

and is therefore constant wherever $r_1 + r_2$ is constant.

But if $r_1 + r_2$, *i.e.* $AP + BP$, is constant, the locus of P in the plane of the paper is an ellipse whose foci are A and B. Hence its locus in space is the surface obtained by rotating this ellipse about AB as axis. The equipotential surfaces are therefore ellipsoids of revolution obtained by rotating confocal ellipses, whose foci are A and B, about AB as axis. Some of these ellipses are shown in the figure.

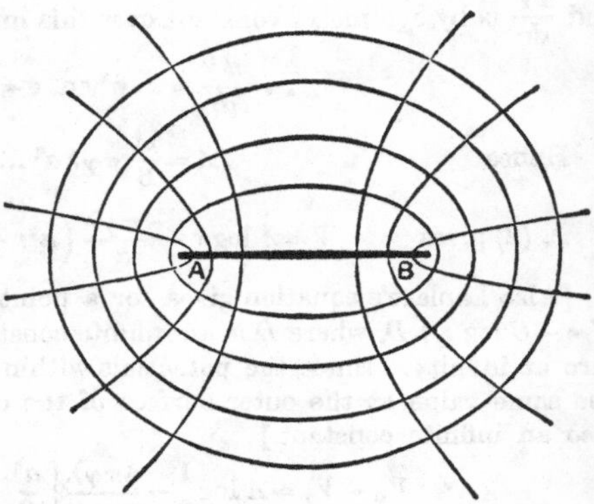

Again, in the case of the spherical shells and sphere of Arts. 300 and 301, the potential must clearly depend only on the distance of the point from the centre, and is thus constant for all points which lie on the surface of a concentric sphere. Thus the equipotential surfaces in these cases are a series of concentric spheres.

311. ***The attracting force at any point P is normal to the equipotential surface which passes through P.***

For if P' be any point close to P on the equipotential surface passing through P, then, by Art. 294, the attracting force in the direction PP'

$$= \text{Lt.} \frac{\text{Potential at } P' - \text{Potential at } P}{PP'}$$

$= 0$, since P and P' are on the same equipotential surface.

Hence for every direction PP', which lies in the tangent plane at P, the attracting force is zero.

The resultant attraction at P must therefore be in the direction of the normal to the equipotential surface through P.

Analytically, any line lying in the tangent plane at P, whose direction cosines are (l, m, n), is perpendicular to the normal to the equipotential surface $\phi(x, y, z) = C$, whose direction cosines are proportional to

$$\frac{d\phi}{dx}, \; \frac{d\phi}{dy}, \text{ and } \frac{d\phi}{dz}.$$

Hence $$\frac{d\phi}{dx}.l+\frac{d\phi}{dy}.m+\frac{d\phi}{dz}.n=0.$$

But since $\frac{d\phi}{dx}$, $\frac{d\phi}{dy}$, and $\frac{d\phi}{dz}$ are the components of the attractive force along the axes, this equation expresses the fact that the component of the attractive force resolved along the line (l, m, n) is zero.

This being true for all lines lying in the tangent plane, it follows that the resultant attractive force must be normal to the equipotential surface, and so no work is done by the attracting mass as the particle is moved from one point of an equipotential surface to any other point of the same equipotential surface.

312. *If δn be the length of the element of the normal drawn from any point P to meet the next consecutive equipotential surface, the resultant attracting force at P varies inversely as δn.*

For if V be the potential at P, and $V+\delta V$ the potential at the point P' where the normal PP' meets the next equipotential surface, the attractive force at P, by Art. 294,

$$=\frac{\text{Potential at } P' - \text{Potential at } P}{PP'}=\frac{(V+\delta V)-V}{\delta n}=\frac{\delta V}{\delta n}.$$

Hence for points P on the same equipotential surface V the resultant attraction varies inversely as δn.

313. The Equipotential Surfaces are often called Level Surfaces, or *Surfaces de niveau*, from an analogy with the Earth. If we consider gravity constant, the equipotential surface at any point of the Earth's surface is a horizontal plane, or rather a portion of a very large sphere concentric with the Earth. Just as no work is done against gravity in moving a particle along a smooth horizontal plane, so no work is done against the attraction of a mass if we move a particle along one of its equipotential surfaces.

314. *Lines of Force.* If from a point P there be drawn an element PP' in the direction of the resulting attraction at P, then an element $P'P''$ in the direction of the resulting attraction at P', and so on, then, in the limit when these elements are taken indefinitely small, they lie on a curve which is called a Line of Force.

In other words, a Line of Force is a curve such that the tangent at any point of it is in the direction of the resultant attractive force at that point.

Since the elements PP', $P'P''$, ... are normal to the equipotential surfaces through P, P', ..., it follows that the Line of Force is at each point of its length perpendicular to the corresponding equipotential surface.

Hence the Lines of Force are lines which cut orthogonally the system of Equipotential Surfaces.

In the case of the spherical shells and sphere of Arts. 300 and 301, the lines of force are any straight lines drawn from the centre O. They cut orthogonally all the Equipotential Surfaces, *i.e.* the spheres whose centres are O.

In the case of a thin rod AB the equipotential curves are ellipses whose foci are A and B (Fig. Art. 310). Now a series of confocal ellipses are cut orthogonally by a series of hyperbolas with the same foci. Hence the lines of forces for a thin rod consist of the hyperbolas whose foci are the ends of the rod. Some of these hyperbolas are shewn in the figure of Art. 310.

315. *Tube of Force.* If through every point of a small closed curve we draw the corresponding line of force, we obtain a Tube of Force.

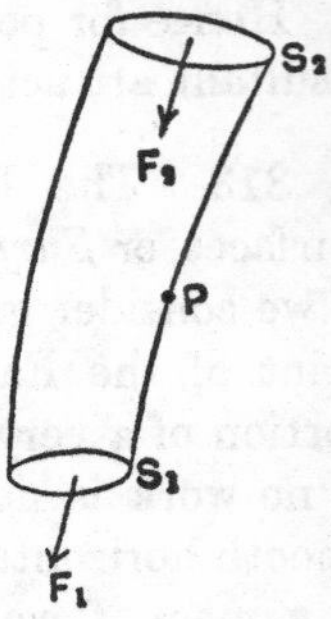

By the definition of a line of force, it is clear that at any point P on the curved part of a tube of force there is no resultant attraction in a direction normal to the curve.

Consider the portion of a Tube of Force bounded by two small normal sections S_1 and S_2, let F_1 and F_2 be the attractive forces normal to the ends S_1 and S_2, and let this portion of the tube contain none of the attracting matter.

Apply the theorem of Art. 303 to the tube. At all points such as P on the curved portion the element of the integral is zero, since there is no normal force there. The only elements of this integral therefore come from the ends, and the theorem therefore gives

$$F_1 . S_1 + (-F_2) S_2 = 0.$$

Hence $F_1 . S_1 = F_2 . S_2$.

The same theorem holds whatever be the length of the tube considered. Hence $F_1 . S_1$ must be the same so long as we keep to the same tube, *i.e.* the attractive force at any point of the same tube of force is inversely proportional to the normal section of the tube.

316. As a particular case consider the attraction of a solid sphere at an external point. The tubes of force are thin cones whose vertices are at the centre O of the sphere. S_1 is then a section of such a small cone, and hence its area varies as the square of the distance from the centre. It follows, as was proved in Art. 287, that the resultant attraction at an external point varies inversely as the square of the distance from the centre.

Again take an infinite solid circular cylinder. By symmetry, the lines of force starting from any point O of its axis are straight lines perpendicular to the axis OA of the cylinder. The tubes of force in this case are wedges and the area of the section S_1 varies as the distance from the axis OA. Hence the resultant attraction of the infinite cylinder, at any point external to itself, varies inversely as the distance of the point from the axis. [Cf. Art. 285, Ex. 2.]

EXAMPLES

1. N and S are the poles of a magnet. If θ and ϕ are the angles that NP and SP make with NS produced, and $NP=r$, $SP=r'$, the lines of force due to the magnet are given by

$$\cos\theta - \cos\phi = \text{const.},$$

and the equipotential surfaces by the revolution about NS of the curves

$$\frac{1}{r} - \frac{1}{r'} = \text{const.}$$

2. Two infinite straight rods, alike in every respect, intersect one another at right angles. Shew that the equipotential curves in their plane are rectangular hyperbolas.

3. If the law of attraction of a thin uniform straight rod is that of the inverse cube, shew that the equipotential surfaces are generated by the revolution of a family of curves whose polar equation is

$$r^2 - a^2 = \pm 2ar \sin\theta \cot(cr \sin\theta)$$

about the initial line, $2a$ being the length of the rod and c a parameter.

4. The equations of two infinite rods of the same density are

$$y = x \tan \alpha, \quad z = c \quad \text{and} \quad y = -x \tan \alpha, \quad z = -c.$$

Find the equations of their equipotential surfaces, and shew that, for a particle placed on the axis of x, the region in which displacements are stable is separated from those in which they are unstable by the surfaces

$$cy \cos \alpha - xz \sin \alpha = \pm (xy \sin \alpha \cos \alpha + cz).$$

[The separating surface is given by $V_{(x,\,y,\,z)} = V_{(x,\,0,\,0)}$, where y and z are small compared with x.]

317. *Work done by the mutual attractive forces of the particles of a self-attracting system, whilst the particles are brought from an infinite distance from one another to the positions they occupy in the given system.*

Let the component particles be $m_1, m_2, \ldots$, and let $A_1, A_2, \ldots$ be their positions in the given system.

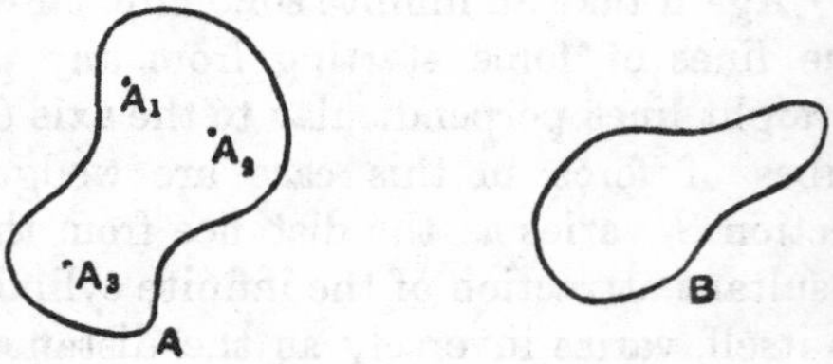

Let the distance between m_1 and m_2 be called r_{12}, and in general let the distance between m_s and m_t be r_{st}.

First, bring m_1 from infinity to its assigned position A_1; the work done is zero; for there are no particles of the system near enough to exert attraction on it.

Next, bring m_2 from infinity to its position A_2; the work done

$$= \text{the potential of } m_1 \text{ at } A_2 \times m_2 = \gamma \cdot \frac{m_1 m_2}{r_{12}}.$$

Next, bring m_3 to its position A_3; the work done on it

$$= m_3 \,(\text{potential of } m_1 \text{ and } m_2 \text{ at } A_3) = \gamma \frac{m_3 m_1}{r_{13}} + \gamma \frac{m_3 m_2}{r_{23}},$$

and so on for the other particles of the system.

Hence the total work done in collecting all the particles from rest at infinite distances from one another to their positions in the configuration A

$$= \gamma \frac{m_1 m_2}{r_{12}} + \gamma m_3 \left[\frac{m_1}{r_{13}} + \frac{m_2}{r_{23}}\right] + \gamma m_4 \left[\frac{m_1}{r_{14}} + \frac{m_2}{r_{24}} + \frac{m_3}{r_{34}}\right] + \ldots + \ldots$$

$$= \gamma \cdot \Sigma \frac{m_1 m_2}{r_{12}} \quad \ldots\ldots\ldots\ldots\ldots\ldots\ldots\ldots\ldots\ldots(1).$$

When the particles have all been brought to their positions in configuration A, let V_1, V_2,... be the potentials of the system at A_1, A_2, ...

Then
$$V_1 = \gamma\frac{m_2}{r_{12}} + \gamma\frac{m_3}{r_{13}} + \gamma\frac{m_4}{r_{14}} + \ldots,$$
$$V_2 = \gamma\frac{m_1}{r_{12}} + \gamma\frac{m_3}{r_{23}} + \gamma\frac{m_4}{r_{24}} + \ldots,$$
....

Hence, clearly, expression (1)
$$= \tfrac{1}{2}\Sigma(V_1 m_1).$$

[For in the expression $\Sigma(V_1 m_1)$ any such term as $\gamma\frac{m_s m_t}{r_{st}}$ is twice repeated, once in the element $V_s \times m_s$ and once in the element $V_t \times m_t$.]

Hence the work done in bringing the particles from infinity to the configuration A
$$= \tfrac{1}{2}\Sigma(V_1 m_1) = \tfrac{1}{2}\int V . dm,$$
where V is the potential of the body A at any element dm of itself, and the integration is taken throughout the configuration A.

Conversely, the work done by the mutual attractive forces of the system as its particles are scattered to an infinite distance from one another is $-\tfrac{1}{2}\int V . dm$.

We can hence find the work done as the body changes from one configuration A to another configuration B.

For this work

= work done in the changing of its form from A to infinity

+ " " " " " " " " " infinity to B

$= -\tfrac{1}{2}\int V . dm + \tfrac{1}{2}\int V' . dm'$,

where the first integral is taken throughout the system in the configuration A, and the second is similarly taken throughout the configuration B.

318. **Ex.** *A self-attracting sphere, of uniform density ρ and radius a, changes to one of uniform density and radius b; shew that the work done by its mutual attractive forces is $\frac{3}{5}\gamma M^2\left(\frac{1}{b} - \frac{1}{a}\right)$, where M is the mass of the sphere.*

In the first shape the potential at a distance x from the centre, by Art. 301,

$$=\pi\gamma\rho\left(2a^2-\frac{2x^2}{3}\right).$$

$$\therefore\ \frac{1}{2}\int V.dm=\frac{1}{2}\int_0^a \pi\gamma\rho\left(2a^2-\frac{2x^2}{3}\right)\times 4\pi x^2\rho\,dx$$

$$=4\pi^2\gamma\rho^2\left[\frac{1}{3}a^5-\frac{1}{15}a^5\right]=\frac{16}{15}\pi^2\gamma\rho^2a^5=\frac{3}{5}\gamma.\frac{M^2}{a}.$$

So $$\frac{1}{2}\int V'.dm'=\frac{3}{5}\gamma.\frac{M^2}{b}.$$

Hence the required work $=\frac{3}{5}\gamma.M^2\left(\frac{1}{b}-\frac{1}{a}\right)$.

EXAMPLES

1. Shew that the work required against gravity to condense the Earth, supposed of uniform density, into a thin spherical shell of the same radius is equal to $\frac{Ma}{10}$ foot-lbs., M being the mass of the Earth in lbs. and a its radius in feet.

2. The radius of a sphere is a and its mass is M; if the density at any point of it varies inversely as the distance from the centre, shew that the work done in bringing the particles of the sphere from infinity under their mutual attraction is $\frac{2}{3}\frac{\gamma M^2}{a}$.

3. Shew that the work done in collecting the particles of a thin circular disc, attracting according to the Newtonian Law, from an infinite distance is $\frac{8}{3}\frac{\gamma M^2}{\pi a}$, where M is the mass of the disc and a its radius.

[The potential V at a distance x from the centre is easily seen to be

$4\gamma\rho\int_0^{\frac{\pi}{2}}\sqrt{a^2-x^2\sin^2\theta}\,d\theta$. Hence the required work

$$=\frac{1}{2}\int_0^a V.2\pi\rho x\,dx=4\pi\gamma\rho^2\int_0^a x\,dx\left[\int_0^{\frac{\pi}{2}}\sqrt{a^2-x^2\sin^2\theta}\,d\theta\right]$$

$$=4\pi\gamma\rho^2\int_0^a\int_0^{\frac{\pi}{2}}x\sqrt{a^2-x^2\sin^2\theta}\,dx\,d\theta=4\pi\gamma\rho^2\int_0^{\frac{\pi}{2}}\left[-\frac{(a^2-x^2\sin^2\theta)^{\frac{3}{2}}}{3\sin^2\theta}\right]_0^a$$

$$=\frac{4\pi\gamma\rho^2a^3}{3}\int_0^{\frac{\pi}{2}}\frac{1-\cos^3\theta}{\sin^2\theta}d\theta=\frac{4\pi\gamma\rho^2a^3}{3}\int_0^{\frac{\pi}{2}}\left[\cos\theta+\frac{1}{1+\cos\theta}\right]$$

$$=\frac{4\pi\gamma\rho^2a^3}{3}\left[\sin\theta+\tan\frac{\theta}{2}\right]_0^{\frac{\pi}{2}}=\frac{8\pi\gamma\rho^2a^3}{3}=\frac{8}{3}\frac{\gamma M^2}{\pi a}.]$$

4. In the case of a homogeneous oblate spheroid of eccentricity $\sin\beta$, semi-major axis a, and mass M, the work done in moving the particles to an infinite distance against their own attractions is $\frac{3}{5}\gamma\frac{M^2}{a}\frac{\beta}{\sin\beta}$.

5. ***Prove that for an attracting mass M, whose potential is V and whose resultant force is R at the point (x, y, z),***

$$\frac{1}{2}\int V\,dM = \frac{1}{8\pi\gamma}\iiint R^2\,dx\,dy\,dz,$$

the latter integral extending throughout all space.

Verify the case for a solid homogeneous sphere.

Consider the integral $\iiint V.\nabla^2 V\,dx\,dy\,dz$ taken through all space bounded by an infinite sphere.

On integration by parts,

$$\left\{\int V\frac{d^2V}{dx^2}dx\right\}\delta y\,\delta z = \left[V\frac{dV}{dx}\right]\delta y\,\delta z - \left[\int\left(\frac{dV}{dx}\right)^2 dx\right]\delta y\,\delta z \quad ...(1).$$

Here $\left[V\frac{dV}{dx}\right]$ must have given to it its values at the points where the parallelopiped, on $\delta y\,\delta z$ as base and with its other sides parallel to Ox, meets the infinite sphere. Hence, if δS be the element of the sphere cut off by this parallelopiped $\delta y\,\delta z = \delta S.\cos\lambda$, where λ is the angle the normal to δS makes with the axis of x.

Hence $$\iint\left[V\frac{dV}{dx}\right]\delta y\,\delta z = \int V\frac{dV}{dx}\cos\lambda\,dS,$$

taken all over the infinite sphere. Now $V\frac{dV}{dx}$ is a quantity of the order of the inverse cube of the distance of δS from any point of the attracting mass, and δS is of the order of the square of this same infinite distance. Hence

$$\int V\frac{dV}{dx}\cos\lambda\,dS = 0.$$

Thus (1) gives

$$\iiint V\frac{d^2V}{dx^2}dx\,dy\,dz = -\iiint\left(\frac{dV}{dx}\right)^2 dx\,dy\,dz,$$

and hence

$$\iiint V.\nabla^2 V\,dx\,dy\,dz = -\iiint\left[\left(\frac{dV}{dx}\right)^2 + \left(\frac{dV}{dy}\right)^2 + \left(\frac{dV}{dz}\right)^2\right]dx\,dy\,dz$$

$$= -\iiint R^2\,dx\,dy\,dz.$$

Now within the attracting mass $\nabla^2 V = -4\pi\gamma\rho$, and, without it, $\nabla^2 V = 0$.

Hence $$\iiint V\rho\,dx\,dy\,dz = \frac{1}{4\pi\gamma}\iiint R^2\,dx\,dy\,dz,$$

i.e. $$\frac{1}{2}\int V\,dM = \frac{1}{8\pi\gamma}\iiint R^2\,dx\,dy\,dz.$$

In the case of a solid sphere the left-hand member $= \frac{3}{5}\frac{\gamma M^2}{a}$ as in Art. 318. Within the sphere $R = \frac{\gamma M}{a^3}r$ and without it $R = \gamma\frac{M}{r^2}$ [Art. 287].

Hence $\frac{1}{8\pi\gamma}\iiint R^2 dx\,dy\,dz = \frac{1}{8\pi\gamma}\iiint R^2 . 4\pi r^2 dr$

$$= \frac{1}{2\gamma}\left[\int_0^a \frac{\gamma^2 M^2 r^2}{a^6} . r^2 dr + \int_a^\infty \frac{\gamma^2 M^2}{r^4} . r^2 dr\right] = \frac{3\gamma M^2}{5a}.$$

319. *The potential of any attracting mass, or masses, cannot be an absolute maximum or an absolute minimum at any point at which there is none of the attracting mass.*

For let P be the point and describe a very small sphere whose centre is P. If the potential V at P were an absolute maximum, it must be greater than at any point Q of the sphere. Hence $\frac{dV}{dn}$ must be negative whatever be the direction in which δn is drawn from P. Hence the normal force N of attraction, which equals $\frac{dV}{dn}$ at any point Q of the sphere, must be always negative, so that $\int N . dS$ taken over the sphere must be a negative quantity.

But, by Gauss' Theorem (Art. 303), this integral must be zero, since the very small sphere with centre P contains none of the attracting mass. It is therefore impossible that the potential at P should be a maximum.

Neither can it be an absolute minimum at P; for then, by a similar reasoning, $\int N . dS$ would be a positive quantity.

Cor. 1. EARNSHAW'S THEOREM. *If a particle be in equilibrium under the action of forces following the law of the inverse square, the equilibrium cannot be stable for all displacements.*

For if it be at rest at P and be displaced through a small distance δn to Q, then, if it is to return towards P, $\frac{dV}{dn}$ must be negative; and if the equilibrium is to be stable this must be true for all such displacements. Hence V must be an absolute maximum which is impossible by the preceding article.

Neither could the equilibrium be unstable for all displacements; for this would require that $\frac{dV}{dn}$ be positive for all directions and therefore V an absolute minimum at P.

Cor. 2. *If the potential has any given constant value V at all points of a closed surface S, which does not contain any of the attracting mass, it has the same value V at all points of the space enclosed by S.*

For if not, there must be some point inside S at which the potential is either greater, or less, than at any other point within S, *i.e.* there must be some point at which the potential is either a maximum or a minimum. But this is impossible by the preceding proposition.

320. If the potential of a distribution is given throughout all space, we can determine the corresponding distribution. For, V being known, we can find $\nabla^2 V$ for every point of space. Wherever it is zero, the corresponding density of the distribution is zero by Poisson's equation, *i.e.* there is no attracting mass at all such points; wherever it is not zero, the corresponding density of the distribution is $-\dfrac{1}{4\pi\gamma}\nabla^2 V$.

If the form of the potential function inside any surface S is different from its form outside, and if there be an abrupt change in the value of $\dfrac{dV}{dn}$ as we pass across this surface, then the surface distribution σ on S is given by Art. 282. For if V_1 be the potential just inside S, and V_2 the potential just outside S, and δn be an element of the outward drawn normal, the result of Art. 282 may be written in the form

$$\left(-\frac{dV_2}{dn}\right)-\left(-\frac{dV_1}{dn}\right)=4\pi\gamma\sigma,$$

i.e.

$$\sigma=\frac{1}{4\pi\gamma}\left[\frac{dV_1}{dn}-\frac{dV_2}{dn}\right] \quad \text{.................(1)},$$

where σ is the superficial density of the stratum on S.

Ex. *What distribution of matter will produce a potential*

$$\frac{\gamma M}{a}\left(1-\frac{x}{3a}\right) \quad \textit{or} \quad \frac{\gamma M}{r}\left(1-\frac{ax}{3r^2}\right),$$

according as $r\ (=\sqrt{x^2+y^2+z^2})$ *is less or greater than* a?

Let V_1 be the potential inside the sphere of radius a, and V_2 the potential outside, so that

$$V_1=\frac{\gamma M}{a}\left(1-\frac{r}{3a}\cos\phi\sin\theta\right),$$

and

$$V_2=\gamma M\left(\frac{1}{r}-\frac{a\cos\phi\sin\theta}{3r^2}\right).$$

Hence, if ρ_1 and ρ_2 are the densities within and without the sphere, we have, by equation (2) of Art. 307, on performing the differentiations,

$$\rho_1 = 0,$$

and $$\rho_2 = 0,$$

so that there is no distribution within or without the spherical surface of radius a.

Also, from the previous article, if σ is the density of the distribution on this spherical surface, we have

$$4\pi\gamma\sigma = \left(\frac{dV_1}{dr}\right)_{r=a} - \left(\frac{dV_2}{dr}\right)_{r=a}$$

$$= \frac{\gamma M}{a}\left(-\frac{\cos\phi\sin\theta}{3a}\right) - \gamma M\left[-\frac{1}{a^2} + \frac{2\cos\phi\sin\theta}{3a^2}\right]$$

$$= \frac{\gamma M}{a^2}(1 - \cos\phi\sin\theta),$$

so that $\sigma = \frac{M}{4\pi a^2}\left(1 - \frac{x}{a}\right)$, giving the distribution of matter on the spherical surface.

321. *If S be a closed equipotential surface, of potential V, which includes a mass M of the attracting matter, and if a thin stratum of attracting matter be placed on S, whose density ρ at any point P is equal to $-\frac{1}{4\pi\gamma}\cdot\frac{dV}{dn}$, where δn is an element of the outward drawn normal at P, then the potential of the stratum at all points external to S is equal to that of M, and the sum of the potentials of the stratum and that of M' is constant throughout S and equal to V, where M' is the part of the attracting mass external to S.*

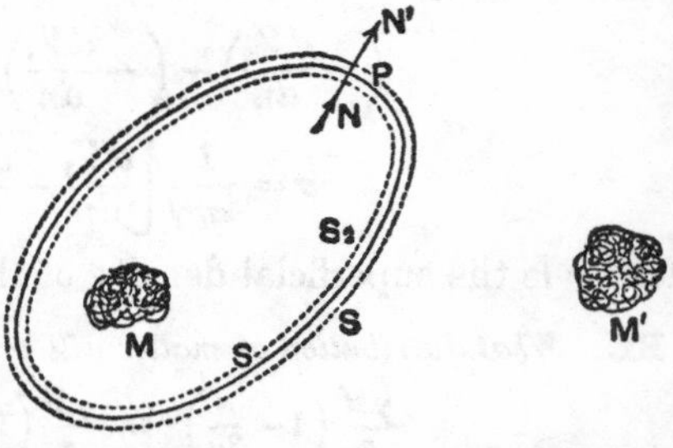

Let us find the density ρ of a stratum at P such that its potential together with that of M' shall $= V$ at each point of S.

Now the potential of M and M' together (which make up the whole attracting mass) $= V$ at each point of S.

Hence the potential of the whole stratum = the potential of M at each point of S.

If therefore we change the sign of ρ (*i.e.* we change the stratum into a repulsive instead of an attractive system), the potential of M and the stratum, of density $-\rho$ at P, is equal to zero all over a surface S_1 just outside S.

But the potential of M and the stratum $(-\rho)$ is also zero all over a sphere of infinite radius.

Hence, by Art. 319, Cor. 2 (since in the space bounded by S_1 and this infinite sphere there is none of the mass M or of the stratum), the potential of M and the stratum $(-\rho)$ is zero throughout all the space between S_1 and the infinite sphere, *i.e.* throughout all space external to S_1,

i.e. the potential of $M=$ that of the stratum of density ρ throughout all space external to S, in the limit ....(1).

Again, by the same corollary to Art. 319, since the potentials of the stratum and M' are together equal to V at all points of a surface S_2 inside S, and indefinitely close to S, it follows that the sum of their potentials inside S_2 is constant and equal to V throughout the interior of the surface S_2 (since S_2 contains none of the attracting system, consisting of the stratum and M'), and hence, in the limit, throughout the interior of the surface S...(2).

Let N and N' be the normal attractions, both measured outwards, of the whole stratum at points just within and just without the surface at P.

Then, by Art. 282,

$$4\pi\gamma\rho = N - N' \qquad\text{.......................(3).}$$

Now, from (2), it follows that

$$N + \text{the normal attraction of } M' \text{ at } P = 0.$$

And, from (1), it follows that

$$N' = \text{the normal attraction of } M \text{ at } P.$$

Hence (3) gives

$$\begin{aligned} -4\pi\gamma\rho &= \text{normal attraction of } M' \text{ at } P + \text{normal attraction of } M \text{ at } P \text{, both measured outwards} \\ &= \text{outward normal attraction of the whole attracting mass at } P \\ &= \frac{dV}{dn}, \end{aligned}$$

so that the density ρ of the stratum at $P = -\dfrac{1}{4\pi\gamma}\dfrac{dV}{dn}$.

Cor. 1. Since the potentials of the stratum and the mass M are equal at all points external to S, they are equal at infinity.

It follows that the mass of the stratum $= M$.

Otherwise thus. The mass of the stratum

$$=\int \rho\, dS = -\frac{1}{4\pi\gamma}\int \frac{dV}{dn}\,.\,dS$$

$$= -\frac{1}{4\pi\gamma} \times \text{surface integral of normal attraction taken over } S$$

= the mass inside S, by Art. 303, $= M$.

Cor. 2. Let there be no attracting mass external to S, *i.e.* let M' be zero; then at each point of space external to S the potential of the stratum = that of M, and within S the potential of the stratum $= V =$ the potential of M at the surface of S.

322. As a simple example of the previous article, let M be a particle O, and M' be zero. Then S is a sphere, whose centre is at O and whose radius is any quantity r.

Hence $$\rho = -\frac{1}{4\pi\gamma}\frac{dV}{dr} = -\frac{1}{4\pi\gamma}\frac{d}{dr}\left(\frac{\gamma M}{r}\right)$$

$$= \frac{M}{4\pi r^2}.$$

Hence, outside S, the potential of this stratum equals that of the particle $M = \dfrac{\gamma M}{\text{distance from } O} = \dfrac{\gamma\,.\,4\pi r^2\rho}{\text{distance from } O}$, and, inside S, the potential of this stratum is constant and = the potential of M at the surface of the sphere $= \dfrac{\gamma M}{r} = \dfrac{\gamma\,.\,4\pi r^2\rho}{r}$.

These are the results of Art. 300.

323. Ex. *If M be the mass of a homogeneous thin straight bar of length $2c$, and $2a$ be the major axis of one of its equipotential curves, then a distribution of mass on the equipotential curve equal to $\dfrac{1}{4\pi}\dfrac{Mp}{a(a^2-c^2)}$ at a point P, where p is the perpendicular from the middle point of the bar upon the tangent at P, will have the same potential as the bar at all points in external space.*

Let AB be the rod and $AP = r_1$, $BP = r_2$, and y = the perpendicular from P on AB.

Then $-\dfrac{dV}{dn}$ = attraction of the rod at $P = \dfrac{2\gamma k\rho}{y}\sin\dfrac{APB}{2}$ (Art. 277)

$$= \frac{2\gamma k\rho\,.\,2c\sin\dfrac{APB}{2}}{r_1 r_2 \sin APB} = \frac{2\gamma k\rho c}{r_1 r_2\cos\dfrac{APB}{2}}.$$

Now $\cos^2 \frac{APB}{2} = \sin^2 BPT$, where PT is the tangent at P to the equipotential curve,

$$= \frac{b^2}{r_1 r_2} = \frac{a^2 - c^2}{r_1 r_2},$$

and $\sqrt{r_1 r_2}$ = semi-diameter conjugate to $CP = \frac{ab}{p} = \frac{a\sqrt{a^2 - c^2}}{p}$.

$$\therefore -\frac{dV}{dn} = \frac{2\gamma k \rho c p}{a(a^2 - c^2)} = \frac{\gamma M p}{a(a^2 - c^2)}.$$

Hence the density of the stratum $= \frac{1}{4\pi} \frac{Mp}{a(a^2 - c^2)}$.

Outside the equipotential curve, the stratum and the bar have equal potentials at all points.

Inside the equipotential curve, by Cor. 2 of Art. 321, the potential of the stratum is the same at all points and is equal to the potential of the bar at all points of the equipotential curve, *i.e.* it

$$= \gamma k \rho \log \frac{r_1 + r_2 + 2c}{r_1 + r_2 - 2c} \quad \text{(Art. 298)}$$

$$= \gamma \frac{M}{2c} \log \frac{a + c}{a - c}.$$

EXAMPLES

1. The values of V at any point at a distance r from a fixed point O are

$$V = 2\pi\gamma\rho (a^2 - b^2), \text{ if } r < b < a;$$

$$V = 2\pi\gamma\rho \left(a^2 - \frac{r^2}{3} - \frac{2}{3}\frac{b^3}{r}\right), \text{ if } b < r < a;$$

and

$$V = \frac{4\pi\gamma\rho}{3} \frac{a^3 - b^3}{r}, \text{ if } b < a < r.$$

Shew that the attracting system is a spherical shell of density ρ, whose boundaries are spheres of centre O and radii a and b.

2. Find whether any distribution of matter will give rise to the following potentials:

$$V = \frac{4\gamma A a z}{r^3}, \text{ when } r > a;$$

$$V = \gamma A \frac{3a^2 + 4az - 3r^2}{a^3}, \text{ when } r < a, \text{ where } r^2 = x^2 + y^2 + z^2.$$

[Inside the sphere $r = a$, the density is $\frac{9A}{2\pi a^3}$; outside, it is zero; upon its surface, the surface density is $\frac{3A}{2\pi a^2}\left(\frac{2z}{a} - 1\right)$.]

3. The potential of a certain distribution of matter at a point (x, y, z) is

$$\frac{4\gamma\mu\pi}{3}\frac{a^4}{r}+\frac{4\gamma\mu\pi}{15}\frac{a^6(2x^2-y^2-z^2)}{r^5},$$

or

$$\frac{4\gamma\mu\pi}{3}a^3+\frac{4\gamma\mu\pi}{15}a(2x^2-y^2-z^2),$$

according as $r\,[=\sqrt{x^2+y^2+z^2}]$ is greater or less than a; find the distribution.

[The surface density on the surface of the sphere $r=a$ is μx^2; inside and outside there is no matter.]

4. The potential outside a certain cylindrical boundary, whose edges are parallel to the axis of z, is zero; inside, it is $V=x^3-3y^2x-ax^2+3ay^2$. Find the distribution of matter.

5. Find the distribution of matter which will produce the following potentials:

$V=1$ within the ellipsoid $\frac{x^2}{\mu^2a^2}+\frac{y^2}{\mu^2b^2}+\frac{z^2}{\mu^2c^2}=1$; $(\mu<1)$;

$V=\frac{1}{1-\mu^2}\left[1-\frac{x^2}{a^2}-\frac{y^2}{b^2}-\frac{z^2}{c^2}\right]$ between the above ellipsoid and $\frac{x^2}{a^2}+\frac{y^2}{b^2}+\frac{z^2}{c^2}=1$;

$V=0$ outside the ellipsoid $\frac{x^2}{a^2}+\frac{y^2}{b^2}+\frac{z^2}{c^2}=1$.

6. If the potential of a given attracting mass at a point (x, y, z) is $\gamma.\frac{a^2(x^2-y^2)+2b^2xy}{(x^2+y^2)^2}$, shew that a surface density $-\frac{1}{2\pi}\sqrt{\frac{a^4+b^4}{(x^2+y^2)^3}}$ on one of the equipotential surfaces will produce at all external points the same potential as the original mass.

7. Find the distribution throughout a sphere, of radius a, and on its surface, which, for a point distant r from its centre, gives potential $\lambda(2a^3-r^3)$ for internal points, and potential $\lambda\frac{a^4}{r}$ for external points.

$\left[\rho=\frac{3\lambda r}{\pi\gamma}\right.$ within the sphere; $\rho=0$ outside it; $\left.\sigma=-\frac{\lambda a^2}{2\pi\gamma}\text{ on its surface.}\right]$

CHAPTER XVI

EQUILIBRIUM OF SLIGHTLY ELASTIC BEAMS

324. IF we want to find the form of a thin rod or beam, loaded in any way, we must have some relation between the Bending Moment and the shape.

We shall here assume that the Bending Moment is proportional to the curvature, *i.e.* that the Bending Moment $= \frac{K}{\rho}$, where ρ is the radius of curvature of the beam. K is called the flexural rigidity.

If the beam before being loaded were of curvature $\frac{1}{\rho_1}$ at the point considered, then this Bending Moment would be

$$K\left(\frac{1}{\rho} - \frac{1}{\rho_1}\right).$$

This assumption, originally due to Bernoulli and Euler, may be looked upon as established by experiment as true approximately in the case of ordinary beams whose lengths are great compared with their transverse dimensions.

A proof, involving some assumptions which are not in the strict sense correct, is given in the next article.

325. *A rectangular beam is bent, without tension; to shew that the bending moment at any point varies as the curvature.*

When a beam naturally straight is bent into the form in the figure, it is clear that the fibres at the upper part near E are in a state of tension, and those at the lower part near F in a state of compression. The line GG' which separates the fibres, which are respectively in tension and compression, is called the neutral line.

As an approximation to the truth, we shall assume that any plane section of the beam perpendicular to its axis continues to be a plane section after the bending.

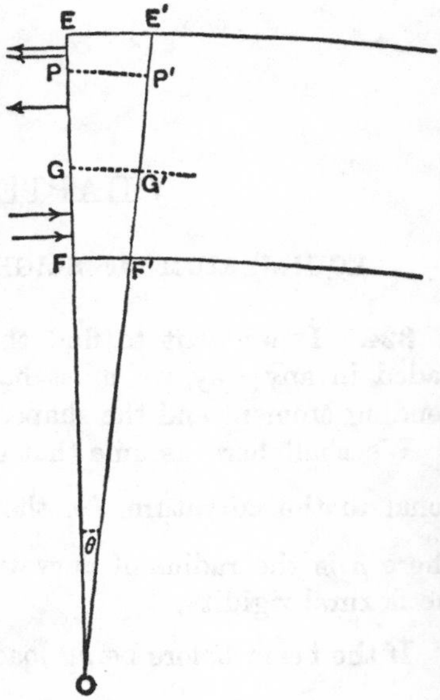

Let the normals at two consecutive points of the neutral line GG' meet in O, and let $OG = \rho$.

Consider any fibre PP' in the plane of the paper on the side of GG' towards E.

If E be Hooke's Modulus of Elasticity for this fibre, its tension per unit of area

$$= E\,\frac{PP' - GG'}{GG'}$$

$$= E\,\frac{(\rho + x) - \rho}{\rho} = E\,\frac{x}{\rho},$$

where $GP = x$.

Hence, if δA be the area of the cross section of the fibre at P, its tension $= \dfrac{Ex}{\rho}\,.\,\delta A$.

[If P be on the side of G towards F, x is negative, and this tension becomes a compression.]

Since the beam has no tension, the sum of the tensions exerted by the fibres perpendicular to OE is zero.

Hence $\Sigma\,\dfrac{Ex}{\rho}\,.\,\delta A = 0$, the summation being taken over the whole section $RSS'R'$ at G perpendicular to the plane of the paper.

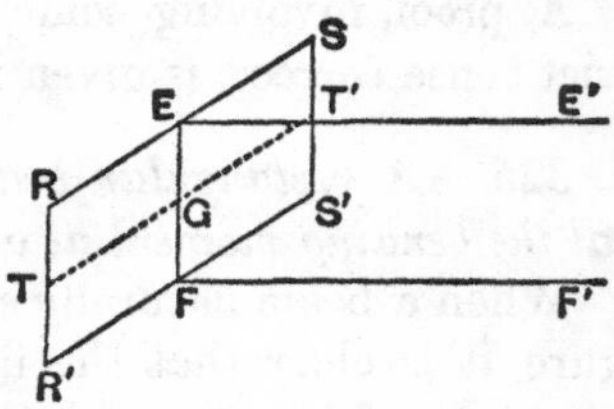

Hence $\qquad \Sigma x\,\delta A = 0.$

It follows, by the formulae giving the position of the centroid of any area, that the line through G perpendicular to the plane of the paper must pass through the centroid of the section $RSS'R'$; thus, if the plane of the paper be a symmetrical section of the beam, G must be the centroid of the section $RSS'R'$.

By taking moments about the line GT perpendicular to the plane of the paper, the resultant couple about it

$$=\Sigma \frac{Ex}{\rho} . \delta A \times x = \frac{E}{\rho} . \Sigma x^2 \delta A.$$

[It is clear that the tensions above G combine with those below G to produce a couple.]

But $\Sigma x^2 \delta A$ is the moment of inertia of the section $RSS'R'$ about the line GT, and is usually denoted by I.

Hence the resultant couple, *i.e.* the bending moment, is equal to $\frac{EI}{\rho}$, where E is Hooke's Modulus of Elasticity, ρ is the radius of curvature of the neutral line, and I is the moment of inertia of the cross section of the beam about a line through its centroid perpendicular to the length of the beam.

326. When the beam is only slightly flexible, *i.e.* when the quantity EI is large, the expression $\frac{EI}{\rho}$ can be simplified.

For
$$\frac{1}{\rho} = \frac{\pm \frac{d^2y}{dx^2}}{\left\{1 + \left(\frac{dy}{dx}\right)^2\right\}^{\frac{3}{2}}}.$$

If the beam differs only a little from a straight line, then $\frac{dy}{dx}$ is small, if x be measured horizontally and y vertically.

In this case $\pm \frac{d^2y}{dx^2}$ is an approximation to the value of $\frac{1}{\rho}$, so that $\pm EI \frac{d^2y}{dx^2}$ is the bending moment. The ambiguity in sign must be determined by the sign of $\frac{d^2y}{dx^2}$.

327. Ex. *A uniform slightly flexible rod AC, of length $2a$, is supported at its two ends, and also at its middle point B; the supports being in the same horizontal line, find the thrusts on them and the equation to the curve in which the rod rests.*

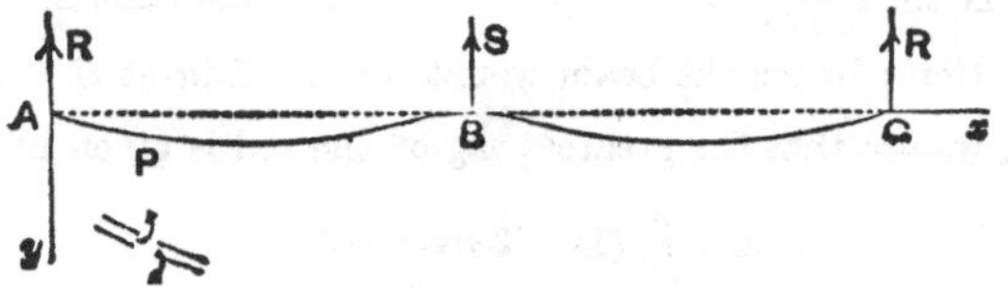

Let P be any point in AB; then the bending moment at it on the part to the left of P is equal to $\frac{EI}{\rho}$ in the sense ↻. Also, if the axis of y be vertically downwards, $\frac{dy}{dx}$ is decreasing at P and hence $\frac{d^2y}{dx^2}$ is negative, so that $\frac{1}{\rho}$ is equal to $-\frac{d^2y}{dx^2}$ approximately.

If R be the reaction at A and w be the weight of the rod per unit of length we have, by taking moments about P for the portion PA,

$$-EI\frac{d^2y}{dx^2}=\frac{EI}{\rho}=R.x-wx.\frac{x}{2} \quad\text{.....................(1).}$$

$$\therefore\ -EI\frac{dy}{dx}=R\frac{x^2}{2}-\frac{wx^3}{6}+D \quad\text{.....................(2).}$$

But, by symmetry, $\frac{dy}{dx}=0$ when $x=a$.

$$\therefore\ D=\frac{wa^3}{6}-R\frac{a^2}{2}.$$

$$\therefore\ -EIy=R\frac{x^3}{6}-\frac{wx^4}{24}+\left(\frac{wa^3}{6}-R\frac{a^2}{2}\right)x \quad\text{............(3),}$$

the constant being zero since x and y vanish together.

Also $y=0$ when $x=a$.

$$\therefore\ 0=R\frac{a^3}{6}-\frac{wa^4}{24}+\left(\frac{wa^4}{6}-\frac{Ra^3}{2}\right).$$

Hence $R=\frac{3}{8}wa=\frac{3}{16}$ of the whole weight.

Hence S, the thrust of the central support, $=\frac{5}{8}$ of the whole weight.

Substituting the value of R in (1), (2) and (3), we have

$$-EI\frac{d^2y}{dx^2}=\frac{E}{\rho}=\frac{w}{2}\left(\frac{3ax}{4}-x^2\right) \quad\text{....................(4),}$$

$$-EI\frac{dy}{dx}=\frac{w}{2}\left(\frac{3ax^2}{8}-\frac{x^3}{3}-\frac{a^3}{24}\right) \quad\text{..................(5),}$$

and
$$-EIy=\frac{w}{2}\left(\frac{ax^3}{8}-\frac{x^4}{12}-\frac{a^3x}{24}\right)=-\frac{w}{48}x(x-a)^2(2x+a) \quad\text{......(6).}$$

From (4), it follows that there is a point of contrary flexure when $x=\frac{3a}{4}=\frac{3}{4}.AB$; that the bending moment is a maximum when $x=\frac{3a}{8}$ and that it is then $\frac{9wa^2}{128}$ ↻; at B the bending moment is $\frac{wa^2}{8}$ in the opposite direction; hence the beam would rupture first at B.

From (5), we see that the greatest sag of the rod is given by

$$x=\frac{a}{16}(1+\sqrt{33})=\cdot 42a.$$

The points of inflexion of a beam are often called its "hinges" or "virtual joints." For, since there is no bending moment at these points, hinges or joints could be introduced at them without altering the equilibrium of the beam.

328. In Art. 129 it was shewn that

$$\text{Load} = -\frac{d}{dx}\,[\text{Shearing Force}]$$

and
$$\text{Shearing Force} = \frac{d}{dx}\,[\text{Bending Moment}],$$

and from Art. 326 we have, in the case of slightly elastic beams,

$$\text{Bending Moment} \propto \frac{d}{dx}\,[\text{Slope}]$$

and
$$\text{Slope} = \frac{d}{dx}\,[\text{Deflection}].$$

Hence the Load Curve, the Shearing Force Curve and the Bending Moment Curve bear to one another the same relations as the Bending Moment Curve, the Slope Curve and the Deflection Curve.

Hence we shall have a theorem similar to that of Art. 133, *viz.* that *Any two tangents to the Deflection Curve intersect in a point which is vertically below the centre of gravity of the corresponding part of the Bending Moment Curve.*

We can verify this directly in the case of the example of the last article.

For the Bending Moment Curve here, for the part AB, is

$$y = M = \frac{w}{2}\left[\frac{3ax}{4} - x^2\right] \;\ldots\ldots\ldots\ldots\ldots\ldots(1),$$

and the Deflection Curve is

$$\frac{2EI}{w}\,y = \frac{x^4}{12} - \frac{ax^3}{8} + \frac{a^3x}{24} \;\ldots\ldots\ldots\ldots\ldots\ldots(2).$$

The tangent to (2) at the point (x_1, y_1) is easily seen to be

$$\frac{2EIy}{w} = x\left[\frac{x_1^3}{3} - \frac{3ax_1^2}{8} + \frac{a^3}{24}\right] - \frac{x_1^4 - ax_1^3}{4}.$$

This intersects the tangent at (x_2, y_2) where

$$x = \frac{6\,[(x_2^4 - x_1^4) - a\,(x_2^3 - x_1^3)]}{8\,(x_2^3 - x_1^3) - 9a\,(x_2^2 - x_1^2)}.$$

Also the abscissa of the centre of gravity of the corresponding part of the curve (1) is given by

$$\bar{x} = \frac{\int_{x_1}^{x_2} y\,dx \,.\, x}{\int_{x_1}^{x_2} y\,dx} = \frac{\int_{x_1}^{x_2} (3ax^2 - 4x^3)\,dx}{\int_{x_1}^{x_2} (3ax - 4x^2)\,dx}$$

$$= \frac{a\,(x_2^3 - x_1^3) - (x_2^4 - x_1^4)}{\frac{3a}{2}(x_2^2 - x_1^2) - \frac{4}{3}(x_2^3 - x_1^3)} = \frac{6\,[(x_2^4 - x_1^4) - a\,(x_2^3 - x_1^3)]}{8\,(x_2^3 - x_1^3) - 9a\,(x_2^2 - x_1^2)}.$$

Hence the result given.

EXAMPLES

1. If a slightly elastic rod rests with its two ends on two supports at the same level, prove that the deflection at a distance x from one end is

$$\frac{w}{24K}\,x\,(a-x)\,\{a^2+ax-x^2\},$$

where K is the flexural rigidity, a the length, and wa the total weight.

2. A uniform bridge, of weight W', formed of a single plank, is supported at its ends; a man, of weight W, stands on the bridge at a point whose distances from the ends are a and b. Shew that the deflection just under the man is $\dfrac{W'\,(a^2+3ab+b^2)+8\,Wab}{24EI\,(a+b)}\,.\,ab.$

3. An elastic rod, clamped at one end so that it is horizontal there, is bent by its own weight; shew that the deflection of its extremity is $\frac{3}{8}$ths of what it would be if the deflection were caused by a weight equal to its own weight hung on at its extremity.

4. A uniform beam AB, of length l, is supported at its ends and loaded with a weight W at a point Q, where $AQ=a$. If the weight of the beam be neglected, shew that the equation of AQ is

$$EIy = \frac{W\,(l-a)}{6l}\,[a\,(2l-a)\,x - x^3],$$

and that of QB is

$$EIy = \frac{Wa}{6l}\,[(l^2-a^2)\,(l-x) - (l-x)^3].$$

Shew also that the deflection at any point P when the load is at Q is equal to the deflection at Q when the same load is at P.

5. A heavy uniform rod rests horizontally on two pegs, one of which is at one end. Shew that the second peg must be placed at a distance from this end equal to two-thirds of the length of the rod, if the bending moment at the middle point is to be zero.

6. A slightly elastic beam AB, of weight W and length $2a$, is supported at its ends and at its middle point C. If the thrusts on the supports are equal, shew that the depth of C below AB is $\frac{7}{144}\frac{Wa^3}{K}$, and is equal to $\frac{7}{15}$ of the depth of C below AB when the ends only are supported.

7. A uniform beam is supported at its two points A, B of trisection, AB being horizontal; shew that the height of the middle point of the beam above AB is to the depth of each end of the beam below AB as 19 : 128.

8. A thin uniform slightly flexible rod is of length $4a$, has a weight W attached to its middle point, and is supported at two points at equal distances a on each side of the middle point. If the tangents at the points of support are horizontal, shew that W must be one-sixth of the weight of the rod.

9. A concentrated load travels from one end of a beam to the other. Shew that the ratio of the slopes of the deflection curve at the two ends commences with the value 2, and is equal to $\frac{5}{4}$ when the weight is at one-third the span.

10. A single-line railway bridge is carried by two main girders, each of 40 feet span. The total weight of a locomotive standing on the bridge is 68 tons distributed upon 4 axles, the leading axle carrying 8 tons and each of the others 20 tons. The distances of the leading axle and of the other axles from one end of the bridge are 6 ft., 13½ ft., 21 ft. and 29 ft. respectively. Shew how to find the maximum deflection of the girder, and where it occurs.

329. Clapeyron's Equation of the three moments.

If M_1, M_2, and M_3 are the bending moments of a uniform loaded beam at three successive points of support, A_1, A_2, and A_3 which are in a horizontal line, to shew that

$$aM_1 + 2(a+b)M_2 + bM_3 = \frac{w}{4}(a^3 + b^3),$$

where w is the load per unit of length, $A_1A_2 = a$, and $A_2A_3 = b$.

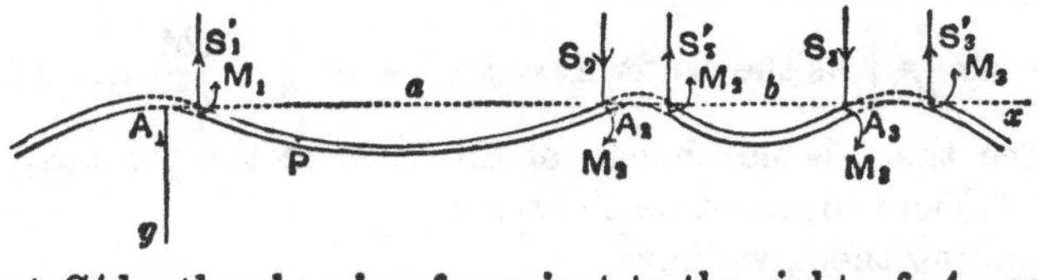

Let S_1' be the shearing force just to the right of A_1, and S_2 and S_2' the shearing forces just to the left and right of A_2.

Take A_1 as origin, $A_1A_2A_3$ as the axis of x and let the axis of y be drawn vertically downwards.

Then for a point P in A_1A_2 distant x from A_1 we have, by taking moments for the part to the left of P, as in Art. 327,

$$-EI\frac{d^2y}{dx^2}=\frac{EI}{\rho}=S_1'.x-\frac{w}{2}x^2-M_1 \text{ ....(1).}$$

Also, by taking moments about A_2 for the part A_1A_2,

$$M_2=\tfrac{1}{2}wa^2-S_1'a+M_1 \text{ ....(2).}$$

Eliminating S_1', we have

$$-EI\frac{d^2y}{dx^2}=\frac{w}{2}(ax-x^2)-M_1\left(1-\frac{x}{a}\right)-M_2\frac{x}{a}\text{ .....(3).}$$

Hence

$$-EI\frac{dy}{dx}=\frac{w}{2}\left(\frac{ax^2}{2}-\frac{x^3}{3}\right)-M_1\left(x-\frac{x^2}{2a}\right)-M_2\frac{x^2}{2a}+C\text{ .....(4),}$$

and

$$-EIy=\frac{w}{2}\left(\frac{ax^3}{6}-\frac{x^4}{12}\right)-M_1\left(\frac{x^2}{2}-\frac{x^3}{6a}\right)-M_2\frac{x^3}{6a}+Cx+D.$$

Now $y=0$ when $x=0$ or a.

Hence $D=0$, and

$$C=M_1\frac{a}{3}+M_2\frac{a}{6}-\frac{wa^3}{24}\text{ ....(5).}$$

Hence, from (4),

$$\left[-EI\frac{dy}{dx}\right] \text{ at the point } A_1=C=M_1\frac{a}{3}+M_2\frac{a}{6}-\frac{wa^3}{24}\text{ ...(6),}$$

and

$$\left[-EI\frac{dy}{dx}\right] \text{ at the point } A_2=\frac{w}{12}a^3-(M_1+M_2)\frac{a}{2}+C$$

$$=\frac{wa^3}{24}-M_1\frac{a}{6}-M_2\frac{a}{3}\text{ .....(7).}$$

So, from the equilibrium of the part A_2A_3, the result similar to (6) would be found to be

$$\left[-EI\frac{dy}{dx}\right] \text{ at the point } A_2=M_2\frac{b}{3}+M_3\frac{b}{6}-\frac{wb^3}{24}\text{ .....(8).}$$

Since there is no change of direction at the point A_2, the results (7) and (8) must be the same.

Equating them, we have

$$M_1a+2M_2(a+b)+M_3b=\frac{w}{4}(a^3+b^3).$$

[It will be noted that in the preceding the moments M_1, M_2, and M_3 are measured positively in the opposite direction from the cases of Arts. 129 and 332.]

Reaction at any support in terms of the bending moments M_1, M_2, and M_3.

From the equilibrium of A_1A_2, we have

$$S_1' - S_2 = wa,$$

so that, from (2),

$$S_2 = \frac{M_1 - M_2}{a} - \frac{wa}{2}.$$

From an equation similar to (2) for the part A_2A_3, we have

$$S_2' = \frac{wb}{2} + \frac{M_2 - M_3}{b}.$$

Hence the reaction at the support A_2

$$= S_2' - S_2 = \frac{w(a+b)}{2} - \frac{M_1 - M_2}{a} + \frac{M_2 - M_3}{b}.$$

330. If A_2, instead of being on the same level with A_1 and A_3, is at distances y_1 and y_2 below them respectively, the equation of the three moments is easily seen to be

$$M_1 a + 2M_2(a+b) + M_3 b = \frac{w}{4}(a^3 + b^3) - 6EI\left(\frac{y_1}{a} + \frac{y_2}{b}\right).$$

331. Ex. 1. Assume that in the question of Art. 329 the beam ends at A_1, A_3, so that we have a beam supported at A_1, A_2, A_3, where $A_1A_2 = a$ and $A_2A_3 = b$.

Then $\qquad M_1 = M_3 = 0$, and $M_2 = \frac{w}{8}(a^2 - ab + b^2)$,

if the beam be of uniform load w per unit of length.

Let R_1, R_2, R_3 be the reactions at A_1, A_2, A_3. Taking moments about A_2, we have

$$M_2 = \frac{wa^2}{2} - R_1 a,$$

and therefore $\qquad R_1 = \frac{w}{8}\left[\frac{3a^2 + ab - b^2}{a}\right].$

So also $\qquad R_3 = \frac{w}{8}\left[\frac{3b^2 + ab - a^2}{b}\right],$

and $\qquad R_2 = w(a+b) - R_1 - R_3 = \frac{w}{8}(a+b)\left[\frac{a^2 + 3ab + b^2}{ab}\right].$

Suppose $a > b$. Then R_3 is negative if

$$a^2 - ab > 3b^2, \quad i.e. \text{ if } a > \frac{b}{2}(1 + \sqrt{13}),$$

i.e. if $a > b \times 2{\cdot}3$ approximately. In this case the end A_3 would have to be weighted if it is to be kept in contact with its support.

Again, the bending moment M ↻ at any point distant x from A_1

$$=R_1 x-\frac{wx^2}{2}=wx\,\frac{3a^2+ab-b^2}{8a}-\frac{wx^2}{2}.$$

It is thus a maximum when $x=\frac{3a^2+ab-b^2}{8a}$, and its value then

$$=\frac{w}{2}\left(\frac{3a^2+ab-b^2}{8a}\right)^2 \quad\text{.................................(1).}$$

Also the bending moment at $A_2=M_2$ ↻ $=\frac{w}{8}(a^2-ab+b^2)$.

The bending moment at A_2 is thus greater than (1) if

$$16a^2(a^2-ab+b^2)>(3a^2+ab-b^2)^2,$$

i.e. if
$$(11a^2+ab-b^2)^2<128a^4,$$

i.e. if
$$ab-b^2<(8\sqrt{2}-11)\,a^2<{\cdot}3136a^2,$$

i.e. if
$$\left(b-\frac{a}{2}\right)^2+{\cdot}0636a^2>0, \text{ which is true.}$$

Hence, if the beam rupture, it will do so at A_2.

[The results found above for R_1, R_2, and R_3 can be verified by experiment for different values of a and b. We thus have a test of the accuracy of the assumption of Art. 324.]

Ex. 2. *A uniform rod A_1A_4 is supported at its ends A_1 and A_4 and at points A_2, A_3 which divide A_1A_4 into three equal parts; the supports being at the same horizontal level, find the thrusts on them and the bending moments at them in terms of the weight W.*

Let the length of the rod be $l\,(=3a)$ so that $W=w.3a$. Applying the formula of Art. 329 to the supports A_1, A_2, A_3 and then to the supports A_2, A_3, A_4, we have, since clearly M_1 and M_4 are both zero at the free ends,

$$4M_2+M_3=\frac{wa^2}{2},\ \text{ and }\ M_2+4M_3=\frac{wa^2}{2}.$$

$$\therefore\quad M_2=M_3=\frac{wa^2}{10}=\frac{W.l}{90}.$$

Also, if R_1 be the reaction at A_1, we have, by taking moments about A_2 for the part A_1A_2,

$$M_2=\frac{1}{2}wa^2-R_1a,$$

so that
$$R_1=\frac{2}{5}wa=\frac{2}{15}W=R_4, \text{ by symmetry.}$$

Hence R_2 and R_3 must each be $\frac{11W}{30}$.

If we take A_1 as origin and the axis of y vertically downwards, the equation to the middle segment A_2A_3 is easily seen to be

$$-EI\frac{d^2y}{dx^2}=-\frac{w}{2}x^2+R_2(x-a)+R_1x,$$

so that
$$EI\frac{d^2y}{dx^2}=\frac{w}{10}[5x^2-15ax+11a^2].$$

On integrating, since $y=0$ when $x=a$ or $2a$,

$$EIy=\frac{w}{120}(x-a)(x-2a)(5x^2-15ax+11a^2).$$

Hence both y and $\frac{d^2y}{dx^2}$ are zero when $x=a\left[\frac{3}{2}\pm\frac{\sqrt{5}}{10}\right]$.

The middle segment has thus two points of inflexion which both lie on the line joining the points of support.

Ex. 3. *A uniform straight rod ABC is clamped at its ends A, C so that the tangents at these ends are horizontal and is supported at a point B; if A, B, and C are in a horizontal line, and $AB=a$, $BC=b$, find the reactions and bending moments at A, B, and C.*

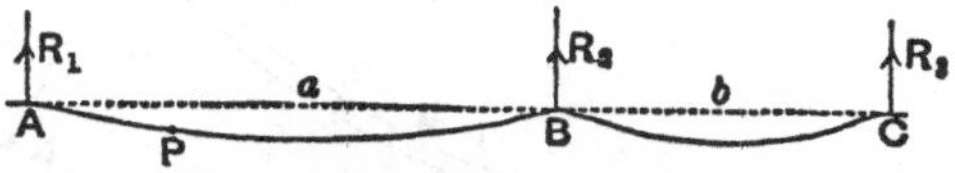

Let M_1, M_2, and M_3 be the bending moments, and R_1, R_2, and R_3 the reactions, at the points A, B, and C.

The theorem of Art. 329 may be applied to this case. For the clamping at A may be supposed to be done by the fixing of two points at it indefinitely close to A. Hence the formula of Art. 329 will apply if we put $A_1A_2=0$ and $A_2A_3=AB=a$, and it gives

$$0+2M_1a+M_2a=\frac{1}{4}wa^3 \quad \text{.....................(1)}.$$

For the three points of support A, B, C we have

$$M_1a+2M_2(a+b)+M_3b=\frac{1}{4}w(a^3+b^3) \quad \text{..............(2)}.$$

Similarly for the clamping at C, which may be supposed to be done by fixing two points at C indefinitely close to C; the formula then gives

$$M_2b+2M_3b+0=\frac{w}{4}b^3 \quad \text{.........................(3)}.$$

Solving (1), (2), and (3), we have

$$M_1=\frac{w}{24}(2a^2+ab-b^2)=\frac{w}{24}(a+b)(2a-b),$$

$$M_2=\frac{w}{12}(a^2-ab+b^2),$$

and
$$M_3=\frac{w}{24}(2b^2+ab-a^2)=\frac{w}{24}(a+b)(2b-a).$$

Again, taking moments about B for the part AB, we have

$$M_2=\frac{wa^2}{2}-R_1a+M_1,$$

so that
$$R_1=\frac{w}{8}\left[4a+b-\frac{b^2}{a}\right].$$

So also
$$R_3=\frac{w}{8}\left[4b+a-\frac{a^2}{b}\right].$$

$$\therefore\ R_2=w(a+b)-R_1-R_3=\frac{w}{8}\frac{(a+b)^3}{ab}=\frac{(a+b)^2}{8ab}\times \text{total weight of the rod.}$$

If the axis of y be drawn vertically downwards, then by taking moments about the section at P, distant x from A, we have, for the part AB, $-EI\frac{d^2y}{dx^2} = R_1 x - M_1 - \frac{wx^2}{2}$.

On substitution, we easily have the equation to the curve AB, and the inclination at B to the horizon.

332. *General equations of equilibrium of a rod, bent in one plane.*

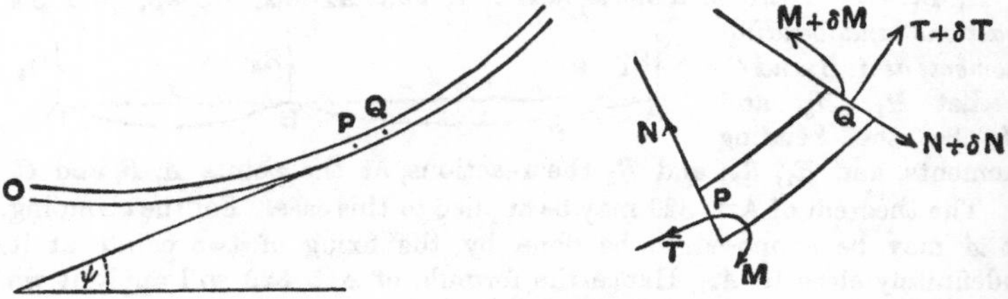

Let PQ be an element of the rod, $OP = s$ where O is a fixed point on the rod, $PQ = \delta s$, and let the tangents at P, Q be inclined at angles ψ and $\psi + \delta\psi$ to the axis of x.

Let T be the tension at P, $T + \delta T$ that at Q.

Let N be the shearing stress at P on the element PQ measured along the inward-drawn normal at P, $N + \delta N$ that at Q on the same element which is therefore along the normal at Q measured outwards.

Let M be the bending moment, or stress-couple, at P on the element PQ and $M + \delta M$ that at Q on the same element in the directions as marked.

Let F and G be the tangential and normal impressed forces on PQ per unit of length. Resolving along the tangent and normal at P, we have

$$-T + (T + \delta T)\cos\delta\psi + (N + \delta N)\sin\delta\psi + F\delta s = 0,$$

and $$N - (N + \delta N)\cos\delta\psi + (T + \delta T)\sin\delta\psi + G\delta s = 0.$$

In the limit when $d\psi$ is indefinitely small, these give

$$\frac{dT}{ds} + \frac{N}{\rho} + F = 0 \quad \text{.....................(1)},$$

and $$\frac{dN}{ds} - \frac{T}{\rho} - G = 0 \quad \text{.....................(2)}.$$

Also, taking moments about P for the element PQ, we have

$$M-(M+\delta M)+(N+\delta N)\,\delta s-G\delta s\,.\,\tfrac{1}{2}\delta s=0,$$

or, in the limit,

$$\frac{dM}{ds}-N=0 \quad \text{.......................(3)}.$$

The equation states that the shear at any point is equal to the differential coefficient of the bending moment with regard to the arc.

If ρ_0 be the radius of curvature of P of the rod when unstrained, we have, in addition, the equation

$$M=K\left[\frac{1}{\rho}-\frac{1}{\rho_0}\right] \quad \text{.......................(4)},$$

where K is the flexural rigidity of the rod.

These four equations give T, N, M and the equation of the curve in which the rod lies.

EXAMPLES

1. A uniform slightly elastic rod, of length $a+b$, rests on three supports in a horizontal line, situated at its ends A, B and at a point C distant a from A; shew that the points of inflexion I and J on the rod are situated in the segments AC and CB respectively, and are such that

$$a\,.\,IC=b\,.\,JC=\tfrac{1}{4}(a^2-ab+b^2).$$

2. A beam, of 40 feet span and carrying a load of 2 tons per foot run, is supported at a distance of 8 feet from one end by a column and at the other end it is built horizontally into a brick pier. Determine the bending moments at the supports, and draw the shearing force and bending moment diagrams to scale. [224 and 64 ft.-tons.]

3. A uniform girder AB, of weight W and length l, is built in firmly at A so as to be horizontal, and the other end B rests on a support in the same horizontal line as A. Shew that the bending moment and the shearing stress at A are $\frac{Wl}{8}$ and $\frac{5W}{8}$ respectively, and draw diagrams for the bending moment and shearing stress for the whole beam, proving that the points of zero and maximum bending moment are at distances $\frac{l}{4}$ and $\frac{5l}{8}$ from A.

Shew also that the beam rests in the form of the curve whose equation is $48EIyl=Wx^2(l-x)(3l-2x)$.

4. A beam AB, of length l, is loaded with a weight W at a point Q where $AQ=a$, and the weight of the beam is neglected. If its ends are built in so that they are horizontal, shew that the equation to AQ is

$$EIy=\frac{W(l-a)^2}{6l^3}x^2[3la-2xa-lx],$$

and that of QB is

$$EIy=\frac{Wa^2}{6l^3}(l-x)^2[3lx-2ax-al].$$

5. A slightly flexible rod, of length $2a$, has one end clamped horizontally; a support is placed under the middle point of the rod so that the free end is in the same horizontal line as the fixed end. Shew that the height of the middle point above the end is $\frac{11\,Wa^3}{240K}$, where K is the flexural constant and W is the weight of the rod. Shew also that the pressure on the support is $\frac{6W}{5}$.

6. A continuous girder, $2l$ feet long, is supported on three piers at equal distances, dividing it into two equal spans. The central pier is of metal and has a vertical motion, due to change of temperature, equal to a feet above and below the horizontal level of the two end piers. Calculate the variation of normal stress, above and below that which occurs when the piers are in line, at a section immediately over the middle pier.

7. A uniform slightly elastic beam, of weight W and length l, has its ends built in so that it is horizontal at both ends. Find its bending moment at any point, and shew that the value at the ends is twice that at the middle point, and that there are points of inflexion at distances $\cdot 211l$ from each end.

8. If a uniform slightly elastic rod be clamped horizontally at each end, and the middle point be pulled upwards by a force through a distance δ above the level of the ends, shew that the magnitude of the force is

$$\frac{24K\delta}{a^3}+\frac{W}{2},$$

and that the bending couples at the ends are equal to $\frac{6K\delta}{a^2}-\frac{1}{24}Wa$, where $2a$ is the length of the rod, W its weight, and K the flexural rigidity.

9. A beam of uniform section, with equal flanges and of span l, is built into walls so that its ends are horizontal and at the same level. One of the walls settles a distance δ without disturbing the horizontality of the ends of the beam. Shew that due to settling the maximum stress induced is $\frac{3Ed\delta}{l^2}$, where d is the depth of the girder and E is Young's Modulus.

10. A continuous girder, of uniform section, rests upon four supports at the same level, forming three equal spans of 100 feet. The girder carries a load of 2 tons per foot uniformly distributed. Draw to scale the bending moment diagram for the whole girder, and calculate the loads carried by each support. [80, 220, 220 and 80 tons wt.]

11. A heavy uniform elastic rod rests on four rigid supports in a horizontal line, two at its extremities and two at points equidistant from them. Find the pressures on the supports when the rod is slightly deflected by its own weight, and shew that the rod ceases to press on the terminal supports when the distance of an intermediate support from the nearer extremity is less than about ·214 of the total length.

12. A continuous beam rests on four supports in the same horizontal line, the width of the middle span being 15 feet and that of each of the side spans 10 feet. The beam carries a uniformly distributed load of 200 pounds per foot. Draw the diagrams of bending moment, and of shearing force and the curve of deflections.

13. A uniform rod (length $6a$, weight $6aw$, flexural rigidity K) is supported symmetrically at the ends A, D and at the points of trisection B, C in such a way that the pressures on the four supports are all equal, and the rod is clamped at A and D so that the tangents at those points are horizontal. Shew that A and D are at a height $\frac{2wa^4}{3K}$ above B and C, that the middle points of AB and CD are points of inflexion, and that the curvature vanishes at B and C without changing sign.

14. A heavy uniform slightly elastic rod rests on five points of support which are all in a horizontal line. Two of the supports are at the ends of the rod, one is at the middle point, and two bisect the distances between the middle point and the ends. Shew that the vertical thrusts on the points of support are in the ratio 11 : 26 : 32.

Prove also that the bending moments at the centre and at each of the supports next it are $\frac{Wl}{56}$ and $\frac{3Wl}{112}$, where W is the weight and $4l$ the length of the rod.

15. A wire, of uniform circular section and originally straight, rests with its ends on two props. Shew that, if T be the maximum fibre-tension at any section, then the bending couple is $\frac{1}{4}\pi r^3 T$, and further, if $2a$ be the span, then at a distance x from the centre of the wire

$$T = W\frac{a^2 - x^2}{ar},$$

where W is what the weight of the wire would have been had its cross section been of unit area.

16. A uniform slightly elastic rod, of weight W, rests with its middle point on a prop; the ends of a uniform string, of weight W', are then tied to its ends so that the string hangs in a catenary; prove that the deflection at each end of the rod is thereby increased in the ratio

$$1 + \frac{8}{3}\frac{W'}{W} : 1.$$

17. A three-hinged arch, having the hinges at the springings and the crown, is of semi-circular shape, the span being 50 feet and the rise 25 feet. It carries a load of 25 tons distributed uniformly (horizontally) over the right-hand half span. Shew how to draw the curves of bending moment on the two halves of the arch.

18. A series of small pegs are fixed into a horizontal floor in a straight line at equal distances, a, apart and a thin uniform rod, naturally straight, is bent in and out between them; shew that, if $R_1, R_2, \ldots$ are the pressures between the rod and successive pegs, then $R_{n-1} - 4R_n + R_{n+1} = 48EIca^{-3}$, where c is the thickness of a peg, E is Young's Modulus and I the moment of inertia of the area of the cross section of the rod about the vertical through the centre of the section.

19. A uniform heavy slightly elastic beam AB, of length $2c$, is supported on three props; shew that it will be strongest, *i.e.* that it will be least likely to break anywhere, if one prop be placed at the centre and the other two at points distant $\frac{6-\sqrt{6}}{5}c$ from the centre.

20. A brittle circular hoop is standing on the ground at rest with its plane vertical. Shew that the breaking moment, due to the weight of the hoop, is greatest at a point whose angular distance θ from the top of the hoop is given by $\tan\theta + \theta = 0$.

333. ***Work done against the stress couples in bending a rod or wire.***

Let PP' be an arc δs of the rod, ψ the angle between the tangents at its ends in its final bent position, so that the stress couple at $P = \frac{EI}{\rho} = \frac{EI}{\delta s} \cdot \psi$.

For any intermediate position between the straight and final bent form, let the angle between the tangents at the end of the same arc δs be ϕ, so that the corresponding stress couple is $\frac{EI}{\delta s} \cdot \phi$.

As ϕ increases to $\phi + \delta\phi$, the work done against this couple $= \frac{EI}{\delta s}\phi \cdot \delta\phi$ [Art. 97]. Hence the whole work done on this arc as ϕ increases from zero to $\psi = \int_0^{\psi} \frac{EI}{\delta s} \cdot \phi\delta\phi = \frac{1}{2}\frac{EI}{\delta s} \cdot \psi^2 = \frac{1}{2}\frac{EI}{\rho^2} \cdot \delta s$.

The whole work done on the rod thus $= \int \frac{1}{2}\frac{EI}{\rho^2}\,ds$, the integral being taken over its whole length.

If the rod, instead of being originally straight, was of curvature $\frac{1}{\rho_1}$ at P, where $\frac{1}{\rho_1}=\frac{\psi_1}{\delta s}$,

$$\text{the stress couple} = \frac{EI}{\delta s}(\phi-\psi_1),$$

and the whole work

$$=\int_0^l\left[\int_{\psi_1}^{\psi}\frac{EI}{\delta s}\cdot(\phi-\psi_1)\,.\,d\phi\right]=\int_0^l\frac{EI}{2\delta s}[\psi-\psi_1]^2$$

$$=\int_0^l\frac{EI}{2}\left(\frac{1}{\rho}-\frac{1}{\rho_1}\right)^2 ds.$$

EXAMPLES

1. *The natural form of a rod, of length a, is an arc of a catenary, of parameter a, having one extremity at the vertex. It is bent into the form of a circular arc of radius a; shew that the work done against the stress couples is $\frac{K}{16a}(10-3\pi)$, where K is the coefficient of flexural rigidity at every point of the rod.*

In a catenary $s=a\tan\psi$, so that $\rho_1=\frac{ds}{d\psi}=a\sec^2\psi=\frac{s^2+a^2}{a}$.

Also $\rho=a$.

Hence the work done

$$=\frac{K}{2}\int_0^a\left[\frac{1}{a}-\frac{a}{s^2+a^2}\right]^2 ds=\frac{K}{2a}\int_0^{\frac{\pi}{4}}[\sec^2\psi-2+\cos^2\psi]\,d\psi,$$

on putting $s=a\tan\psi$,

$$=\frac{K}{2a}\left[\tan\psi-\frac{3\psi}{2}+\frac{1}{4}\sin 2\psi\right]_0^{\frac{\pi}{4}}=\frac{K}{16a}(10-3\pi).$$

2. A uniform beam, of length $2a$ and weight W, rests on a smooth horizontal table and is raised by a force applied at its middle cross section; shew that, when its ends leave the plane, its centre is at a height $\frac{Wa^3}{16EI}$ above them and that the work then done is $\frac{W^2a^3}{20EI}$.

3. A uniform heavy rod, whose length is l and whose weight is W, is supported at its two ends so as to be initially horizontal, and bends slightly under its own weight. Shew that the work done by gravity in bending it is $\frac{W^2l^3}{240EI}$.

4. Shew that the work required to bend a straight wire, of length $2\pi a$, round the rim of a penny, of radius a, is $\frac{\pi EI}{a}$.

5. Shew that the work done in bending a wire, of length $2l$, into the form of a catenary, whose parameter is c, is $\frac{EI}{2c}\left(\tan^{-1}\frac{l}{c}+\frac{lc}{l^2+c^2}\right)$.

6. One end of a heavy slightly flexible wire in the form of a circular quadrant is fixed into a vertical wall, so that the plane of the wire is vertical and the tangent at the fixed end is horizontal. Assuming that the change of curvature at any point is proportional to the moment of the bending couple there, shew that the horizontal deflection at the free end is $\frac{\pi}{8}\frac{wa^4}{K}$, where K is the flexural rigidity, w is the weight of a unit length, and a is the radius of the circle.

7. A uniform stiff wire, of weight πwa and flexural rigidity K, whose natural shape is a semi-circle of radius a, is placed in a vertical plane with its ends on a smooth horizontal plane. Shew that the intrinsic equation to the form assumed by the wire is approximately

$$s = a\phi + wa^4 K^{-1}\{\tfrac{1}{2}\pi\phi + \phi\cos\phi - 2\sin\phi\},$$

s being measured from the highest point.

8. A stiff wire, whose natural shape is a semi-circle of radius a, rests in a vertical plane with the middle point on a horizontal table. Shew that the intrinsic equation to the curve formed is approximately

$$s = a\phi + \frac{wa^4}{K}\left[\left(\frac{\pi}{2} - \phi\right)\cos\phi + 2\sin\phi - \frac{\pi}{2}\right],$$

where w is the weight per unit of arc, K is the flexural rigidity, and it is assumed that $\frac{wa^3}{K}$ is very small.

334. Bending of Long Columns.

Suppose we have a column, or strut, whose length is great compared with the dimensions of its cross section, and which is of uniform strength and was originally straight.

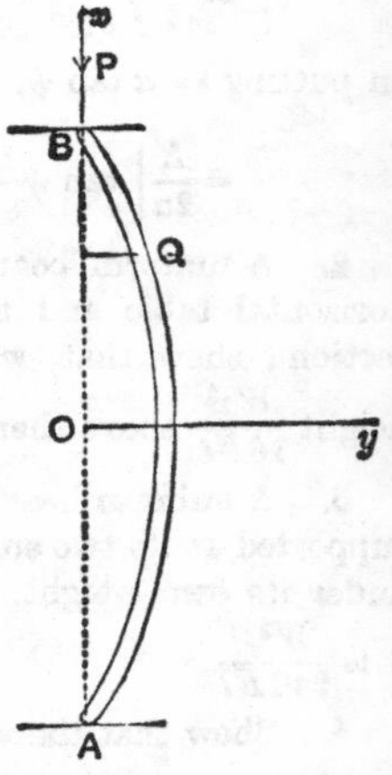

Let it be set up vertically and bear on its upper end a load P; it is required to find the shape it assumes, its deflection being assumed to be small.

Take as origin the middle point between the ground and the point of application of the load, and let Ox be vertical and Oy horizontal.

If the weight of the column be small compared with the load P, the equation of equilibrium is

$$-EI\frac{d^2y}{dx^2} = \frac{EI}{\rho} = Py, \quad i.e. \ \frac{d^2y}{dx^2} = -\frac{P}{EI}y.$$

$$\therefore \ y = A\cos\left[x\sqrt{\frac{P}{EI}}\right] + B\sin\left[x\sqrt{\frac{P}{EI}}\right] \quad \text{......(1)},$$

where A and B are arbitrary constants.

By symmetry, $\frac{dy}{dx}=0$ when $x=0$. Hence $B=0$.

Let the ends of the beam be rounded, so that the tangents at the ends may assume any direction. But, since no couple is applied at the ends, $\frac{EI}{\rho}$, and hence $\frac{d^2y}{dx^2}$, is zero at each end, *i.e.* $\frac{d^2y}{dx^2}=0$ when $x=\pm\frac{l}{2}$, if l be the length of the column.

$$\therefore A\cos\left[\frac{l}{2}\sqrt{\frac{P}{EI}}\right]=0.$$

Hence $$\frac{l}{2}\sqrt{\frac{P}{EI}}=\frac{\pi}{2}, \text{ so that } P=\frac{\pi^2 . EI}{l^2} \quad \text{.........(2).}$$

This gives the end-load which is sufficient to keep the column bent when the curvature has been produced.

If P exceeds this value, the column would give way.

Since I for a circular column varies as the fourth power of the diameter, it follows from (2) that, for columns of the same material, the greatest possible end-load varies directly as the fourth power of the diameter and inversely as the square of the length of the column.

This is known as Euler's Law for the bending of long Columns or Struts.

335. In the previous article if the ends A and B are fixed, so that the tangents at them are vertical, the solution is different. At the end B there must act a couple G ↻; the resultant of this couple and the load P acting at B is a parallel force P acting as in the figure. Its line of action being the axis of x, the equation of equilibrium is as in the last article and the solution is the same, *viz.* equation (1).

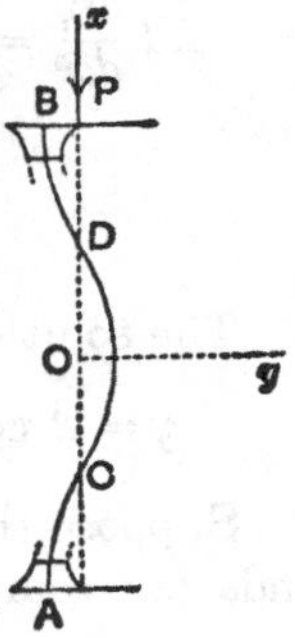

In this case $\frac{dy}{dx}=0$ when $x=0$ or $\pm\frac{l}{2}$.

Hence $B=0$, and $0=-A\sin\left[\frac{l}{2}\sqrt{\frac{P}{EI}}\right]$.

$$\therefore \frac{l}{2}\sqrt{\frac{P}{EI}}=\pi, \text{ so that } P=\frac{4\pi^2EI}{l^2}.$$

The equation to the curve into which the neutral line of the column is bent is now

$$y = A \cos \frac{2\pi x}{l}.$$

Hence there are points of inflexion where $\frac{d^2y}{dx^2} = 0$, *i.e.* where $x = \pm \frac{l}{4}$. These are the points C and D and they lie on the line of action of the resultant of the load and the couple at B.

336. Centrifugal Whirling of Shafts.

Suppose a thin vertical cylindrical shaft to revolve in bearings; it will tend to bend laterally under its rotation if the angular velocity of the rotation is sufficient.

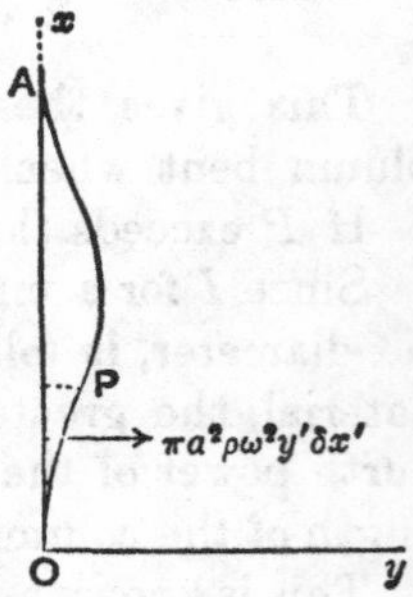

Suppose G↺ to be the bending moment, and S the horizontal thrust, at the lower bearing O. Then, since the "centrifugal force" at any point (x', y') is $\pi a^2 \rho \omega^2 y' \delta x$, where a is the radius and ρ the density of the shaft, the deflection being assumed to be small, and hence $\delta x'$ and $\delta s'$ very nearly the same, the equation of equilibrium is

$$EI \frac{d^2y}{dx^2} = G - S \cdot x + \pi a^2 \rho \omega^2 \int_0^x y' \, dx' \, (x - x') \quad \text{...(1)}.$$

Differentiating twice with respect to x, we have

$$EI \frac{d^3y}{dx^3} = \pi a^2 \rho \omega^2 \int_0^x y' \, dx' - S, \quad \text{and} \quad EI \frac{d^4y}{dx^4} = \pi a^2 \rho \omega^2 y.$$

$$\therefore \frac{d^4y}{dx^4} = \frac{\pi a^2 \rho \omega^2}{E \cdot \pi a^2 \cdot \frac{a^2}{4}} y = \frac{4\rho\omega^2}{Ea^2} y = \frac{\mu^4}{l^4} y, \text{ say, } \quad \text{...(2)}.$$

The solution of this equation is

$$y = A \cos \frac{\mu x}{l} + B \sin \frac{\mu x}{l} + C \cosh \frac{\mu x}{l} + D \sinh \frac{\mu x}{l} \quad \text{...(3)}.$$

Suppose that the shaft is so supported at O and A that the ends are compelled to be vertical there, so that when $x = 0$ we have $y = 0$ and $\frac{dy}{dx} = 0$.

$$\therefore y = A \left[\cos \frac{\mu x}{l} - \cosh \frac{\mu x}{l} \right] + B \left[\sin \frac{\mu x}{l} - \sinh \frac{\mu x}{l} \right] \quad \text{...(4)}.$$

Also, when $x=l$, y and $\frac{dy}{dx}$ both vanish.

$$\therefore\quad 0=A\,[\cos\mu-\cosh\mu]+B\,[\sin\mu-\sinh\mu],$$

and $$0=A\,[-\sin\mu-\sinh\mu]+B\,[\cos\mu-\cosh\mu].$$

$$\therefore\quad \frac{\cos\mu-\cosh\mu}{\sin\mu-\sinh\mu}=\frac{\sin\mu+\sinh\mu}{\cosh\mu-\cos\mu}=-\frac{B}{A}\quad\text{......(5).}$$

$$\therefore\quad (\cos\mu-\cosh\mu)^2+\sin^2\mu-\sinh^2\mu=0.$$

$$\therefore\quad \cos\mu\cosh\mu=1\quad\text{.................(6).}$$

This equation gives μ, and hence, from (5), the value of $\frac{B}{A}$, so that the form of the curve (4) is known.

But, from (2), $\frac{4\rho\omega^2}{Ea^2}=\frac{\mu^4}{l^4}$, so that $\omega=\frac{1}{2}\frac{\mu^2}{l^2}\,.\,a\,.\sqrt{\frac{E}{\rho}}$, giving the required angular velocity. If ω is greater than this value, the shaft bends more and more.

On tracing the graphs of $\cos\mu$ and $\operatorname{sech}\mu$, we easily see that the solution of (6) is very nearly $\frac{3\pi}{2}$. Put then $\mu=\frac{3\pi}{2}+\lambda$, where λ is small, and (6) gives, for a second approximation,

$$\lambda=\frac{1}{\cosh\frac{3\pi}{2}}=\frac{1}{55{\cdot}7}=\cdot 018\text{, from the Tables.}$$

$\therefore\quad \mu=\frac{3\pi}{2}+\cdot 018=4{\cdot}73$, as a second approximation. Substituting in (5), we have $\frac{B}{A}=-\cdot 98$ approximately. From (4), we now have the equation to the curve assumed by the shaft.

For steel, $E=$ about 3×10^7 lbs. wt. per sq. in., and $\rho=$ about 480 lbs. per cubic ft.

337. In the previous question suppose that the ends of the beam are not compelled to be vertical, but that at the ends it is freely supported only, so that $\frac{dy}{dx}$ is not zero there but $\frac{d^2y}{dx^2}$ does vanish there, since there is no bending moment at either of the ends.

In this case G is zero. The equation (2) is as before, and its solution is

$$y = A_1 \cos\frac{\mu x}{l} + B_1 \sin\frac{\mu x}{l} + C_1 \cosh\frac{\mu x}{l} + D_1 \sinh\frac{\mu x}{l}.$$

When $x=0$, $y=0$ and $\frac{d^2y}{dx^2}=0$.

$$\therefore \quad A_1 + C_1 = 0, \quad \text{and} \quad -\frac{\mu^2}{l^2}A_1 + \frac{\mu^2}{l^2}C_1 = 0,$$

so that

$$A_1 = C_1 = 0.$$

So when $x=l$, $y=0$ and $\frac{d^2y}{dx^2}=0$.

$$\therefore \quad 0 = B_1 \sin\mu + D_1 \sinh\mu,$$

and

$$0 = -B_1 \sin\mu + D_1 \sinh\mu.$$

$$\therefore \quad B_1 \sin\mu = D_1 \sinh\mu = 0.$$

$$\therefore \quad \mu = \pi \text{ and } D_1 = 0.$$

Hence

$$\pi^4 = \mu^4 = l^4 \cdot \frac{4\rho\omega^2}{Ea^2},$$

giving $\omega = \frac{1}{2}\frac{\pi^2}{l^2}\cdot a\sqrt{\frac{E}{\rho}}$, as the smallest value of ω.

Also the shape of the curve in this case is

$$y = B_1 \sin\frac{\pi x}{l}.$$

338. *A thin uniform column is set up vertically; to find the greatest height it can have so that it shall not give way under its own weight.*

The origin O being at the upper end of the column, Ox being drawn vertically downwards, and Oy being horizontal, the equation of equilibrium is

$$-EI\frac{d^2y}{dx^2} = \int_0^x w\,d\xi\,(y-\eta),$$

where w is the weight of the column per unit of length, and the deflection is supposed to be very small.

Differentiating, and putting $\frac{w}{EI}=m$ and $\frac{dy}{dx}=p$, this gives

$$\frac{d^2p}{dx^2} = -m\int_0^x p\,d\xi = -mpx.$$

Put $x = u^{\frac{2}{3}}$, and $p = u^{\frac{1}{3}}t$, and this equation becomes

$$\frac{d^2t}{du^2} + \frac{1}{u}\frac{dt}{du} + t\left[\frac{4m}{9} - \frac{1}{9u^2}\right] = 0.$$

Let $$q^2 = \frac{4m}{9} = \frac{4}{9}\frac{w}{EI}, \text{ and } qu = v.$$

Then $$\frac{d^2t}{dv^2} + \frac{1}{v}\frac{dt}{dv} + t\left[1 - \frac{1}{9v^2}\right] = 0$$

$$\therefore \quad t = AJ_{\frac{1}{3}}(v) + BJ_{-\frac{1}{3}}(v),$$

where $J_n(v)$ is the Bessel's function of order n.

$$\therefore \quad p = x^{\frac{1}{2}}\left[AJ_{\frac{1}{3}}(qx^{\frac{3}{2}}) + BJ_{-\frac{1}{3}}(qx^{\frac{3}{2}})\right].$$

Now $\dfrac{dp}{dx} = 0$ when $x = 0$, since the bending moment is zero at the highest point.

Also $$J_{\frac{1}{3}}(qx^{\frac{3}{2}}) = q^{\frac{1}{3}}x^{\frac{1}{2}}\left[1 - \frac{q^2x^3}{2^2 . \frac{4}{3}} + \frac{q^4x^6}{\underline{|2} . 2^4 . \frac{4}{3} . \frac{7}{3}} - \dots\right],$$

and $$J_{-\frac{1}{3}}(qx^{\frac{3}{2}}) = q^{-\frac{1}{3}}x^{-\frac{1}{2}}\left[1 - \frac{q^2x^3}{2^2 . \frac{2}{3}} + \frac{q^4x^6}{\underline{|2} . 2^4 . \frac{2}{3} . \frac{5}{3}} - \dots\right],$$

so that $$\frac{d}{dx}\left[Ax^{\frac{1}{2}}J_{\frac{1}{3}}(qx^{\frac{3}{2}})\right] = Aq^{\frac{1}{3}}, \text{ when } x = 0,$$

and $$\frac{d}{dx}\left[Bx^{\frac{1}{2}}J_{-\frac{1}{3}}(qx^{\frac{3}{2}})\right] = 0, \text{ when } x = 0.$$

Hence $A = 0$, since $\dfrac{dp}{dx} = 0$, when $x = 0$.

$$\therefore \quad p = Bx^{\frac{1}{2}}J_{-\frac{1}{3}}(qx^{\frac{3}{2}}).$$

But $p = 0$ when $x = l$, the height of the column, since the column is fixed vertically in the ground.

Hence $J_{-\frac{1}{3}}(ql^{\frac{3}{2}}) = 0$, an equation to give l.

This gives

$$1 - \frac{3}{2}\cdot\frac{1}{2^2}\cdot\frac{4m}{9}l^3 + \frac{1}{\underline{|2}}\frac{1}{2^4}\cdot\frac{3}{2}\cdot\frac{3}{5}\left(\frac{4m}{9}\right)^2 l^6 - \dots = 0,$$

i.e. $$1 - \frac{ml^3}{2.3} + \frac{m^2l^6}{2.3.5.6} - \frac{m^3l^9}{2.3.5.6.8.9} + \dots = 0.$$

An approximate solution of this equation is found to be $ml^3 = 7{\cdot}84$, so that $l^3 = 7{\cdot}84 \times \dfrac{EI}{w}$, giving the greatest height of the column.

If we take a solid steel column of one foot radius, whose density is 480 lbs. per cubic foot and for which $E = 3 \times 10^7$ lbs. wt. per square inch, this formula gives $l =$ about 260 feet.

EXAMPLES

1. A straight steel rod, of length 20 feet and diameter one inch, is set up vertically, and its ends fixed so that the tangents at them are both vertical; shew that the greatest load it will bear is about 1010 lbs.

2. A straight steel rod, of uniform circular section and 5 feet long, is found to deflect one inch under a central load of 20 pounds when tested as a beam simply supported at its ends. Determine the critical load for the same beam when used as a vertical strut with rounded ends.

[About 247 lbs.]

3. A steel spindle, $\frac{3}{4}$ inch in diameter, is to be supported in bearings having spherical seatings, and is required to rotate 3000 times per minute. Calculate the maximum distance permissible between the centres of the bearings so that there shall be no whirling of the shaft. [1·21 ft. nearly.]

4. A long thin rod AB is set up vertically and loaded with a weight W at B, the lower end A being compelled to remain vertical. If it be slightly elastic and be of length l, shew that the equation of the curve it assumes is

$$y = y_1\left[1 - \cos\frac{\pi x}{2l}\right],$$

where A is the origin and y_1 is the horizontal displacement of B, and $\dfrac{EI}{W} = \dfrac{4l^2}{\pi^2}$.

Shew also that the beam does not bend if $W < \dfrac{\pi^2 EI}{4l^2}$.

[From this and Art. 335 it follows that a rod, with both ends fixed so that it is vertical at both ends, will support a weight sixteen times as great as it will if the upper end be free to move sideways and the lower end only be fixed in a vertical position.]

5. If in the question of Art. 334 the lower end is fixed so that the tangent at it is vertical, and the upper end is compelled, by a horizontal force applied to it, to always remain in the vertical line through the fixed end, prove that

$$\sqrt{\frac{P}{EI}}\,l = \tan\left[\sqrt{\frac{P}{EI}}\,l\right], \text{ so that } \sqrt{\frac{P}{EI}}\,l = 4{\cdot}493,$$

and hence that $\qquad P = 20{\cdot}187 \times \dfrac{EI}{l^2} = 2{\cdot}045 \times \dfrac{\pi^2 EI}{l^2}$ approx.

6. In the question of Art. 336, if the shaft have one end compelled to be vertical and the other end freely supported, shew that μ is given by the equation $\tan\mu = \tanh\mu$, and hence that it equals 3·93 nearly.

If the second end be quite free, shew that μ is given by the equation $\cos\mu\cosh\mu = -1$, and hence that it equals 1·875 nearly. [In this case the shearing force is zero at the second end as well as the bending moment, so that both $\frac{d^3y}{dx^3}$ and $\frac{d^2y}{dx^2}$ are zero there.]

7. If a bow be constructed of uniform material, shew that its intrinsic equation when strung is $\frac{ds}{d\phi}\sqrt{\sin^2\frac{a}{2} - \sin^2\frac{\phi}{2}} = a\sin\frac{a}{2}$.

If the bow be only slightly flexible, and if its length be $2l$ and the length of the string be $2a$, where l and a are nearly equal, shew that the equation to the curve it assumes is $y = \frac{4\sqrt{a(l-a)}}{\pi} \cdot \sin\frac{\pi x}{2a}$, and that the tension of the string is $\frac{\pi^2 EI}{4a^2}$.

8. A horizontal bracket, of length a, is attached to the upper end of a vertical pillar, of length l, which has its lower end built in. When carrying a load W at the extremity of the bracket, the pillar bends slightly. Shew that, on account of the flexure of the pillar, the bending moment at its base is increased, whatever be the length of the bracket, in the ratio $\sec\left(\sqrt{\frac{W}{EI}}\,l\right)$, where E is Young's Modulus and I is the moment of inertia of the cross section of the pillar about the line through its centre perpendicular to the plane of bending.

9. A straight girder, of uniform section, is laid upon a horizontal bed of compressible material. At a point in the girder where the depression is y inches, the pressure between the girder and the bed is found to be Ky tons per inch run. If the girder is subjected to a vertical load of W tons concentrated at a point equidistant from its ends, shew that the distribution of pressure is $\frac{1}{2}aWe^{-ax}[\cos ax + \sin ax]$ tons per inch run, where x is measured from the loaded section and $a = \sqrt[4]{\frac{K}{4EI}}$. I is the moment of inertia of the section of the girder about its neutral axis and E is Young's Modulus for the material.

Printed in the United States
By Bookmasters

Printed in the United States
By Bookmasters